INTRODUCTORY
MODERN
ALGEBRA

INTRODUCTORY MODERN ALGEBRA

A Historical Approach

Saul Stahl
University of Kansas

A Wiley-Interscience Publication
JOHN WILEY & SONS, INC.
New York ■ Chichester ■ Brisbane ■ Toronto ■ Singapore ■ Weinheim

This text is printed on acid-free paper.

Copyright © 1997 by John Wiley & Sons, Inc.

All rights reserved. Published simultaneously in Canada.

Reproduction or translation of any part of this work beyond that permitted by Section 107 or 108 of the 1976 United States Copyright Act without the permission of the copyright owner is unlawful. Requests for permission or further information should be addressed to the Permissions Department, John Wiley & Sons, Inc., 605 Third Avenue, New York, NY 10158-0012.

Library of Congress Cataloging in Publication Data:
Stahl, Saul.
    Introductory modern algebra / Saul Stahl.
      p.   cm.
    "A Wiley-Interscience publication."
    Includes bibliographical references (p. -) and index.
    ISBN 0-471-16288-4 (cloth : alk. paper)
    1. Algebra, Abstract.   I. Title.
QA162.S73   1997
512'.2--dc20   96-19469

Printed in the United States of America

10 9 8 7 6 5 4 3

Contents

Preface   ix

1. The Early History   1
   1.1. The Breakthrough, 1

2. Complex Numbers   9
   2.1. Rational Functions of Complex Numbers, 9
   2.2. Complex Roots, 17
   2.3. Solvability by Radicals (I), 24
   2.4. Ruler and Compass Constructibility of Regular Polygons, 27
   2.5. Orders of Roots of Unity, 39
   2.6. The Existence of Complex Numbers*, 41

3. Solutions of Equations   47
   3.1. The Cubic Formula, 47
   3.2. Solvability by Radicals (II), 51
   3.3. Other Types of Solutions*, 52

4. Modular Arithmetic   57
   4.1. Modular Addition, Subtraction, and Multiplication, 57
   4.2. The Euclidean Algorithm and Modular Inverses, 62
   4.3. Radicals in Modular Arithmetic*, 68
   4.4. The Fundamental Theorem of Arithmetic*, 69

5. The Binomial Theorem and Modular Powers   74
   5.1. The Binomial Theorem, 74
   5.2. Fermat's Theorem and Modular Powers, 85
   5.3. The Multinomial Theorem*, 91
   5.4. The Euler ϕ-Function*, 93

*Optional.

6. Polynomials Over a Field — 98
6.1. Fields and Their Polynomials, 98
6.2. The Factorization of Polynomials, 106
6.3. The Euclidean Algorithm for Polynomials, 111
6.4. Elementary Symmetric Polynomials, 118
6.5. Lagrange's Solution of the Quartic Equation*, 124

7. Galois Fields — 129
7.1. Galois's Construction of His Fields, 129
7.2. The Galois Polynomial, 138
7.3. The Primitive Element Theorem, 142
7.4. On the Variety of Galois Fields*, 145

8. Permutations — 152
8.1. Permuting the Variables of a Function (I), 152
8.2. Permutations, 155
8.3. Permuting the Variables of a Function (II), 163
8.4. The Parity of a Permutation, 166

9. Groups — 181
9.1. Permutation Groups, 181
9.2. Abstract Groups, 191
9.3. Isomorphisms of Groups and Orders of Elements, 197
9.4. Subgroups and Their Orders, 203
9.5. Cyclic Groups and Subgroups, 212
9.6. Cayley's Theorem, 215

10. Quotient Groups and Their Uses — 221
10.1. Quotient Groups, 221
10.2. The Rigorous Construction of Fields, 231
10.3. Galois Groups and the Resolvability of Equations (An informal discussion), 243

11. Topics in Elementary Group Theory — 250
11.1. The Direct Product of Groups, 250
11.2. Some Classifications, 254

Appendix A. Excerpts from Al-Khwarizmi's *Solution of the Quadratic Equation* — 261

Appendix B. Excerpts from Cardano's *Ars Magna* — 266

*Optional.

Appendix C. Excerpts from *A Demonstration of the Impossibility of the Algebraic Resolution of General Equations whose Degree Exceeds Four* by Niels H. Abel   271

Appendix D. *On the Theory of Numbers* by Évariste Galois   277

Appendix E. *The Theory of Groups* by Arthur Cayley   285

Appendix F. Mathematical Induction   288

Biographies   297

Bibliography   301

Solutions to Selected Odd Exercises   303

Symbols   315

Pronounciation Guide   317

Index   319

Preface

A CONCRETE AND HISTORICAL APPROACH

It is common knowledge among mathematicians that much of modern algebra has its roots in the issue of solvability of equations by radicals. The purpose of this text is to provide the undergraduate mathematics majors and the prospective high school mathematics teachers with a one semester introduction to modern algebra that keeps this relationship in view at all times.

Most modern algebra texts employ an axiomatic strategy that begins with abstract groups and ends with fields, ignoring the issue of solvability of equations by radicals. By contrast, we follow the paper trail from the Renaissance solution of the cubic equation to Galois's description of his ideas. In the process, all the important concepts are encountered, each in a well-motivated manner.

One year of calculus provides all the information required for the comprehension of all the topics in this text.

DISTINGUISHING FEATURES

Historical Development

Students prefer to know the real reasons that underlie the creation of the mathematical structures they encounter. They also enjoy being placed in direct contact with the works of the prime movers of mathematics. This text tries to bring them as close to the source as possible.

Finite groups and fields are rooted in some specific investigations of Lagrange, Gauss, Cauchy, Abel, and Galois regarding the solvability of equations by radicals. This text makes these connections explicit. Gauss's proof of the constructibility of the regular 17-sided polygon is incorporated into the development, and the argument given is merely a paraphrase of that which appears in the *Disquisitiones*. Similarly the proof of Theorem 8.8 is just a reorganization of that given by Abel in his paper on the *quintic equation*. The construction of *Galois fields* is accomplished in the form of a commentary on the opening pages of Galois's *On the Theory of Numbers* which are quoted verbatim in the text. Several important documents are also included as appendixes. A considerable amount of historical discussion is integrated into the development of the subject matter.

Cohesive Organization

The historical development of the material allows for very little flexibility. Each chapter elucidates some of the preceding material and motivates ideas that come later. The advantage of this approach is the same as that of good motivation in general: It aids comprehension by providing the students with a framework in which to fit the various concepts they encounter. A one-semester course can be constructed on the basis of Sections 1.1, 2.1-5, 3.1-2, 4.1-2, 5.1-2, 6.1-3, 7.1-3, 8.1-4, 9.1-5, and 10.1.

The figure on the opposite page illustrates the author's perception of the evolution of abstract group theory (ignoring all the geometric and much of the number theoretic contributions). The number at the right in each box denotes the chapter in which the topic is discussed. Solid arrows correspond to connections that are treated in some depth, whereas those that are displayed by dashed arrows are touched on only informally.

Chapters 1-3 are dedicated to the formalization of the notion of solvability by radicals. Gauss's proof of the constructibility of the regular 17-sided polygon is the capstone theorem of this part of the course. Field theory is developed in Chapters 4-7. The *Primitive Element Theorem* of Section 7.3 unifies many of the important concepts that precede it and motivates the notion of cyclicity that comes later. Group theory is developed in Chapters 8-10. This begins with an explanation of the relevance of permutations to solvability by radicals, goes on to the discussion of permutation groups and abstract groups, and concludes with the description of quotient groups. The final Chapter 11 acquaints students with some of the standard tools of elementary group theory.

Pedagogy

Each section is followed by its own set of exercises. These range from the routine to the challenging. Each chapter has an additional set of easy review exercises added to remind the students of the chapter's main points. There are over 1000 of these end of section and chapter review exercises. The answers to selected odd exercises appear at the end of the book. Each chapter is also accompanied by a collection of supplementary computer and/or mathematical projects. Some of the latter inolve open questions.

Each chapter begins with an introduction and concludes with a summary. The purposes of both the introduction and the summary are to provide the student with an overview of the chapter, and sometimes to comment on its

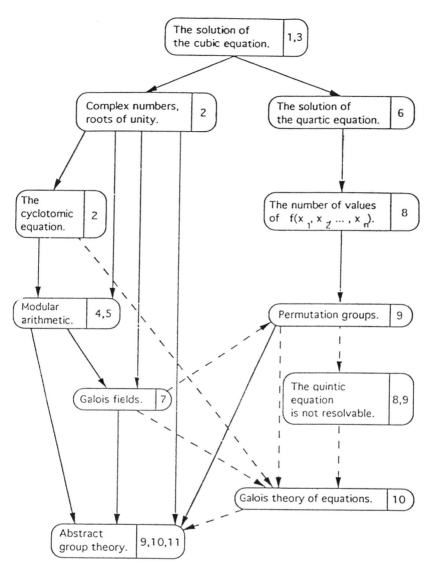

The genesis of the theory of finite groups.

relationship to the previous chapters. The examples are integrated into the exposition. Each chapter's new terms are listed, together with the page on which they are defined, following that chapter's summary.

An instructor's manual is available. It contains the answers to all the end of section and chapter review exercise. Some suggested homework assignments and tests are also included.

Acknowledgments

First and foremost I wish to acknowledge the substantial contributions made by Fred Galvin who rooted out several inaccuracies in the original development, improved and/or corrected many of the proofs, both in the text and the manual, suggested new exercises, and used the manuscript in his class. Thanks are also due to Andy Magid, Phil Montgomery, and Todd Eisworth who also tested the manuscript in their classes and made valuable suggestions and to my colleague Paul J. McCarthy who was kind enough to lend me both an ear and his algebraic expertise. It remains to gratefully acknowledge the efforts of Jessica Downey, Steve Quigley, Rosalyn Farkas, and Lisa Van Horn of John Wiley & Sons on behalf of this book.

SAUL STAHL

June 1996

INTRODUCTORY MODERN ALGEBRA

CHAPTER 1

The Early History

This chapter contains an informal account of the early history of the issue of solvability of equations of degrees 1, 2, and 3 in a single unknown. The formulas that provide the solutions lead in a natural way to the discussion of the origins of complex numbers. We also take this opportunity to review some well-known information about the quadratic equation.

1.1 THE BREAKTHROUGH

There is a general agreement among historians of mathematics that modern mathematics came into being in the midsixteenth century when the combined efforts of the Italian mathematicians Scipione del Ferro, Niccolo Tartaglia, and Gerolamo Cardano produced a formula for the solution of cubic equations. For the first time ever western mathematicians succeeded in cracking a problem whose solution eluded the best mathematical minds of antiquity. Archimedes, one of the greatest mathematicians, scientists, and engineers of all times, had solved some cubic equations in terms of the intersections of a suitable parabola and hyperbola. Omar Khayyam, one of the most prominent of the Arab mathematicians and poets, also expended much effort on his geometrical solutions of special cases of the cubic equation but could not find the general formula. However, the significance of this accomplishment of the Renaissance mathematicians is not limited to the difficulty of the problem that was solved. We will try to show how the issues raised by this solution eventually led to the creation of modern algebra and the discovery of mathematical landscapes that were undreamed of, even by such imaginative investigators as Archmides and Khayyam.

The interest in algebraic equations goes back to the beginnings of written history. The Rhind Mathematical Papyrus, found in Egypt circa 1856, is a list of mathematical problems compiled some time during the second half of the nineteenth century BC or even earlier. The twenty-fourth of these problems reads: "A quantity and its 1/7 added become 19. What is the quantity?" In other words, what is the solution to the equation

$$x + \frac{x}{7} = 19?$$

The method employed by the scribes has come to be known as the *method of false position*. He replaces the unknown by 7 and observes that

$$7 + \frac{7}{7} = 8$$

From this he concludes that the correct answer is obtained upon multiplying the first guess of 7 by 19/8:

$$x = 7 \cdot \frac{19}{8} = \frac{133}{8}$$

Interestingly enough, the scribe double-checks his solution by substituting it into the original problem and verifying that

$$\frac{133}{8} + \frac{133/8}{7} = 19$$

We will not discuss the merits and limitations of the method of false position except to note that the idea of obtaining a correct solution to an equation by starting out with a possibly false guess and then modifying that guess has been refined into powerful techniques for finding numerical solutions, one of which will be described in Section 3.3. We do, however, wish to point out that the general first-degree equation is today defined as

$$ax + b = 0, \quad a \neq 0$$

and that the rules of algebra yield

$$x = -\frac{b}{a}$$

as its unique solution.

The Mesopotamian mathematicians of that time could solve much more intricate equations. They had in fact already developed techniques for solving what we nowadays call quadratic equations. These techniques employed the geometrical method of "completing the square." The Greeks, Indians, and Arabs all were aware of this method, having either derived them independently or perhaps learnt from their predecessors and/or neighbors. In the ninth century the Persian mathematician al-Khwarizmi wrote the book *Hisab al-jabr w'al-muqa-balah* in which he carefully explained a compendium techniques learned from several past civilizations. The clarity of his exposition won both him and his book immortality in that the portion *al-jabr* of the title evolved into the word *algebra,* and the author's name is the source of the word *algorithm.* An excerpt from this book expounding the solution to the quadratic equation

$$x^2 + 10x = 39$$

appears in Appendix A. The modern solution of the quadratic also relies on the completion of the square. The general quadratic equation has the form

$$ax^2 + bx + c = 0, \quad a \neq 0 \qquad (1)$$

and its solutions are found by first factoring out the coefficient a and then completing the rest to a perfect square. Thus we first divide (1) through by a to obtain the equation

$$x^2 + \frac{b}{a}x + \frac{c}{a} = 0 \qquad (2)$$

The left side of (2) is then transformed to a near perfect square:

$$x^2 + \frac{b}{a}x + \frac{c}{a} = \left(x + \frac{b}{2a}\right)^2 - \frac{b^2}{4a^2} + \frac{c}{a}$$

$$= \left(x + \frac{b}{2a}\right)^2 - \frac{b^2 - 4ac}{4a^2}$$

The original quadratic equation has thus been transformed to

$$\left(x + \frac{b}{2a}\right)^2 - \frac{b^2 - 4ac}{4a^2} = 0$$

or

$$\left(x + \frac{b}{2a}\right)^2 = \frac{b^2 - 4ac}{4a^2}$$

4 The Early History

or

$$x + \frac{b}{2a} = \pm\sqrt{\frac{b^2 - 4ac}{4a^2}} = \frac{\pm\sqrt{b^2 - 4ac}}{2a}$$

Hence the general quadratic equation (1) has the two solutions

$$x_{1,2} = \frac{-b \pm \sqrt{b^2 - 4ac}}{2a}$$

It is clear that if a, b, and c are real numbers, then these two solutions are real and distinct when $b^2 - 4ac > 0$, they are real and identical when $b^2 - 4ac = 0$, and they are imaginary and distinct when $b^2 - 4ac < 0$. Another important fact to bear in mind (Exercises 5, 6) is that

$$x_1 + x_2 = -\frac{b}{a} \quad \text{and} \quad x_1 x_2 = \frac{c}{a}$$

from which it follows that it is easy to construct a quadratic equation whose roots are prespecified. As we will have several occasions to refer to these identities later, they are stated as a proposition whose proof is relegated to Exercise 14.

■ **PROPOSITION 1.1.** *For any two numbers r and s the quadratic equation*

$$x^2 - (r + s)x + rs = 0$$

has r and s as its roots.

It is reasonable at this point to raise the ante and ask for a formula that will yield the solution of the general cubic equation

$$ax^3 + bx^2 + cx + d = 0 \tag{3}$$

There are indications that the Mesopotamians already tried to systematize the search for solutions of cubic equations, and we know for a fact that the Greeks attempted the same. As was mentioned above, the final breakthrough did not occur until the middle of the sixteenth century, when it was shown that a solution of the equation

$$x^3 + px + q = 0$$

is given by the expression

$$x = \sqrt[3]{-\frac{q}{2} + \sqrt{\frac{q^2}{4} + \frac{p^3}{27}}} - \sqrt[3]{\frac{q}{2} + \sqrt{\frac{q^2}{4} + \frac{p^3}{27}}} \qquad (4)$$

As we will see later, very little additional work is required to pass from this formula on to a formula for the general cubic equation (3), and so formula (4) can be considered as the crucial step, even though it does not yield the solution to the most general cubic equation.

In analogy with the ancient solutions of the quadratic, this solution was obtained by a geometrical process of completing the cube. Excerpts from Cardano's description of the solution are contained in Appendix B. A modern derivation of this formula appears in Chapter 3, and we restrict ourselves here to the examination of some instructive applications of formula (4). Surprisingly this formula raises some very interesting questions.

Consider the cubic equation $x^3 - 1 = 0$. Here $p = 0$ and $q = -1$, so formula (4) yields

$$x = \sqrt[3]{\frac{1}{2} + \sqrt{\frac{1}{4} + 0}} - \sqrt[3]{-\frac{1}{2} + \sqrt{\frac{1}{4} + 0}}$$

$$= \sqrt[3]{\frac{1}{2} + \frac{1}{2}} - \sqrt[3]{-\frac{1}{2} + \frac{1}{2}} = 1$$

which is as it should be. However, for the equation $x^3 + 6x - 20 = 0$, which Cardano uses as an illustration in his *Ars Magna*, the same formula yields the solution

$$x = \sqrt[3]{10 + \sqrt{100 + 8}} - \sqrt[3]{-10 + \sqrt{100 + 8}}$$

$$= \sqrt[3]{\sqrt{108} + 10} - \sqrt[3]{\sqrt{108} - 10}$$

It can be easily verified with the aid of a calculator that the above solution agrees with 2 to at least 8 decimal places, and the mathematical verification that the agreement is absolute is left to Exercise 1. Our purpose in presenting this example was to draw attention to the possibility that formula (4) may present a correct solution in an unnecessarily complicated form. This obfuscation becomes much more disturbing in the case of the equation $x^3 - 15x - 4 = 0$, treated by Rafael Bombelli in his *Algebra* (1572). Formula (4) yields the solution

$$x = \sqrt[3]{2 + \sqrt{-121}} - \sqrt[3]{-2 + \sqrt{-121}} \qquad (5)$$

6   The Early History

However, it is easily verified by inspection that $x = 4$ is also a solution of this cubic, and since

$$x^3 - 15x - 4 = (x - 4)(x^2 + 4x + 1)$$

two more solutions of the original equation are obtained by solving the quadratic

$$x^2 + 4x + 1 = 0$$

Since the solutions of this quadratic are $-2 \pm \sqrt{3}$, we are faced with the question of which of the three numbers 4 or $-2 \pm \sqrt{3}$ is disguised as expression (5)? Moreover this complicated expression involves square roots of negative numbers, in other words, imaginary quantities, whereas 4 and $-2 \pm \sqrt{3}$ are all real numbers. This apparent paradox was resolved by Bombelli who simplified expression (5) by setting

$$\sqrt[3]{2 \pm \sqrt{-121}} = a \pm b\sqrt{-1}$$

cubing both sides and deriving $a = 2$ and $b = 1$ from the resulting simultaneous equations. Rather than exhibit the details of his solution, we simply point out that indeed

$$(2 + \sqrt{-1})^3 = 2^3 + 3 \cdot 2^2 \cdot \sqrt{-1} + 3 \cdot 2 \cdot (\sqrt{-1})^2 + (\sqrt{-1})^3$$
$$= 8 + 12\sqrt{-1} - 6 - \sqrt{-1}$$
$$= 2 + 11\sqrt{-1} = 2 + \sqrt{-121}$$

and similarly

$$(-2 + \sqrt{-1})^3 = -2 + \sqrt{-121}$$

Consequently

$$\sqrt[3]{2 + \sqrt{-121}} - \sqrt[3]{-2 + \sqrt{-121}} = 2 + \sqrt{-1} - (-2 + \sqrt{-1})$$
$$= 4$$

Thus users of the cubic formula ignore the so-called imaginary numbers at their pe..i... Such prejudices come at the cost of losing some real solutions to real equations. This is further born out by the innocent-looking equation $x^3 - 3x = 0$. Formula (4) yields the solution

$$x = \sqrt[3]{\sqrt{-1}} - \sqrt[3]{\sqrt{-1}}$$

Even if one is very skeptical about the existence of imaginary quantities, it is very tempting to believe in them..just long enough for the above radicals to cancel out and to yield the root $x = 0$, which we know to be correct.

The solution to the cubic equation is the context within which imaginary numbers were first discussed by mathematicians. Cardano toyed with them and then rejected them as useless. Bombelli gave them more credence, but it was not until about two hundred years later that the work of L. Euler, and later that of C. F. Gauss, A. L. Cauchy, and N. H. Abel, turned the complex number system, consisting of both the real and imaginary numbers, into an indispensible tool for all mathematical researchers.

The Ferro-Tartaglia-Cardano formula suffers from a serious deficiency. This formula yields at most one solution for any cubic equation, even when such an equation is known to have three distinct real roots, as is the case for $x^3 - x = 0$ whose roots are $0, \pm 1$. In view of the fact that the quadratic formula of (1) does succeed in incorporating all the solutions into one expression, it would not seem unreasonable to expect the same of the cubic counterpart. As we will see in the next chapter, the complex numbers will enable us to find just such an expression.

EXERCISES 1.1

1. Prove that $\sqrt[3]{\sqrt{108} + 10} - \sqrt[3]{\sqrt{108} - 10} = 2$.
2. Prove that $\sqrt{28 - 10\sqrt{3}} - \sqrt{7 - 4\sqrt{3}} = 3$.
3. Solve the equation $3x^2 - 2x - 2 = 0$.
4. Solve the equation $x^4 - 3x^2 + 2 = 0$.

If r and s are the roots of the quadratic equation $ax^2 + bx + c = 0$, prove the identities in Exercises 5–7.

5. $r + s = -\dfrac{b}{a}$.
6. $rs = \dfrac{c}{a}$.
7. $r^2 + s^2 = \dfrac{b^2 - 2ac}{a^2}$.

If r and s are the roots of the quadratic equation $ax^2 + bx + c = 0$, rewrite the expressions in Exercises 8–13 in terms of a, b, and c. (Wherever necessary, you may assume that the denominators are not zero.)

8. $\dfrac{1}{r} + \dfrac{1}{s}$
9. $r^3 + s^3$
10. $r^2 s + r s^2$
11. $(r - s)^2$
12. $\dfrac{1}{r^2} + \dfrac{1}{s^2}$
13. $\dfrac{1}{r^2 s} + \dfrac{1}{r s^2}$

14. Prove Proposition 1.1.
15. If r and s are the roots of the equation $x^2 + px + q = 0$, what is the quadratic equation whose roots are $r + s$ and rs.

8 The Early History

16. If $r, s \neq 0$ are the roots of the equation $x^2 + px + q = 0$, what is the quadratic equation whose roots are $1/r$ and $1/s$.
17. For what real values of α are the roots of the equation $x^2 + \alpha x + \alpha = 0$ real?
18. For what values of m will the equation $x^2 - 2x(1 + 3m) + (3 + 2m) = 0$ have equal roots?

CHAPTER SUMMARY

This introductory chapter was used to briefly review the solutions of the first- and second-degree equations in a single unknown. The history of the solution of the cubic equation was also discussed, and the relationship of this formula to the complex number system was examined.

Chapter Review Exercises

Mark the following true or false.

1. Every real number is the solution of some equation.
2. Every pair of real numbers is the solution set of some quadratic equation.
3. Every equation has at least one solution.

New Terms

Complex numbers	7
Cubic equation	4
First-degree equation	2
Imaginary numbers	6
Method of false position	2
Quadratic equation	3

CHAPTER 2

Complex Numbers

Throughout history the introduction of new numbers has been greeted with considerable resistance on the part of mathematicians. Legend has it that the discoverer of irrational numbers was rewarded by being drowned by his fellow Greeks. Be that as it may, the fact is that these numbers have been tagged with the pejorative label of *irrational*, a word that, when used in nonmathematical contexts, has definite derogatory connotations. The same of course applies to the *negative* numbers. The *imaginary* numbers have been cursed with what is arguably the worst nomenclature in mathematics. Given the considerable difficulties that the average students face in learning the rigorous discipline of mathematics, can they be blamed for balking at having to contend with quantities that mathematicians themselves admit are *imaginary*?

The best way to overcome people's resistance to a new concept is to convince them of its utility. Accordingly it will be shown that the widening of our field of operations to include the complex numbers greatly enhances the power of the Ferro-Tartaglia-Cardan cubic formula. Next the complex numbers will be used to solve some ruler and compass construction problems of plane geometry. Only in this chapter's last section will the issue of the existence of the complex numbers be addressed.

2.1 RATIONAL FUNCTIONS OF COMPLEX NUMBERS

Just as was done by the mathematicians of the eighteenth and nineteenth centuries, we assume here the existence of a number i, which has the property that

$$i^2 = -1$$

The rigorous proof of i's existence is deferred to Section 2.6. In the meantime the number i is to be treated just like a variable, with the sole additional stipulation that whenever i^2 occurs within an algebraic expression, it can be replaced by -1. A *complex number* is an expression of the form $a + bi$ where a and b are any real numbers. When $b \neq 0$ such a number is called *imaginary*, and when $b = 0$ it is said to be *real*. These complex numbers can be added and subtracted as polynomials. Thus

$$(5 - 3i) + (-2 + 5i) = 5 - 3i - 2 + 5i = 3 + 2i$$
$$(5 - 3i) - (-2 + 5i) = 5 - 3i + 2 - 5i = 7 - 8i$$

The multiplication of complex numbers also resembles that of polynomials, except that each occurrence of i^2 is replaced by -1. Thus

$$(5 - 3i)(-2 + 5i) = -10 + 25i + 6i - 15i^2$$
$$= -10 + 31i - 15(-1)$$
$$= -10 + 31i + 15 = 5 + 31i$$

The division of complex numbers mimics the well-known process of rationalizing denominators. Thus

$$\frac{5 - 3i}{-2 + 5i} = \frac{5 - 3i}{-2 + 5i} \cdot \frac{-2 - 5i}{-2 - 5i}$$
$$= \frac{-10 - 25i + 6i - 15}{(-2)^2 - (5i)^2}$$
$$= \frac{-25 - 19i}{4 + 25} = -\frac{25}{29} - \frac{19}{29}i$$

Surprisingly all of these arithmetical operations can be given very interesting visual, or geometric, interpretations. To accomplish this, we represent each complex number $a + bi$ by the point (a, b) of the Cartesian plane. The point (a, b) is called the *Cartesian representation* of the complex number $a + bi$. Given two complex numbers $a + bi$ and $c + di$, let their Cartesian representations be $P = (a, b)$ and $Q = (c, d)$ (Figure 2.1). Their sum

$$(a + bi) + (c + di) = (a + c) + (b + d)i$$

is represented by the point $R = (a + c, b + d)$. However,

$$\text{slope of } PR = \frac{(b + d) - b}{(a + c) - a} = \frac{d}{c} = \text{slope of } OQ$$

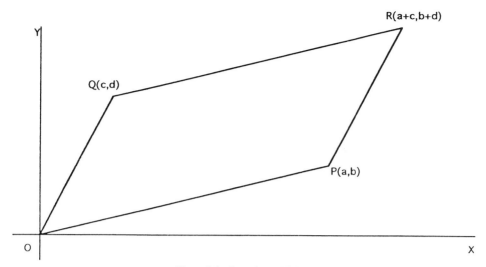

Figure 2.1. Complex addition.

and

$$\text{slope of } QR = \frac{(b+d)-d}{(a+c)-c} = \frac{b}{a} = \text{slope of } OP$$

Consequently $PR \parallel OQ$ and $QR \parallel OP$, and so $OPRQ$ is a parallelogram. Thus we see that the addition of complex numbers resembles that of vectors. These considerations are summarized as follows:

■ **PROPOSITION 2.1.** *Let O denote the origin of the Cartesian plane, and let P and Q be the Cartesian representations of the complex numbers $a + bi$ and $c + di$, respectively. If the sum of the two complex numbers is represented by the point R, then the quadrilateral OPRQ is a parallelogram.*

To give the multiplication of complex numbers a visual interpretation, it is convenient to begin by establishing some conventions. In the sequel the general complex number $a + bi$ will frequently be abbreviated as z. If either a or b is 0, it is omitted from $a + bi$. Thus

$$3 + 0i = 3 \quad \text{and} \quad 0 - 5i = -5i$$

Let $P = (a, b)$ be the Cartesian representation of the complex number $z = a + bi$ (Figure 2.2). The *modulus* of z, denoted by $|z|$, is the length of the line

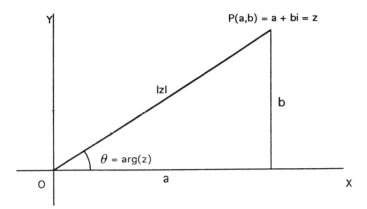

Figure 2.2. The argument and the modulus.

segment OP. In other words,

$$|a + bi| = \sqrt{a^2 + b^2}$$

Thus

$$|2 + 3i| = \sqrt{2^2 + 3^2} = \sqrt{13}$$
$$|3 - 4i| = \sqrt{3^2 + (-4)^2} = 5$$
$$|-2i| = \sqrt{0^2 + (-2)^2} = 2$$

It is clear that the modulus of the real number $a + 0i$ is just its absolute value. Thus the modulus should be regarded as the extension of the notion of absolute value to the complex numbers. The *argument* of $z = a + bi$, denoted $arg(z)$, is the counterclockwise angle from the positive x-axis to the ray OP where P is the Cartesian representation of z. As can be seen in Figure 2.3, the arguments of 3, $1 + i$, $3i$, -2, $-2 - 2i$, and $3 - 3i$ are 0, $\pi/4$, $\pi/2$, π, $5\pi/4$, and $7\pi/4$, respectively. For our purposes here, it is convenient to identify angles whose measures differ by the full angle of 2π. Thus it will be convenient sometimes to regard 1 as having argument 2π or 4π rather than 0. The reasons for this will become clear after we have discussed the geometrical interpretation of the multiplication of complex numbers.

The argument θ of the general complex number $z = a + bi$ is easily computed from the reltation (Figure 2.2)

$$\tan \theta = \frac{b}{a}$$

2.1 Rational Functions of Complex Numbers 13

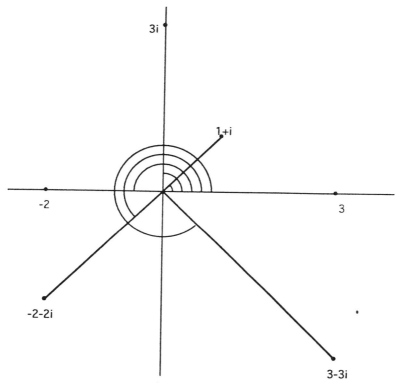

Figure 2.3. Some complex numbers.

but the quadrant in which z lies must be taken into account. Thus

$$\arg(1+i) = \arctan\left(\frac{1}{1}\right) = \frac{\pi}{4}$$

whereas

$$\arg(-2-2i) = \pi + \arctan\left(\frac{-2}{-2}\right) = \frac{5\pi}{4}$$

Observe that if the complex number $z = a + bi$ has argument θ, then, by Figure 2.2,

$$z = a + bi = |z|\left(\frac{a}{|z|} + \frac{b}{|z|}i\right) = |z|(\cos\theta + i\sin\theta)$$

We refer to $|z|(\cos\theta + i\sin\theta)$ as the *polar form* of z. The complex numbers $1 + i$, 5, i, $-2i$ have polar forms $\sqrt{2}(\cos\pi/4 + i\sin\pi/4)$, $5(\cos 0 + i\sin 0)$, $\cos\pi/2 + i\sin\pi/2$, and $2(\cos 3\pi/2 + i\sin 3\pi/2)$, respectively. On the other hand, the number $3 + 4i$ has polar form $5(\cos\alpha + i\sin\alpha)$, where $\alpha = \arctan(4/3) \approx 53.13°$.

Just as the addition of complex numbers has a geometrical interpretation in terms of Cartesian coordinates, so their multiplication can be easily visualized in terms of their polar forms. Let the complex numbers z and w be given in terms of their polar forms

$$z = |z|(\cos\theta + i\sin\theta) \quad \text{and} \quad w = |w|(\cos\phi + i\sin\phi)$$

The trigonometric formulas for the functions of the sums of two angles yield

$$zw = |z||w|(\cos\theta + i\sin\theta)(\cos\phi + i\sin\phi)$$
$$= |z||w|[(\cos\theta\cos\phi - \sin\theta\sin\phi) + i(\cos\theta\sin\phi + \sin\theta\cos\phi)]$$
$$= |z||w|[\cos(\theta + \phi) + i\sin(\theta + \phi)]$$

Thus the product zw has polar form $|z||w|[\cos(\theta + \phi) + i\sin(\theta + \phi)]$. Since it follows from Figure 2.2 that every complex number is completely determined by its argument and modulus, we conclude that

$$\arg(zw) = \theta + \phi \quad \text{and} \quad |zw| = |z||w|$$

Hence we have proved the following theorem:

■ **THEOREM 2.2.** *Let z and w be any two complex numbers. Then*

$$\arg(zw) = \arg(z) + \arg(w) \quad \text{and} \quad |zw| = |z||w|$$

If $z = i$ and $w = 1 + i$, then $zw = i(1 + i) = -1 + i$. Consequently

$$\arg(z) + \arg(w) = \frac{\pi}{2} + \frac{\pi}{4} = \frac{3\pi}{4} = \arg(zw)$$

and

$$|z||w| = 1 \cdot \sqrt{2} = \sqrt{2} = |zw|$$

Angles whose measures differ by integer multiples of 2π are considered to be identical. Thus the number i has all the following as its arguments

$$\ldots, \frac{-7\pi}{2}, \frac{-3\pi}{2}, \frac{\pi}{2}, \frac{5\pi}{2}, \frac{9\pi}{2}, \ldots$$

This is necessitated by such observations as the fact that

$$\frac{\pi}{2} = \arg(i) = \arg[(-1)(-i)] = \arg(-1) + \arg(-i)$$

$$= \pi + \frac{3\pi}{2} = \frac{5\pi}{2}$$

This fact, which appears to be a nuisance at this point, will in fact turn out to be very useful in the next section. Some more light will be shed on this in Section 9.4.

Theorem 2.2 is commonly referred to as the *argument principle*. It has many interesting and useful consequences. For example, it clearly implies that

$$\arg(z^2) = 2\arg(z) \quad \text{and} \quad |z^2| = |z|^2$$

Similarly

$$\arg(z^3) = 3\arg(z) \quad \text{and} \quad |z^3| = |z|^3$$

In fact a simple induction procedure yields the following observation:

■ **COROLLARY 2.3.** *If z is any nonzero complex number and k is any positive integer, then*

$$\arg(z^k) = k\arg(z) \quad \text{and} \quad |z^k| = |z|^k$$

This observation can be put to good use in computing large powers of complex numbers. Consider the problem of computing $(1 + i)^{100}$. By Corollary 2.3,

$$\arg[(1+i)^{100}] = 100\arg(1+i) = 100 \cdot \frac{\pi}{4} = 25\pi = \pi$$

and

$$|(1+i)^{100}| = |1+i|^{100} = (\sqrt{2})^{100} = 2^{50}$$

Hence

$$(1+i)^{100} = 2^{50}(\cos\pi + i\sin\pi) = -2^{50}$$

Corollary 2.3 also holds for nonpositive exponents if we define

$$z^0 = 1 \quad \text{for all } z$$

16 Complex Numbers

and

$$z^{-k} = \left(\frac{1}{z}\right)^k \quad \text{for } z \neq 0, \, k = 1, 2, 3, \ldots$$

The proof of this fact is relegated to Exercise 29. Exercise 28 calls for proving that integer powers of complex numbers obey the same rules as do the more familiar powers of real numbers, to wit,

$$z^m z^n = z^{m+n} \quad \text{and} \quad (z^m)^n = z^{mn}$$

EXERCISES 2.1

Find the argument and modulus of each of the complex numbers in Exercises 1–4.

1. $2 + 3i$ 2. $3 - 2i$ 3. $-3 - 4i$ 4. $-1 + 7i$

Express the complex quantities in Exercises 5–21 in the form $a + bi$, where a and b are real numbers.

5. $(2 + 3i) + (5 - i)$
6. $(17 - 3i) + (2 + 3i)$
7. $(2 + 3i) - (5 - i)$
8. $(17 - 3i) - (2 + 3i)$
9. $(2 + 3i)(5 - i)$
10. $(17 - 3i)(2 + 3i)$
11. $\dfrac{2 + 3i}{5 - i}$
12. $\dfrac{17 - 3i}{2 + 3i}$
13. $\dfrac{\sqrt{3} + 5i}{2 - \sqrt{3}i}$
14. $\dfrac{a + bi}{a - bi} - \dfrac{a - bi}{a + bi}$
15. $\dfrac{(2 - i)^2}{1 + i}$
16. $(1 + i)^4$
17. $(1 - 2i)^4$
18. $(1 - i)^{63}$
19. i^{4321}
20. $\left(\dfrac{1 - i}{1 + i}\right)^{127}$
21. i^{4n+3} (n is an integer)

Solve the following equations in Exercises 22–25 for z and w:

22. $(1 + 2i)z + 5 = 0$
23. $(1 + i)z + 5i = \dfrac{z}{1 - i} - 2$
24. $iz - w = 1 + i$ and $(1 + i)z + iw = 1$
25. $(1 - i)z + iw = i$ and $2z - (2 + i)w = 1$
26. Prove that if z and w are any two complex numbers, then

$$|z + w| \leq |z| + |w|$$

27. If $z = a + bi$ is any complex number, where a and b are real, we define $\bar{z}$, the *conjugate* of z, to be $a - bi$.
   (a) Find the relation between the moduli and arguments of z and $\bar{z}$.
   (b) Prove that $\overline{z + w} = \bar{z} + \bar{w}$, $\overline{zw} = \bar{z}\bar{w}$, and $\overline{z^{-1}} = \bar{z}^{-1}$.

(c) If a, b, c, and d are any real numbers, prove that z is a root of the equation $ax^3 + bx^2 + cx + d = 0$ if and only if $\bar{z}$ is also a root of the same equation.

28. Prove that if z is any complex number and m and n are any integers, then $z^m z^n = z^{m+n}$ and $(z^m)^n = z^{mn}$.

29. Prove that Corollary 2.3 holds for every integer k.

30. Prove that the three distinct complex numbers z_1, z_2, z_3 are collinear if and only if there exists a real number λ such that $z_2 = (1 - \lambda)z_1 + \lambda z_3$. [Hint: Examine the expression $(z_2 - z_1)/(z_3 - z_1)$.]

31. Prove that if z and w are two complex numbers, then the distance between them equals $|z - w|$.

32. Prove that the midpoint of the line segment joining the complex numbers z and w is

$$\frac{z + w}{2}$$

33. Prove that the four complex numbers z_1, z_2, z_3, z_4 lie on either a common straight line or a common circle if and only if the following number is real:

$$\frac{(z_1 - z_3)/(z_1 - z_2)}{(z_4 - z_3)/(z_4 - z_2)}$$

34. Let z_1, z_2, z_3, z_4 be four complex numbers such that $|z_1| = |z_2| = |z_3| = |z_4| = 1$. Prove that z_1, z_2, z_3, z_4 form a rectangle if and only if $z_1 + z_2 + z_3 + z_4 = 0$.

35. Prove that the center of gravity of the triangle whose vertices are the complex numbers z_1, z_2, z_3 is

$$\frac{z_1 + z_2 + z_3}{3}$$

(Hint: Recall that the center of gravity of a triangle coincides with the intersection of its three medians.)

36. Prove that if $|\zeta| = 1$, then there is a real number b such that

$$\frac{1 + \zeta}{1 - \zeta} = bi$$

37. Prove that if $z = a + bi$, where a and b are real, then

$$\frac{|a| + |b|}{\sqrt{2}} \leqslant |z| \leqslant |a| + |b|$$

2.2 COMPLEX ROOTS

In the previous section the four arithmetical operations were extended to complex numbers. Next we examine the process of finding roots of complex numbers. What, for example, is $\sqrt{i}$? Before addressing this question, it behooves us to recall that even $\sqrt{1}$ involves some ambiguities. Sometimes it is 1 and sometimes it is -1, or both ± 1. We therefore define $\sqrt[n]{z}$, for any complex number z and for any positive integer n, to be the set of all the complex numbers w such that

$$w^n = z$$

Returning to $\sqrt{i}$, let j be any complex number such that $j^2 = i$. Then $2\arg(j) = \arg(i)$ and $|j|^2 = |i| = 1$. Consequently, using $\arg(i) = \pi/2$,

$$\arg(j) = \frac{1}{2}\arg(i) = \frac{\pi}{4}$$

and since $|j|$ is, by definition, positive,

$$|j| = 1$$

Thus

$$j = 1\left(\cos\frac{\pi}{4} + i\sin\frac{\pi}{4}\right) = \frac{\sqrt{2}}{2} + \frac{\sqrt{2}}{2}i$$

Another square root of i is of course

$$-j = -\frac{\sqrt{2}}{2} - \frac{\sqrt{2}}{2}i$$

An alternate method for arriving at $-j$ is to recall that $\arg(i)$ could also have been taken as $5\pi/2$, in which case we obtain the square root

$$1\left(\cos\frac{5\pi}{4} + i\sin\frac{5\pi}{4}\right) = -\frac{\sqrt{2}}{2} - \frac{\sqrt{2}}{2}i = -j$$

It can be easily verified by direct calculations that

$$\left(\frac{\sqrt{2}}{2} + \frac{\sqrt{2}}{2}i\right)^2 = i = \left(-\frac{\sqrt{2}}{2} - \frac{\sqrt{2}}{2}i\right)^2$$

This procedure yields three different values for $\sqrt[3]{1}$. Taking as the argument of 1 the successive values $0, 2\pi, 4\pi, 6\pi, \ldots$, each of the elements of $\sqrt[3]{1}$ must have as its argument one of the values $0, 2\pi/3, 4\pi/3, 2\pi, \ldots$. The modulus of 1 being 1, it follows that the modulus of $\sqrt[3]{1}$ must be the real cube root of 1, which is

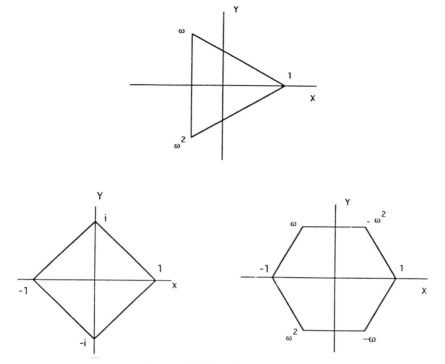

Figure 2.4. Complex roots of unity.

also 1. Hence we get as our cube roots of 1 the following numbers:

$$1(\cos 0 + i \sin 0) = 1 \cdot 1 + i \cdot 0 = 1$$

$$1\left(\cos \frac{2\pi}{3} + i \sin \frac{2\pi}{3}\right) = -\frac{1}{2} + \frac{\sqrt{3}}{2} i$$

$$1\left(\cos \frac{4\pi}{3} + i \sin \frac{4\pi}{3}\right) = -\frac{1}{2} - \frac{\sqrt{3}}{2} i$$

$$1(\cos 2\pi + i \sin 2\pi) = 1 \ldots$$

It is clear that this list of cube roots of 1 will cycle through the same three values. The second root, the one with argument $2\pi/3$, is denoted by ω. By Corollary 2.3, the third root, having double the argument of ω and the same modulus of 1, equals ω^2. Note that the Cartesian representations of these three complex cube roots of 1 form an equilateral triangle in the Cartesian plane (Figure 2.4).

The same procedure yields all the nth roots of 1 for each positive integer n.

20 Complex Numbers

■ **THEOREM 2.4.** *Let n be a positive integer, and let*

$$\zeta = \cos\frac{2\pi}{n} + i\sin\frac{2\pi}{n}$$

Then

$$\sqrt[n]{1} = \{1, \zeta, \zeta^2, \zeta^3, \ldots, \zeta^{n-1}\}$$

Proof. By Corollary 2.3,

$$\zeta^n = \cos 2\pi + i\sin 2\pi = 1$$

and so ζ is indeed one of the elements of $\sqrt[n]{1}$. Moreover, for any integer k,

$$(\zeta^k)^n = (\zeta^n)^k = 1^k = 1$$

and hence each ζ^k is indeed an nth root of unity. Since $\arg(\zeta^k) = 2\pi k/n$, it follows that the numbers $1, \zeta, \zeta^2, \ldots, \zeta^{n-1}$ all have distinct arguments, and so they are all distinct complex numbers. The remainder of the proof, that there are no other roots of unity, is relegated to Exercise 29. We note that this also follows from the fact (see Corollary 6.7 below) that a polynomial equation of degree n has at most n roots. ■

The nth root of unity

$$\zeta = \cos\frac{2\pi}{n} + i\sin\frac{2\pi}{n}$$

will be referred to as the *first nth root of unity*. Since the angle subtended by the consecutive roots ζ^{k+1} and ζ^k at the origin equals

$$\arg\left(\frac{\zeta^{k+1}}{\zeta^k}\right) = \arg(\zeta) = \frac{2\pi}{n}$$

and is independent of k, it follows that the elements of $\sqrt[n]{1}$ form a regular n-gon that is centered at the origin. This fact is crucial for the next section, and so we state it as a proposition.

■ **PROPOSITION 2.5.** *For any fixed integer $n \geq 3$, the Cartesian representations of the elements of $\sqrt[n]{1}$ form the vertices of a regular n-gon.*

Since $2\pi/4 = \pi/2$ and

$$\cos\frac{\pi}{2} + i\sin\frac{\pi}{2} = 0 + i = i$$

it follows that

$$\sqrt[4]{1} = \{1, i, i^2, i^3\} = \{1, i, -1, -i\}$$

which form the vertices of a square (Figure 2.4).
Similarly, since $2\pi/6 = \pi/3$ and

$$\cos\frac{\pi}{3} + i\sin\frac{\pi}{3} = \frac{1}{2} + i\frac{\sqrt{3}}{2} = -\omega^2$$

it follows that

$$\begin{aligned}\sqrt[6]{1} &= \{1, -\omega^2, (-\omega^2)^2, (-\omega^2)^3, (-\omega^2)^4, (-\omega^2)^5\} \\ &= \{1, -\omega^2, \omega, -1, \omega^2, -\omega\}\end{aligned}$$

which form the vertices of the regular hexagon of Figure 2.4.

The following proposition is both a natural extension and a corollary of Theorem 2.4. Since our main interest lies in the roots of unity, its proof is omitted and relegated to Exercise 18. The subsequent example clarifies the purport of the proposition.

■ **PROPOSITION 2.6.** *Let n be a positive integer and z any nonzero complex number with argument θ. If*

$$\zeta = \cos\frac{2\pi}{n} + i\sin\frac{2\pi}{n}$$

then

$$\sqrt[n]{z} = \left\{\sqrt[n]{|z|}\left(\cos\frac{\theta}{n} + i\sin\frac{\theta}{n}\right)\zeta^k \,\bigg|\, k = 0, 1, 2, \ldots, n-1\right\}$$

where $|\sqrt[n]{z}|$ denotes the common modulus of all the elements of $\sqrt[n]{z}$.

Since $-1 + i$ has modulus $\sqrt{2}$ and argument $3\pi/4$, and since

$$\cos\frac{\pi}{4} + i\sin\frac{\pi}{4} = \frac{1}{\sqrt{2}}(1 + i)$$

it follows that

$$\sqrt[3]{-1+i} = \left\{ \sqrt[3]{\sqrt{2}} \left(\frac{1+i}{\sqrt{2}} \right), \sqrt[3]{\sqrt{2}} \left(\frac{1+i}{\sqrt{2}} \right) \omega, \sqrt[3]{\sqrt{2}} \left(\frac{1+i}{\sqrt{2}} \right) \omega^2 \right\}$$

$$= \left\{ \frac{1+i}{2^{1/3}}, \frac{1+i}{2^{1/3}} \omega, \frac{1+i}{2^{1/3}} \omega^2 \right\}$$

It should be noted here that calculators are very handy in computing complex roots too. Thus, to compute $\sqrt[4]{2+3i}$, we note that

$$\sqrt[4]{|2+3i|} = \sqrt[4]{\sqrt{2^2+3^2}} = \sqrt[8]{13} \approx 1.378$$

and

$$\arg(\sqrt[4]{2+3i}) = \tfrac{1}{4} \arg(2+3i) = \tfrac{1}{4} \arctan(\tfrac{3}{2}) \approx 14.077°$$

Hence, if we set

$$w = \cos[\tfrac{1}{4} \arctan(\tfrac{3}{2})] + i \sin[\tfrac{1}{4} \arctan(\tfrac{3}{2})] \approx .970 + .243i$$

then

$$\sqrt[4]{2+3i} \approx 1.378 \{w, wi, -w, -wi\}$$
$$\approx 1.378\{.970 + .243i, -.243 + .970i, -.970 - .243i, .243 - .970i\}$$
$$\approx \{1.336 + .335i, -.335 + 1.336i, -1.336 - .335i, .335 - 1.336i\}$$

The solution of the quadratic equation detailed in Chapter 1 works for complex coefficients as well. Accordingly the roots of the equation $iz^2 + 2z - 2i = 0$ are

$$\frac{-2 \pm \sqrt{2^2 - 4 \cdot i \cdot (-2i)}}{2i} = \frac{-2 \pm \sqrt{-4}}{2i} = \frac{-2 \pm 2i}{2i} = i \pm 1$$

We conclude this section with a curious fact that will shortly prove unexpectedly useful.

■ **PROPOSITION 2.7.** *For any fixed integer $n > 1$, the sum of the elements of $\sqrt[n]{1}$ is 0.*

Proof. Let ζ be the first nth root of unity. By Theorem 2.4 the elements of $\sqrt[n]{1}$ can be listed as

$$1, \zeta, \zeta^2, \ldots, \zeta^{n-1}$$

The formula for the geometric progression now yields

$$1 + \zeta + \zeta^2 + \zeta^3 + \cdots + \zeta^{n-1} = \frac{1 - \zeta^n}{1 - \zeta} = \frac{1 - 1}{1 - \zeta} = 0 \qquad \blacksquare$$

EXERCISES 2.2

Express each of the elements of the sets in Exercises 1–12 in the form $a + bi$, where a and b are real numbers.

1. $\sqrt[8]{1}$
2. $\sqrt[12]{1}$
3. $\sqrt[6]{-1}$
4. $\sqrt[4]{-i}$
5. $\sqrt{3 - 4i}$
6. $\sqrt{8 - 30i}$
7. $\sqrt[3]{1 + i}$
8. $\sqrt[3]{i}$
9. $\sqrt[3]{-i}$
10. $\sqrt[5]{100 - 37i}$
11. $\sqrt{c^2 - 1 + 2ci}$
12. $\sqrt{4cd - 2(c^2 - d^2)i}$

13. Resolve the following paradox:

$$1 = \sqrt{1} = \sqrt{(-1)(-1)} = \sqrt{-1}\sqrt{-1} = i \cdot i = i^2 = -1$$

Find the complex solutions of the equations in Exercises 14–17.

14. $z^2 - 6z + 9 + 2i = 0$
15. $(1 - 2i)z^2 + 2z + 1 = 0$
16. $z^2 - (1 + i)z + 5i = 0$
17. $z^2 + 3(1 + i)z - (2 - 3i) = 0$
18. Prove Proposition 2.6.

Prove the identities in Exercises 19–22.

19. $(1 + \omega^2)^{16} = \omega$.
20. $(3 + 5\omega + 3\omega^2)^9 = 512$.
21. $a^3 + b^3 = (a + b)(a + b\omega)(a + b\omega^2)$.
22. $(a + b + c)(a + b\omega + c\omega^2)(a + b\omega^2 + c\omega) = a^3 + b^3 + c^3 - 3abc$.

If ζ is an nth root of unity, simplify ζ^k for the values of n and k specified in Exercises 23–26.

23. $n = 10, k = 135$
24. $n = 135, k = 999$
25. $n = 999, k = 12{,}345$
26. $n = 12{,}345, k = 10^6$
27. Prove that $z \in \sqrt[n]{1}$ if and only if $\bar{z} \in \sqrt[n]{1}$.
28. Prove that for every positive integer $n > 1$,

$$\sum_{k=1}^{n} \cos \frac{2\pi k}{n} = 0 = \sum_{k=1}^{n} \sin \frac{2\pi k}{n}$$

29. Complete the proof of Theorem 2.4 by showing that the given list contains all the nth roots of 1.

24 Complex Numbers

30. Prove that the three complex numbers A, B, C form a counterclockwise equilateral triangle if and only if $A + B\omega + C\omega^2 = 0$. What equation characterizes clockwise equilateral triangles?
31. Suppose that ABC is any triangle in the plane. Let A', B', C', be three points such that $A'BC$, $AB'C$, ABC', are all clockwise (or all counterclockwise) equilateral triangles. Prove that the centers of triangles $A'BC$, $AB'C$, ABC', also form an equilateral triangle.

2.3 SOLVABILITY BY RADICALS (I)

We now have sufficient tools at our disposal to formalize the notion of an *algebraic solution* of an equation

$$a_0 x^n + a_1 x^{n-1} + \cdots + a_{n-1} x + a_n = 0$$

where $a_0, a_1, \ldots, a_n$ are any complex numbers. A *solution* of this equation is of course another complex number r such that

$$a_0 r^n + a_1 r^{n-1} + \cdots + a_{n-1} r + a_n = 0$$

The value of the solution r clearly depends on the coefficients $a_0, a_1 \ldots, a_n$, and the solution is said to be *algebraic* if this dependence involves only radicals and the four arithmetic operations. More precisely, let $\mathbb{Z}$ denote the set of integers, and let V be any set of complex numbers. The complex number z is said to have an *algebraic expression in* V if there exists a sequence of complex numbers $z_1, z_2, \ldots, z_n = z$ such that for each $i = 1, 2, \ldots, n$, either

$$z_i \in \sqrt[m]{z_{i-1}} \qquad \text{for some positive integer } m > 1$$

or else the number z_i is obtained by adding, subtracting, multiplying, or dividing some two elements of $\mathbb{Z} \cup V \cup \{z_1, z_2, \ldots, z_{i-1}\}$. Thus each of the solutions of the quadratic equation $ax^2 + bx + c = 0$ $(a \neq 0)$, namely

$$z = \frac{-b \pm \sqrt{b^2 - 4ac}}{2a}$$

has an algebraic expression in the quantities $\{a, b, c\}$ because it is possible to choose

$$z_1 = b^2, \quad z_2 = ac, \quad z_3 = 4z_2, \quad z_4 = z_1 - z_3$$
$$z_5 = \text{any element of } \sqrt{z_4}, \quad z_6 = -b + z_5,$$
$$z_7 = z_6/a, \quad z = z_8 = z_7/2$$

If the complex number z has an algebraic expression in V whose only radicals are square roots, then we say that z has a *degree 2 algebraic expression in V*. Thus the above solutions of the quadratic equation clearly have degree 2 radical expressions in $\{a, b, c\}$. If no radicals appear in an algebraic expression of z in V, then z is said to have a *rational expression in V*. For example, if $c \neq -1$, then

$$\frac{2 - ab}{1 + c}$$

has a rational expression in $\{a, b, c\}$ with $n = 4$, where

$$z_1 = ab, \quad z_2 = 2 - z_1, \quad z_3 = 1 + c, \quad \text{and} \quad z = z_4 = \frac{z_2}{z_1}$$

If z has an algebraic expression in V, and V happens to be empty, we say that z has an *algebraic expression in the integers*. Finally, we say that an equation

$$a_0 x^n + a_1 x^{n-1} + a_2 x^{n-2} + \cdots + a_n = 0$$

is *solvable by radicals* or *algebraically resolvable* if each of its roots has an algebraic expression in the coefficients $\{a_0, a_1, a_2, \ldots, a_n\}$. Thus the quadratic equation $ax^2 + bx + c = 0$ is solvable by radicals, as is easily verified by examining the quadratic formula above. The equation

$$x^4 + ax^3 + bx^2 + ax + 1 = 0 \tag{1}$$

is also solvable by radicals. To see this we observe that 0 is not a root of this equation so that it can be divided by x^2 and its terms can be regrouped as

$$x^2 + \frac{1}{x^2} + a\left(x + \frac{1}{x}\right) + b = 0$$

or

$$\left(x + \frac{1}{x}\right)^2 + a\left(x + \frac{1}{x}\right) + b - 2 = 0$$

Setting $u = x + 1/x$, we note that

$$u^2 + au + b - 2 = 0 \quad \text{and} \quad x^2 - ux + 1 = 0$$

Thus x has an algebraic expression in $\{u\}$, and u in turn has an algebraic expression in $\{a, b\}$. This of course means that every solution of equation (1) has an algebraic expression in its coefficients.

EXERCISES 2.3

Decide whether the expressions in Exercises 1–11 are rational or (degree 2) algebraic expressions in $\{x, y, z\}$.

1. $\dfrac{x + y - 1}{x - z + 2}$

2. $\dfrac{x + \sqrt{y - 1}}{x - z + 2}$

3. $\dfrac{x + y - 1}{x - z + \sqrt{2}}$

4. $(x + y)^{10^{10}}$

5. $10^{10^{10}}$

6. x^x

7. 10^x

8. $\dfrac{i}{\sqrt{x + \sqrt{2y} - \sqrt{3z}}}$

9. $\omega + \sqrt[16]{x + 123}$ (where ω is a cube root of unity)

10. $\sqrt[x]{y + z}$

11. $\sin x$

12. Explain why the solutions of the simultaneous equations

$$ax + by + c = 0$$
$$dx + ey + f = 0$$

have rational expressions in $\{a, b, c, d, e, f\}$ whenever $ae \neq bd$.

13. Explain why the solutions of the simultaneous equations

$$ax + by + c = 0$$
$$dx + ey + f = 0$$

need not have rational expressions in $\{a, b, c, d, e, f\}$ when $ae = bd$.

14. Explain why the solutions of the simultaneous equations

$$x^2 + y^2 + x + y + a = 0$$
$$x^2 + y^2 + 2x + 3y + b = 0$$

have degree 2 algebraic expressions in $\{a, b\}$.

15. Prove that the equation $x^4 + 3x^3 - x^2 + 3x + 1 = 0$ is algebraically resolvable.
16. Prove that the equation $x^6 - 2x^5 - 4x^3 - 2x + 1 = 0$ is algebraically resolvable.
17. Explain why the solutions of the equation $x^8 + ax^4 + 1 = 0$ have degree 2 algebraic expressions in $\{a\}$.
18. Explain why the solutions of the simultaneous equations

$$x^8 + y^8 = a \quad \text{and} \quad x^8 y^8 = b$$

have degree 2 algebraic expressions in $\{a, b\}$.

19. Explain why the solutions of the simultaneous equations

$$x^8 + y^8 = a \quad \text{and} \quad x^4 y^4 = b$$

have degree 2 algebraic expressions in $\{a, b\}$.

20. Explain why the solutions of the simultaneous equations

$$x^9 + y^9 = a \quad \text{and} \quad x^3 y^3 = b$$

have algebraic expressions in $\{a, b\}$.

21. Show that the solution of the equation

$$x^6 + x^5 + x^4 + x^3 + x^2 + x + 1 = 0$$

can be reduced to the solution of some quadratic and cubic equations.

22. Prove that the roots of the equation $(A + C - B)x^2 + 2Cx + (B + C - A) = 0$ have rational expressions in A, B, C.

23. Prove that the roots of the equation

$$ABC^2 x^2 + (3A^2 + B^2)Cx - (6A^2 + AB - 2B^2) = 0$$

have rational expressions in A, B, C.

2.4 RULER AND COMPASS CONSTRUCTIBILITY OF REGULAR POLYGONS

The ancient Greek mathematicians, who invented what we have come to call Euclidean geometry and the notion of a rigorous proof, bequeathed their successors a host of unsolved mathematical problems. Best known among these are the questions of whether it is possible to trisect an angle, double a cube, or square a circle by means of a compass and an unmarked ruler alone. Here we treat a lesser known, but equally natural, construction problem, namely, What regular polygons are constructible by ruler and compass alone? The other three problems are discussed informally at the end of the section.

The ruler and compass constructions of the equilateral triangle, the square, and the regular hexagon are standard fare in the high school curriculum. That the regular pentagon is also so constructible is true, but not so widely known. This is proved below in Proposition 2.12. A regular octagon is easily constructed by inscribing a square in a circle and then drawing the two diameters that are perpendicular to the sides of the square (Figure 2.5). In general, it is clear that given any regular n-gon it is possible to derive from it a regular $2n$-gon by drawing radii perpendicular to its sides. Hence the regular n-gon is construcible for $n = 2^{m+2}, 3 \cdot 2^m, 5 \cdot 2^m$ for $m = 0, 1, 2, \ldots$. If a regular pentagon and an equilateral triangle are inscribed in a circle so that they share a vertex, as in Figure 2.6, then arc AB is $2/5 - 1/3 = 1/15$ of the total circumference of the circle. It follows that the regular 15-sided polygon is also constructible by

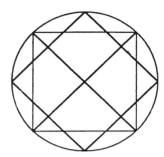

Figure 2.5. A square and a regular octagon.

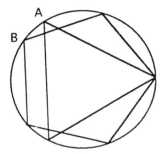

Figure 2.6. An equilateral triangle and a regular pentagon.

ruler and compass. This information is summarized as the following proposition:

■ **PROPOSITION 2.8.** *The regular n-sided polygon is constructible by ruler and compass for* $n = 3, 4, 5, 6, 8, 10, 12, 15, 16$.

In 1796 C. F. Gauss proved that the regular 17-sided polygon is also constructible by ruler and compass alone, and we will discuss in detail a proof that appears in his *Disquisitiones Arithmeticae* of 1801. Before that, however, it is necessary to make some general remarks about ruler and compass constructibility. We will work in the Cartesian plane, so it will be assumed that a certain line segment has been designated as the *unit segment*. For the sake of brevity it will henceforth be said that a configuration is *constructible* when it is constructible by ruler and compass alone. A real number α is said to be *constructible* if it is possible to construct a line segment whose length is $|\alpha|$ times the length of the unit segment.

2.4 Ruler and Compass Constructibility of Regular Polygons 29

■ **PROPOSITION 2.9.** *The point (x, y) is constructible if and only if its coordinates x and y are constructible real numbers.*

Proof. The nature of Cartesian coordinates is such that it is possible to pass from a point to its coordinates, and vice versa, by means of straight lines that are perpendicular to the axes. Since such perpendiculars are well known to be constructible, we are done. ■

It is obvious that the number 1 is constructible, and it follows from elementary Euclidean geometry that if α and β are constructible real numbers, so are $\alpha \pm \beta$. In particular, every integer is constructible. The next lemma will provide us with a host of real constructible numbers.

■ **LEMMA 2.10.** *If α and β are nonzero constructible real numbers, then so are*

1. *their product $\alpha\beta$,*
2. *their quotient β/α,*
3. *the square root $\sqrt{|\alpha|}$.*

Proof. It is clear that we may restrict attention to positive α and β.

1. On the positive x and y axes (Figure 2.7), let B, C, D be points such that the lengths of the segments OB, OC, OD are 1, α, and β, respectively. Using a standard ruler and compass construction, draw through D a straight line that is parallel to BC and intersects OY at E. Since $\triangle OBC$ and $\triangle ODE$ are similar, it follows that

$$\frac{OD}{OB} = \frac{OE}{OC} \quad \text{or} \quad \frac{\beta}{1} = \frac{OE}{\alpha}$$

and so the constructible line segment OE has length $\alpha\beta$.

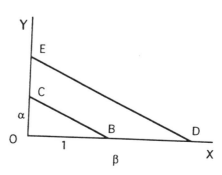

Figure 2.7. The multiplication of constructible numbers.

30 Complex Numbers

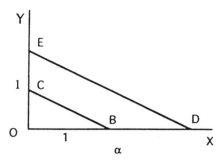

Figure 2.8. The reciprocal of a constructible number.

2. In view of part 1 above, it suffices to show that $1/\alpha$ is constructible. On the positive x and y axes (Figure 2.8), let B, D, E be points such that the lengths of the segments OB, OD, OE are 1, α, 1, respectively. Join the points DE, and draw a line through B that is parallel to DE. If this line intersects OY at the point C, then, because of the similarity of $\triangle OBC$ and $\triangle ODE$, it follows that

$$\frac{OD}{OB} = \frac{OE}{OC} \quad \text{or} \quad \frac{\alpha}{1} = \frac{1}{OC}$$

and so the constructible line segment OC has length $1/\alpha$.

3. Let ADB be three collinear points such that the line segments AD and DB have lengths α and 1, respectively (Figure 2.9). Let C be the intersection of the line perpendicular to AB at the point D with the semicircle that has AB as its diameter (all these are well known to be constructible). The triangles ACD and CBD are right triangles, each of which also shares an acute angle with the right $\triangle ABC$. Thus $\triangle ACD$ and $\triangle CBD$ are both similar to $\triangle ABC$, and hence they are also similar to each other. Consequently

$$\frac{AD}{CD} = \frac{CD}{BD} \quad \text{or} \quad \frac{\alpha}{CD} = \frac{CD}{1}$$

and hence the constructible segment CD has length $\sqrt{\alpha}$. ∎

It follows from Lemma 2.10 that every rational number is constructible, as is every real number that has a degree 2 algebraic expression in the integers. A complex number is said to be *constructible* if its Cartesian representation is a constructible point. The above geometric proposition results in an algebraic description of some constructible complex numbers.

2.4 Ruler and Compass Constructibility of Regular Polygons

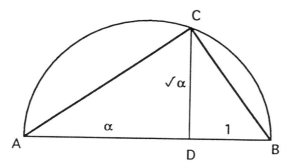

Figure 2.9. The square root of a constructible number.

■ **COROLLARY 2.11.** *If the complex number $z = x + iy$ has a degree 2 algebraic expression in the integers, then it is constructible.*

Proof. Because of the recursive nature of the definition of algebraic expressions, it suffices to show that if $z = x + iy$ and $w = u + iv$ are constructible complex numbers, then so are $z + w$, $z - w$, zw, z/w, and $\sqrt{z}$. However, if z and w are constructible complex numbers, then, by Proposition 2.9 and Lemma 2.10, so are

$$z \pm w = (x \pm u) + i(y \pm v)$$

$$zw = (xu - yv) + i(xv + yu)$$

$$\frac{1}{z} = \frac{x}{x^2 + y^2} - i\frac{y}{x^2 + y^2}$$

and

$$\frac{z}{w} = z\left(\frac{1}{w}\right)$$

To argue the constructibility of $\sqrt{z}$, we note that it consists of the intersections of the circle centered at the origin and of radius $\sqrt{|z|}$ with the straight line that bisects the argument of z. Since $\sqrt{|z|}$ is constructible by Lemma 2.10.3, and angle bisection is a well-known ruler and compass construction, we are done. ■

The converse of this corollary also holds, and its proof is relegated to Exercise 28.

Our approach to proving the constructibility of the regular pentagon and the regular 17-sided polygon is based on the observation that the elements of $\sqrt[n]{1}$ form a regular n-gon centered at the origin of the Cartesian plane. By

Corollary 2.11 it suffices to show that each of the vertices of these polygons, when regarded as a complex number, has a degree 2 algebraic expression in the integers.

■ **PROPOSITION 2.12.** *The regular pentagon is constructible by ruler and compass alone.*

Proof. Let ε denote the first fifth root of unity. Since the remaining roots are ε^2, ε^3, ε^4, and 1, it follows from Lemma 2.10.1 that it suffices to prove that ε has a degree 2 algebraic expression in the integers. By Proposition 2.7, ε satisfies the equation

$$\varepsilon + \varepsilon^2 + \varepsilon^3 + \varepsilon^4 = -1$$

Set

$$A = \varepsilon + \varepsilon^4 \quad \text{and} \quad B = \varepsilon^2 + \varepsilon^3$$

and note that

$$A + B = \varepsilon + \varepsilon^4 + \varepsilon^2 + \varepsilon^3 = -1$$

and

$$AB = (\varepsilon + \varepsilon^4)(\varepsilon^2 + \varepsilon^3) = \varepsilon^3 + \varepsilon^4 + \varepsilon^6 + \varepsilon^7 = \varepsilon^3 + \varepsilon^4 + \varepsilon + \varepsilon^2$$
$$= -1$$

Hence, by Proposition 1.1, A and B are the solutions of the quadratic equation

$$x^2 + x - 1 = 0$$

and therefore A has a degree 2 algebraic expression in the integers. On the other hand,

$$A = \varepsilon + \varepsilon^4 = \varepsilon + \frac{1}{\varepsilon}$$

so that $\varepsilon^2 - A\varepsilon + 1 = 0$, and hence ε is a solution of the quadratic equation

$$x^2 - Ax + 1 = 0$$

It follows that ε has a degree 2 algebraic expression in A, which in turn has a degree 2 algebraic expression in the integers. Hence, by Corollary 2.11, ε is constructible. ■

2.4 Ruler and Compass Constructibility of Regular Polygons 33

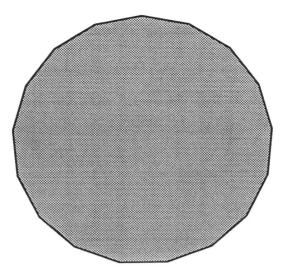

Figure 2.10. The regular 17-gon.

Exercise 1 calls for the explicit description of the coordinates of ε as degree 2 algebraic expressions in the integers. Exercise 29 calls for a ruler and compass construction of the regular pentagon. We now turn to the regular 17-sided polygon (Figure 2.10).

■ **THEOREM 2.13.** *The regular 17-sided polygon is ruler and compass constructible.*

Proof. Let

$$\zeta = \cos\frac{2\pi}{17} + i\sin\frac{2\pi}{17}$$

be the first 17th root of unity. Just as was the case for the regular pentagon, it suffices to show that ζ has a degree 2 algebraic expression in the integers. This will be accomplished by listing a sequence of elements, the last being ζ, such that each member of the sequence has a degree 2 algebraic expression in the previous ones, and the first has a degree 2 algebraic expression in the integers.

We begin with the not at all natural observation that by Proposition 2.7,

$$\zeta + \zeta^3 + \zeta^9 + \zeta^{10} + \zeta^{13} + \zeta^5 + \zeta^{15} + \zeta^{11} + \zeta^{16}$$
$$+ \zeta^{14} + \zeta^8 + \zeta^7 + \zeta^4 + \zeta^{12} + \zeta^2 + \zeta^6$$
$$= -1$$

Next set

$$A = \zeta + \zeta^9 + \zeta^{13} + \zeta^{15} + \zeta^{16} + \zeta^8 + \zeta^4 + \zeta^2$$
$$B = \zeta^3 + \zeta^{10} + \zeta^5 + \zeta^{11} + \zeta^{14} + \zeta^7 + \zeta^{12} + \zeta^6$$
$$C = \zeta + \zeta^{13} + \zeta^{16} + \zeta^4$$
$$D = \zeta^9 + \zeta^{15} + \zeta^8 + \zeta^2$$
$$E = \zeta^3 + \zeta^5 + \zeta^{14} + \zeta^{12}$$
$$F = \zeta^{10} + \zeta^{11} + \zeta^7 + \zeta^6$$
$$G = \zeta + \zeta^{16} = \zeta + \zeta^{-1}$$
$$H = \zeta^{13} + \zeta^4$$

It is not at all clear at this point what the pattern is for writing out the imaginary 17th roots of unity in the first of the above equations, but once that is taken for granted, the pattern for forming the subsequent sequences is quite obvious.

Since $G = \zeta + \zeta^{-1}$, it follows that

$$\zeta^2 - G\zeta + 1 = 0$$

and hence ζ has a degree 2 algebraic expression in G. Next

$$G + H = \zeta + \zeta^{16} + \zeta^{13} + \zeta^4 = C$$

and

$$GH = (\zeta + \zeta^{16})(\zeta^{13} + \zeta^4) = \zeta^{14} + \zeta^5 + \zeta^{29} + \zeta^{20}$$
$$= \zeta^{14} + \zeta^5 + \zeta^{12} + \zeta^3 = E$$

Hence

$$G, H \text{ are the roots of } x^2 - Cx + E = 0$$

Again, by making use of Proposition 2.7, it is easily verified (Exercise 3) that

$$C + D = A, \quad CD = -1$$

and that

$$E + F = B, \quad EF = -1$$

2.4 Ruler and Compass Constructibility of Regular Polygons

From this it follows that

$$C, D \text{ are the roots of } x^2 - Ax - 1 = 0$$

and

$$E, F \text{ are the roots of } x^2 - Bx - 1 = 0$$

Finally, leaving some details to Exercise 4, we see that

$$A + B = -1 \quad \text{and} \quad AB = -4$$

Therefore

$$A, B \text{ are the roots of } x^2 + x - 4 = 0$$

At this point we can conclude that

A and B have degree 2 algebraic expressions in the integers,
C and E have degree 2 algebraic expressions in A and B,
G has a degree 2 algebraic expression in C and E,
ζ has a degree 2 algebraic expression in G.

Consequently ζ has a degree 2 algebraic expression in the integers. Thus, the Cartesian representations of ζ and all of its powers are constructible. ∎

While the above proof demonstrates the ruler and compass constructibility of the regular 17-sided polygon, it sidesteps the issue of actually constructing the vertices. The fact that the abscissa of the point ζ that is used in the proof of the above theorem is known to be

$$-\frac{1}{16} + \frac{1}{16}\sqrt{17} + \frac{1}{16}\sqrt{34 - 2\sqrt{17}}$$
$$-\frac{1}{8}\sqrt{17 + 3\sqrt{17} - \sqrt{34 - 2\sqrt{17}} - 2\sqrt{34 + 2\sqrt{17}}}$$

clarifies the reasons behind this avoidance.

Gauss did not limit himself to the 17-gons. He went on to prove that for each prime number p the equation

$$x^p - 1 = 0 \tag{2}$$

has a resolution of the type exhibited in Proposition 2.12 and Theorem 2.13. More specifically, Gauss demonstrated that the roots of $x^p - 1 = 0$ have algebraic expressions in the integers that call only for radicals of the type

$$\sqrt[q]{x}$$

where q is any prime factor of $p - 1$. In particular, if $p - 1$ is a power of 2, as is the case for $p = 5, 17$, then these roots all have degree 2 algebraic expressions in the integers, as illustrated by Proposition 2.12 and Theorem 2.13.

This of course means that for such p the regular p-sided polygon is constructible by ruler and compass. It so happens (Exercise 5) that if p is a prime such that $p - 1$ is a power of 2, then

$$p = 2^{2^m} + 1$$

for some integer m. Unfortunately, only five such values of p are known, namely

$$3 = 2^{2^0} + 1, \quad 5 = 2^{2^1} + 1, \quad 17 = 2^{2^2} + 1$$
$$257 = 2^{2^3} + 1, \quad 65{,}537 = 2^{2^4} + 1$$

The next number in this sequence, $2^{2^5} + 1 = 4{,}294{,}967{,}297$ is not a prime, since it equals $641 \cdot 6{,}700{,}417$, as discovered by Euler in the eighteenth century. In fact, all of the numbers $2^{2^m} + 1$ for $m = 5, 6, \ldots, 16$ are now known to be composite.

Gauss also claimed to have proved rigorously that when p is a prime that is not of the form $2^n + 1$ (e.g., 7 or 11), then the regular p-sided polygon is not constructible by ruler and compass alone. While he never wrote down his proof, this fact is now known to be true.

The strange ordering of the 17th roots of unity that was used in the proof of Theorem 2.13 merits some discussion. The series

$$\zeta + \zeta^3 + \zeta^9 + \zeta^{10} + \zeta^{13} + \zeta^5 + \zeta^{15} + \zeta^{11} + \zeta^{16} + \zeta^{14}$$
$$+ \zeta^8 + \zeta^7 + \zeta^4 + \zeta^{12} + \zeta^2 + \zeta^6$$

is such that each root is the cube of the previous one. Thus

$$(\zeta^3)^3 = \zeta^9, \ (\zeta^9)^3 = \zeta^{27} = \zeta^{10}, \ldots, (\zeta^2)^3 = \zeta^6$$

The general implications of the existence of such a sequence, those which Gauss used to treat equation (2) above, lie beyond the scope of this text. However, it is necessary to point out that the existence of such a sequence cannot be taken for granted. For example, if instead of cubing each term we only squared them, we would obtain the series

$$\zeta + \zeta^2 + \zeta^4 + \zeta^8 + \zeta^{16} + \zeta^{15} + \zeta^{13} + \zeta^9 + \zeta + \cdots$$

2.4 Ruler and Compass Constructibility of Regular Polygons

which fails to contain all the imaginary 17th roots of unity. That some such power does cycle through all the imaginary pth roots of unity will eventually be established in Theorem 7.8. At this point we leave the determination of such powers to trial and error. For example, if $p = 11$, then squaring works because

$$\zeta + \zeta^2 + \zeta^4 + \zeta^8 + \zeta^5 + \zeta^{10} + \zeta^9 + \zeta^7 + \zeta^3 + \zeta^6$$

contains all the imaginary 11th roots of unity, whereas cubing does not work because the series

$$\zeta + \zeta^3 + \zeta^9 + \zeta^5 + \zeta^4 + \zeta + \cdots$$

comes up short.

Three other constructibility problems were mentioned in this section's opening paragraph. Using tools similar to those that were introduced here, together with some linear algebra, it can be shown that the ruler and compass constructions in question do not exist. Specifically, *the squaring of a circle* calls for the ruler and compass construction of a square whose area equals that of a given circle. In particular, squaring a circle of radius 1 is tantamount to solving the equation

$$\pi \cdot 1^2 = x^2$$

and so, if successful, such a construction would imply that $x = \sqrt{\pi}$ has a degree 2 expression in the integers. However, in 1882 Ferdinand Lindemann (1852–1939) proved that π has no algebraic expressions in the integers whatsoever. Consequently neither does $\sqrt{\pi}$ have such an expression, and hence there is no general ruler and compass construction for squaring arbitrary circles. Other numbers, like π that have no algebraic expressions in the integers are $e = 2.718281\ldots$ and

$$.1234567891011121314151617181920212223$$

The doubling of a cube calls for the ruler and compass construction of the side of a cube that has double the volume of a given cube. In particular, the doubling of the unit cube, if successful, would prove that $\sqrt[3]{2}$ is a constructible number. Again, it is known that $\sqrt[3]{2}$ has no degree 2 algebraic expressions in the integers, and so this construction too is impossible.

Finally, the trisection of the angle calls for the ruler and compass construction of an angle that is one-third of a given angle. In particular, if successful, such a construction would yield an angle of

$$20° = \frac{60°}{3}$$

38 Complex Numbers

This in turn would imply that $\cos 20°$ is a constructible number. However,

$$\tfrac{1}{2} = \cos 60° = \cos(3 \cdot 20°) = 4\cos^3 20° - 3\cos 20°$$

and so $\cos 20°$ is a solution of the cubic equation

$$4x^3 - 3x - \tfrac{1}{2} = 0$$

The solutions of this equation can be shown to have no degree 2 algebraic expressions in the integers. Consequently there is no general ruler and compass construction for trisecting angles.

The credit for hammering in the final nail into the coffins of the angle trisection and cube doubling problems in 1837 goes to the little known French mathematician Pierre Wantzel (1814–1848).

EXERCISES 2.4

1. Show that if $\varepsilon = \cos 2\pi/5 + i \sin 2\pi/5$, then

$$\varepsilon = \frac{\sqrt{5}-1}{4} + \frac{\sqrt{10+2\sqrt{5}}}{4} i$$

2. With ε as in Exercise 1, express ε^2, ε^3, ε^4 in the form $a + bi$.
3. If C, D, E, F are defined in the proof of Theorem 2.13, prove that $CD = EF = -1$.
4. If A and B are as defined in the proof of Theorem 2.13, prove that $AB = -4$.
5. Prove that if p is a prime integer such that $p-1$ is a power of 2, then there exists an integer m such that $p = 2^{2^m} + 1$.
6. For which of the integers $n = 3, 4, \ldots, 100$ can you assert that a regular n-gon is constructible?
7. For which of the integers $n = 101, 102, \ldots, 200$ can you assert that a regular n-gon is constructible?

Sketch diagrams that demonstrate the constructibility of the real numbers in Exercises 8–11.

8. $\sqrt{2}$
9. $\sqrt{5}$
10. $\sqrt{\sqrt{2}+\sqrt{5}}$
11. $\sqrt[4]{2}$

Suppose that you are given a triangle with constructible sides. Which of the following aspects of this triangle listed in Exercises 12–20 can you assert to be constructible? Justify your answers.

12. The perimeter
13. The square of the perimeter
14. The sum of the lengths of the medians
15. The square root of the sum of the lengths of the medians
16. The area

17. The square of its area
18. The cube of the area
19. The cube root of its area
20. The product of its area with its perimeter
21. Prove that the area of the regular hexagon, whose side is constructible, is a constructible real number.
22. Prove that the area of the regular pentagon whose side is constructible is a constructible real number.
23. Explain why the equation $x^8 + x^4 + 1 = 0$ has no real roots and why each of its complex roots is constructible.
24. Explain why the equation $x^{1024} + x^{512} + 1 = 0$ has no real roots and why each of its complex roots is constructible.
25. Let η be the first 7th root of unity. Prove that the quantities $\eta + \eta^2 + \eta^4$ and $\eta^3 + \eta^6 + \eta^5$ are constructible.
26. Let α be the first 11th root of unity. Prove that the sum of α, α^3, α^4, α^5, α^9 is a constructible complex number.
27. Find three distinct 13th roots of unity whose sum is a constructible number.
28. State and prove the converse of Corollary 2.11.
29. Construct a regular pentagon by means of ruler and compass.

2.5 ORDERS OF ROOTS OF UNITY

We have seen that the fourth roots of unity are $\{1, i, -1, -i\}$ and that the sixth roots of unity are $\{1, -\omega^2, \omega, -1, \omega^2, -\omega\}$. However, -1 is already a square root of 1, and ω and ω^2 are also cube roots of 1. If ζ is any root of unity, then the *order* of ζ, denoted by $o(\zeta)$ is the least positive integer m such that

$$\zeta^m = 1$$

Thus

$$o(1) = 1, \quad o(-1) = 2. \quad o(\omega) = 3$$
$$o(-\omega) = 6, \quad o(i) = 4, \quad o(-\omega^2) = 6$$

The following proposition on the order of roots may seem obvious, but it does require formal proof. The integer m is said to be a *divisor* of the integer n (and n is said to be a *multiple* of m) if there is an integer k such that $n = km$. An integer that is greater than 1 and whose only positive divisors are 1 and itself is said to be *prime*. An integer that is greater than 1 and is not a prime is said to be *composite*.

■ **PROPOSITION 2.14.** *If ζ is any complex root of unity and n is any integer, then $\zeta^n = 1$ if and only if n is a multiple of $o(\zeta)$.*

Proof. If n is a multiple of $o(\zeta)$, then there exists an integer m such that $n = o(\zeta)m$, and hence

$$\zeta^n = (\zeta^{o(\zeta)})^m = 1^m = 1$$

Conversely, suppose that n is an integer such that $\zeta^n = 1$. If n is positive, then the process of long division yields integers q and r such that

$$q \geqslant 0, \quad o(\zeta) > r \geqslant 0, \quad \text{and} \quad n = o(\zeta)q + r$$

But then

$$\zeta^r = \zeta^{n - o(\zeta)q} = \frac{\zeta^n}{(\zeta^{o(\zeta)})^q} = \frac{1}{1^q} = 1$$

Since $0 \leqslant r < o(\zeta)$ and $o(\zeta)$ is the least *positive* integer m such that $\zeta^m = 1$, it follows that $r = 0$ and hence that $n = o(\zeta)q$.

If n is zero, then it is trivially a multiple of $o(\zeta)$. If n is negative, then $-n$ is a positive integer such that

$$\zeta^{-n} = \frac{1}{\zeta^n} = 1$$

Thus, by the above consicderations, $o(\zeta)$ is a divisor of $-n$ and therefore also of n. ∎

The following corollary is an immediate consequence of the above proposition. Its proof is relegated to Exercise 17.

■ **COROLLARY 2.15.** *Suppose that ζ is a root of unity, and that a and b are any two integers. Then*

1. $\zeta^a = \zeta^b$ *if and only if $o(\zeta)$ is a divisor of $a - b$,*
2. $1, \zeta, \zeta^2, \ldots, \zeta^{o(\zeta)-1}$ *are all distinct.*

A *primitive* nth root of unity is one which is not an mth root for any $m < n$. Thus i and $-i$ are the only primitive fourth roots of unity, and $-\omega$ and $-\omega^2$ are the only primitive sixth roots of unity. On the other hand, ω and ω^2 are primitive cube roots of unity, and similarly every fifth root of unity except 1 is also a primitive fifth root of unity. It is clear that ζ is a primitive nth root of unity if and only if $n = o(\zeta)$. It also follows from Corollary 2.15.2 that ζ is a primitive nth root of unity if and only if $\zeta^n = 1$ and the numbers $1, \zeta, \zeta^2, \ldots, \zeta^{n-1}$ are all distinct. In particular, the first nth root of unity is always primitive.

We will state and prove several more important facts regarding the orders of roots of unity in Section 5.2, after some necessary tools have been acquired.

EXERCISES 2.5

For each of the values of n in Exercises 1–8 list the elements of $\sqrt[n]{1}$ with their orders.

1. 4
2. 5
3. 6
4. 7
5. 8
6. 9
7. 10
8. 12

For each of the values of n in Exercises 9–12 list the primitive elements of $\sqrt[n]{1}$.

9. 6
10. 11
11. 12
12. 24

13. Prove that if ζ is a root of unity, then $o(\zeta^{-1}) = o(\zeta)$.
14. Prove that if ζ is a primitive nth root of unity, then so is ζ^{-1}.
15. Prove that if ζ is a primitive nth root of unity, then $1, \zeta, \zeta^2, \ldots, \zeta^{n-1}$ are all distinct.
16. Prove that if n is a positive odd integer and ζ is a primitive nth root of unity, then so is ζ^2. Is this also true for even n? Justify your answer.
17. Prove Corollary 2.15.
18. Prove that if ζ is any root of unity, then

$$1 + \zeta + \zeta^2 + \cdots + \zeta^{o(\zeta)-1} = 0$$

19. Prove that if ζ is any root of unity, then

$$1 \cdot \zeta \cdot \zeta^2 \cdot \ldots \cdot \zeta^{o(\zeta)-1} = (-1)^{o(\zeta)-1}$$

20. Prove that if p is a prime number, then every imaginary pth root of unity is necessarily a primitive pth root of unity.
21. Prove that if $n > 4$ is composite, then at least three of the nth roots of unity are not primitive nth roots.

2.6 THE EXISTENCE OF COMPLEX NUMBERS*

This section is devoted to the construction of a number system whose ontological credentials are impeccable and which is indistinguishable from the complex number system. An alternate proof of the existence of complex numbers is offered in Section 10.2 in a much wider and more useful setting.

We begin by defining a *Cartesian number* as an ordered pair (a, b) of real numbers. The two Cartesian numbers $z = (a, b)$ and $w = (c, d)$ are considered to be the same, or *equal*, if and only if $a = c$ and $b = d$. Thus

$$(2, 3) \neq (3, 2) \quad \text{and} \quad (2^2, 3^3) = (4, 27)$$

These Cartesian numbers can be thought of as either pairs of real numbers or as points of the Cartesian plane.

*Optional.

The *addition* and *multiplication* of Cartesian numbers are defined as follows:

$$(a, b) + (c, d) = (a + c, b + d) \tag{3}$$

$$(a, b) \cdot (c, d) = (ac - bd, ad + bc) \tag{4}$$

These definitions are motivated by the fact that the Cartesian number (a, b) is supposed to be a logical construct that mimics the behavior of the intuitive quantity $a + bi$. Thus definitions (3) and (4) mimic the facts that

$$(a + bi) + (c + di) = (a + c) + (b + d)i$$

and

$$(a + bi)(c + di) = (ac - bd) + (ad + bc)i$$

This of course is only the motivation for these definitions. From the purely logical stance, these definitions need no justification. The addition and multiplication of Cartesian numbers can be demonstrated to possess all the usual properties that they have in the context of real numbers (these well known properties will be formalized later in Section 6.1). Thus the addition of Cartesian numbers is commutative because it inherits this property from the addition of real numbers in the following manner:

$$(a, b) + (c, d) = (a + c, b + d) = (c + a, d + b) = (c, d) + (a, b)$$

Similarly the multiplication of Cartesian numbers is associative because

$$\begin{aligned}
[(a, b) \cdot (c, d)] \cdot (e, f) &= (ac - bd, ad + bc) \cdot (e, f) \\
&= [(ac - bd)e - (ad + bc)f, (ac - bd)f + (ad + bc)e] \\
&= (ace - bde - adf - bcf, acf - bdf + ade + bce) \\
&= (ace - adf - bcf - bde, acf + ade + bce - bdf) \\
&= [a(ce - df) - b(cf + de), a(cf + de) + b(ce - df)] \\
&= (a, b) \cdot (ce - df, cf + de) = (a, b) \cdot [(c, d) \cdot (e, f)]
\end{aligned}$$

The remaining proofs of the commutativity, associativity, and distributivity of the addition and multiplication of the Cartesian numbers are relegated to Exercises 1–3. We define the *Cartesian zero* to be the pair $(0, 0)$ and denote it by 0_c. It is clear that if $z = (a, b)$ is any Cartesian number, then

$$z + 0_c = (a, b) + (0, 0) = (a, b) = z$$

2.6 The Existence of Complex Numbers

We define the *Cartesian unity* to be the pair $(1,0)$, and denote it by 1_c. Note that for any Cartesian number $z = (a, b)$,

$$z \cdot 1_c = (a, b) \cdot (1, 0) = (a \cdot 1 - b \cdot 0, a \cdot 0 + b \cdot 1) = (a, b) = z$$

Finally, we address the existence of inverses. It is clear that $(-a, -b)$ is the additive inverse of (a, b) in the sense that

$$(a, b) + (-a, -b) = 0_c$$

If $(a, b) \neq 0_c$, then $a^2 + b^2 \neq 0$, and so

$$(c, d) = \left(\frac{a}{a^2 + b^2}, \frac{-b}{a^2 + b^2} \right)$$

is a well-defined Cartesian number. Finally, it can be verified (Exercise 4) that (c, d) is the multiplicative inverse of (a, b) in the sense that

$$(a, b) \cdot (c, d) = 1_c$$

The foregoing discussion establishes that the set of Cartesian numbers, together with the operations of addition and multiplication constitutes a bona fide number system. We now demonstrate that these numbers are just the complex numbers in disguise by identifying among them a copy of the real number system together with a quantity that behaves just as the imaginary number i is expected to behave.

Observe that

$$(a, 0) + (c, 0) = (a + c, 0 + 0) = (a + c, 0)$$

and

$$(a, 0) \cdot (b, 0) = (a \cdot b - 0 \cdot 0, a \cdot 0 + b \cdot 0) = (ab, 0)$$

In other words, the Cartesian number $(a, 0)$ behaves with respect to Cartesian addition and multiplication just as the real number a behaves with respect to real addition and multiplication. Thus the set of Cartesian numbers whose second coordinate is 0 is indistinguishable from the real number system.

Let i_c denote the Cartesian number $(0, 1)$. Note that

$$i_c^2 = (0, 1) \cdot (0, 1) = (0 \cdot 0 - 1 \cdot 1, 0 \cdot 1 + 1 \cdot 0) = (-1, 0)$$

and that $(-1, 0)$ is the Cartesian number that corresponds to the real number -1. Moreover, if z is any Cartesian number (a, b), then

$$\begin{aligned}(a, 0) + (b, 0)i_c &= (a, 0) + (b, 0) \cdot (0, 1) \\ &= (a, 0) + (b \cdot 0 - 0 \cdot 1, b \cdot 1 + 0 \cdot 0) \\ &= (a, 0) + (0, b) = (a, b) = z\end{aligned}$$

This of course is the Cartesian analog of the fact that the arbitrary complex number z can be written in the form $a + bi$ where a and b are real numbers and i is a square root of -1.

Thus we have shown that the Cartesian numbers, whose existence is justified by definition, behave just like the complex numbers.

EXERCISES 2.6

1. Prove that the addition of Cartesian numbers is associative.
2. Prove that the multiplication of Cartesian numbers is commutative.
3. Prove the distributivity of Cartesian numbers. That is, prove that

$$(a, b)[(c, d) + (e, f)] = (a, b)(c, d) + (a, b)(e, f)$$

4. Prove that if $(a, b) \neq 0_c$, then

$$(a, b) \cdot \left(\frac{a}{a^2 + b^2}, \frac{-b}{a^2 + b^2} \right) = 1_c$$

5. Prove that if $(a, b) \neq 0_c$ and $(a, b) \cdot (x, y) = (a, b)$ for some (x, y), then $(x, y) = 1_c$.
6. Prove that $(a, b) \cdot 0_c = 0_c$.
7. Prove that if $(a, b) \neq 0_c$ and $(a, b) \cdot (c, d) = 0_c$, then $(c, d) = 0_c$.

CHAPTER SUMMARY

The chapter began with an informal definition of the complex numbers and went on to a discussion of the four arithmetical operations and the extraction of roots in this new context, with special emphasis being given to the roots of unity. These operations were used to give a formal definition of the concept of solvability by radicals. The geometry of the roots of unity was then used to prove the ruler and compass constructibility of the regular 17-sided polygon. This application relied on some surprising subtleties inherent in the powers of the roots of unity. The related notion of the order of a root of unity and some of its properties were expounded in the next section. Finally, a rationale justifying the existence of the so-called imaginary numbers was offered in the last section.

Chapter Summary   45

New Terms

Algebraic expression	24
Algebraic resolvability	25
Algebraic solution	24
Argument	12
Argument principle	15
Cartesian numbers	41
Cartesian representation	10
Complex numbers	10
Conjugate	16
Constructible configuration	28
Constructible number	28
Degree 2 algebraic expression	25
First nth root of unity	20
i	9
Modulus	11
$\sqrt[n]{z}$	18
Order of root of unity	39
Polar form of complex number	14
Prime number	39
Primitive root of unity	40
Rational expression	25
Regular 17-sided polygon	28
Ruler and compass constructibility	27
Solvability by radicals	25

Chapter Review Exercises

Mark the following true or false.

1. The sum of two complex numbers is a complex number.
2. The sum of two imaginary numbers is never a real number.
3. The numbers $0, 1 + i, 2 - i, i$ form a parallelogram.
4. $\arg[(1 + 2i)(1 - i)] = \arg(1 - i) + \arg(1 + 2i)$.
5. $|(1 + 2i)^{123}| = |(1 + 2i)|^{123}$.
6. The number 1 has 20 20th roots.
7. The number 1 has 19 20th roots of order 20.
8. The elements of $\sqrt[7]{1}$ form a regular heptagon.
9. If $a \neq 0$, then the solutions of $ax^2 + bx + c = 0$ have a degree 2 algebraic expression in $\{a, b, c\}$.

46   Complex Numbers

10. The solutions of $3x^2 - 7x + 11 = 0$ are constructible.
11. The regular pentagon and the regular 17-sided polygon are the only regular polygons that are constructible.
12. The order of every element of $\sqrt[24]{1}$ is a divisor of 72.
13. If α is a primitive 7th root of unity, then α, α^2, α^3, and α^4 are all distinct.

Supplementary Exercises

1. Write a computer script that will verify that the regular 257-sided polygon is ruler and compass constructible.
2. Write a computer script that will compute the nth roots of any complex number.
3. Is there an analog of Exercise 2.30 for squares and other regular polygons?
4. Investigate the number system obtained by replacing the product rule (4) with

$$(a, b) \cdot (c, d) = (ac + bd, ad + bc)$$

Which numbers have multiplicative inverses? What are the roots of unity like?

5. Make up your own number system and investigate it.

CHAPTER 3

Solutions of Equations

It is our purpose here to discuss solutions of equations from several different points of view. We will touch on the issues of existence of solutions, existence of formulas, solvability by radicals, and computation of solutions.

3.1 THE CUBIC FORMULA

We are now in position to present the modern version of the Ferro–Tartaglia–Cardano solution to the general cubic equation.

■ **THEOREM 3.1.** *Every cubic equation is solvable by radicals.*

Proof. For simplicity we will assume that the cubic equation we want to solve has the form

$$x^3 + ax^2 + bx + c = 0 \qquad (1)$$

It is clear that every cubic equation can be reduced to this form. Next the problem is further simplified by transforming it to a form that is free of the x^2 term. This is accomplished by a transformation of the type $x = \alpha + y$, where the value of α will shortly be specified. Substituting $x = \alpha + y$ into (1), we get

$$(\alpha + y)^3 + a(\alpha + y)^2 + b(\alpha + y) + c = 0$$

or

$$y^3 + (3\alpha + a)y^2 + (3\alpha^2 + 2a\alpha + b)y + (\alpha^3 + a\alpha^2 + b\alpha + c) = 0$$

The choice of $\alpha = -a/3$ will clearly make the above coefficient of y^2 vanish.

48 Solutions of Equations

The equation now reduces to

$$y^3 + py + q = 0 \tag{2}$$

where

$$p = \frac{3b - a^2}{3} \quad \text{and} \quad q = \frac{27c + 2a^3 - 9ab}{27}$$

To solve the reduced cubic of (2), we can make another transformation

$$y = z + \frac{\beta}{z}$$

where the value of β will be chosen so that the resulting equation is solvable. When this value of y is substituted into (2), we obtain

$$\left(z + \frac{\beta}{z}\right)^3 + p\left(z + \frac{\beta}{z}\right) + q = 0$$

or

$$z^3 + (3\beta + p)z \qquad \frac{\cdot \beta p}{z} + \frac{\beta^3}{z^3} = 0$$

Miraculously the choice of $\beta = -p/3$ causes the coefficients of both z and $1/z$ to vanish, leaving us with the equation

$$z^3 + q - \frac{p^3}{27z^3} = 0 \quad \text{or} \quad z^6 + qz^3 - \frac{p^3}{27} = 0 \tag{3}$$

This is a quadratic equation in z^3 with solutions

$$z^3 = \frac{-q \pm \sqrt{q^2 + 4p^3/27}}{2} \tag{4}$$

It is clear that z has an algebraic expression in p and q. Hence, if $z \neq 0$, each of the terms of the sequence p, q, z, y, x has an algebraic expression in $\{a, b, c\}$. If $z = 0$, then $y = z + \beta/z$ fails to be an algebraic expression. However, in that case, by Exercise 10, $x^3 + ax^2 + bx + c = (x + a/3)^3$, and the corresponding equation is clearly algebraically (in fact rationally) resolvable. ∎

The proof of Theorem 3.1 yields six possible values for z from which one can obtain six possible values for x. We will later see (Corollary 6.7) that a cubic equation can have at most three roots, and we now set out to select three of the above six values that always yield all the solutions of the given cubic equation. The proof that these are the "correct" three values is deferred to Section 6.4. If all of the values of z that arise from (4) are 0, then (Exercise 10)

$$x^3 + ax^2 + bx + c = \left(x + \frac{a}{3}\right)^3$$

and so $x = -a/3$ is the triple solution of equation (1). Otherwise, let z_1 be any nonzero value of z obtained from (4), and set $z_2 = \omega z_1$ and $z_3 = \omega^2 z_1$. For each $i = 1, 2, 3$ set

$$y_i = z_i - \frac{p}{3z_i} \quad \text{and} \quad x_i = y_i - \frac{a}{3} \tag{5}$$

Then x_1, x_2, x_3 is the complete solution set to equation (1).

Consider the cubic equation $x^3 - 3x + 2 = 0$. Here $p = -3$, $q = 2$, and

$$\frac{-q \pm \sqrt{q^2 + 4p^3/27}}{2} = \frac{-2 \pm \sqrt{4 + 4(-3)^3/27}}{2} = -1$$

Hence we can choose $z_1 = -1$. This gives

$$x_1 = y_1 = -1 - \frac{-3}{3(-1)} = -2$$

$$x_2 = y_2 = \omega(-1) - \frac{-3}{3\omega(-1)} = -\omega - \omega^2 = 1$$

$$x_3 = y_3 = \omega^2(-1) - \frac{-3}{3\omega^2(-1)} = -\omega^2 - \omega = 1$$

We next turn to an example with complex coefficients — the equation $x^3 + 3ix - (1 + i) = 0$. Here $p = 3i$ $q = -(1 + i)$, and

$$\frac{-q + \sqrt{q^2 + 4p^3/27}}{2} = \frac{1 + i + \sqrt{(1+i)^2 + 4(3i)^3/27}}{2}$$

$$= \frac{1 + i + \sqrt{1 + 2i - 1 - 4i}}{2}$$

$$= \frac{1 + i + 1 - i}{2} = 1$$

50 Solutions of Equations

Consequently we can choose $z_1 = 1$ and obtain

$$x_1 = y_1 = 1 - \frac{3i}{3 \cdot 1} = 1 - i$$

$$x_2 = y_2 = 1 \cdot \omega - \frac{3i}{3 \cdot \omega \cdot 1} = \omega - \omega^2 i$$

$$x_3 = y_3 = 1 \cdot \omega^2 - \frac{3i}{3 \cdot \omega^2 \cdot 1} = \omega^2 - \omega i$$

Finally, we examine an equation in which the coefficient of x^2 is not zero, namely

$$x^3 - 6x^2 - 4 = 0$$

Here $a = -6$, $b = 0$, and $c = -4$ so that $p = -12$ and $q = -20$. Substitution of these values into (4) yields $z^3 = 10 \pm 6$. Choosing $z_1 = \sqrt[3]{4}$, we then obtain

$$x_1 = y_1 - \frac{a}{3} = \sqrt[3]{4} - \frac{-12}{3\sqrt[3]{4}} - \frac{-6}{3} = 2 + \sqrt[3]{4} + \sqrt[3]{16}$$

$$x_2 = y_2 - \frac{a}{3} = \sqrt[3]{4}\,\omega - \frac{-12}{3\sqrt[3]{4}\,\omega} - \frac{-6}{3} = 2 + \sqrt[3]{4}\,\omega + \sqrt[3]{16}\,\omega^2$$

$$x_3 = y_3 - \frac{a}{3} = \sqrt[3]{4}\,\omega^2 - \frac{-12}{3\sqrt[3]{4}\,\omega^2} - \frac{-6}{3} = 2 + \sqrt[3]{4}\,\omega^2 + \sqrt[3]{16}\,\omega$$

In conclusion we note that while the solution of the cubic equation given in (5) above is formally different from that which appears as (4) in Chapter 1, it is not too difficult to show that this latter expression actually equals one of the three roots that are given by (5). The details are relegated to Exercise 17.

EXERCISES 3.1

Use the cubic formula to find all the complex roots of each of the equations in Exercises 1–9.

1. $x^3 + 9x - 6 = 0$
2. $x^3 + 12x - 12 = 0$
3. $x^3 + 18x - 30 = 0$
4. $x^3 - 15x - 126 = 0$
5. $x^3 + 3x^2 + 9x + 5 = 0$
6. $x^3 - 6x^2 + 24x - 44 = 0$
7. $x^3 + 3x^2 - 3x + 2i - 5 = 0$
8. $x^3 + 6ix - 1 - 8i = 0$ [*Hint:* $\sqrt{-63 - 16i} = \pm(1 - 8i)$.]
9. $2x^3 + 3x^2 + 3x + 1 = 0$

10. Prove that if all the values of z in equation (4) are 0, then $p = q = 0$ and $x^3 + ax^2 + bx + c = (x + a/3)^3$.

Use Exercise 10 to solve Exercises 11–12.

11. $x^3 + 3(1 + i)x^2 + 6ix - 2(1 - i) = 0$
12. $x^3 + 6ix^2 - 12x - 8i = 0$

If x_1, x_2, x_3 are the solutions of the cubic equation $x^3 + ax^2 + bx + c = 0$, then the quantity $[(x_1 - x_2)(x_2 - x_3)(x_3 - x_1)]^2$ is called the discriminant *of the given equation.*

13. Prove that the discriminant of the equation $y^3 + py + q = 0$ is $-4p^3 - 27q^2$.
14. Prove that the discriminant of the equation $x^3 + ax^2 + bx + c = 0$ is $18abc - 4a^3c + a^2b^2 - 4b^3 - 27c^2$.
15. Prove that a cubic equation with real coefficients has three distinct roots if its discriminant D is positive, a single real root if D is negative, and at least two equal real roots if D is zero.
16. Show that if a is real, then the equation $x^3 + ax + 2 = 0$ has three real roots if and only if $a \leq -3$.
17. Show that root of the cubic equation given in (4) of Chapter 1 is also one of those given by (5) of this section.

3.2 SOLVABILITY BY RADICALS (II)

Whereas thousands of years elapsed between the solutions of the quadratic and the general cubic equation, only a few more years passed before Cardano assigned the problem of finding a formula for the general fourth-degree, or *quartic*, equation to his disciple Ferrari and the latter succeeded in solving it. Subsequently several other mathematicians presented their own solutions of the quartic. From our perspective the most significant of these is Lagrange's solution, which is presented in detail in Section 6.5.

Quite naturally mathematicians next turned their attention to the general *quintic*, or fifth-degree, equation. This equation, however, presented new difficulties, and no substantial progress was made for another 250 years. During the second half of the eighteenth century some mathematicians recognized the fact that even the innocent-looking equation

$$x^n - 1 = 0 \qquad (6)$$

presented them with challenges. They were aware that one of the solutions of the equation $x^{11} - 1 = 0$ is the complex number

$$\cos \frac{2\pi}{11} + i \sin \frac{2\pi}{11}$$

and consequently this number can be expressed in terms of a radical of the 11th order (i.e., $\sqrt[11]{1}$). However, in 1771 Vandermonde and Lagrange showed that the same number could also be expressed in terms of radicals of the 2nd and 5th orders, thus uncovering some surprising relationships between radicals of the 2nd, 5th, and 11th orders.

Equation (6) came to be known as the *cyclotomic equation* because its zeros, as stated in Proposition 2.5, lie on a common circle. In 1801, in the concluding chapter of his *Disquisitiones Arithmeticae*, Gauss delved into the subtleties of the radicals that are required for the solution of the cyclotomic equation. The proof of Theorem 2.13 was meant to provide a sample of the algebraic manipulations that this task required. The very special nature of the cyclotomic equation notwithstanding, this work turned out to be seminal. Évariste Galois, who completely settled the issue of algebraic resolvability of equations some 30 years later, made repeated references to Gauss's techniques and results in explaining the motivation for his own work.

3.3 OTHER TYPES OF SOLUTIONS*

While this book's main theme is the issue of solvability of polynomial equations by means of algebraic operations, it might be pedagogically advantageous at this point to discuss some other senses in which an equation could be solvable. Consider the equation

$$x^5 - 6x + 3 = 0 \qquad (6)$$

If we set $f(x) = x^5 - 6x + 3$, then

$$\lim_{x \to \infty} f(x) = \infty \quad \text{and} \quad \lim_{x \to -\infty} f(x) = -\infty$$

Since $f(x)$ is a continuous function, it follows that its graph must cross the x-axis at some point and that point clearly yields a solution to equation (6). This argument generalizes easily to the following observation:

If n is an odd positive integer, and $a_1, a_2, \ldots, a_n$ are real numbers, then the equation

$$x^n + a_1 x^{n-1} + a_2 x^{n-2} + \cdots + a_{n-1} x + a_n = 0$$

has a real solution.

Thus it is possible to argue the existence of a solution to an equation without having any information at all about the value of the solution. In fact the mathematicians of the seventeenth and eighteenth centuries became convinced

*Optional.

of the validity of the following sweeping statement:

> *Every equation of the form* $x^n + a_1 x^{n-1} + a_2 x^{n-2} + \cdots + a_{n-1} x + a_n = 0$ *has a solution.*

The solution whose existence is guaranteed here may be complex, but it always exists. The first valid proof of this fact was provided by Gauss in 1796, and it eventually became known as the *Fundamental Theorem of Algebra*. Neither this proof nor any of Gauss's several subsequent alternate proofs provided information about the actual value of the solution; they were concerned only with its existence. In 1826 Niels Abel proved that no analogs of the Ferro–Tartaglia–Cardano cubic formula could exist for the general fifth-degree equation

$$x^5 + ax^4 + bx^3 + cx^2 + dx + e = 0$$

Shortly thereafter Galois constructed a general theory for determining just which general equations have formulas and which specific equations are solvable by radicals. Using that theory, it is possible, for example, to show that the roots of equation (6), the existence of at least one of which is easily demonstrated, do *not* have algebraic expressions in the integers.

There is yet another aspect to solving equations that is essentially different from both the issue of existence and from that of expressibility by algebraic operations, and that is the problem of evaluating a root and writing it down as, say, a decimal number. Just because it is known that a certain number is an algebraic expression in the integers does not mean that we have any idea of its size. The root of equation (6) whose existence was proved by a theoretical argument of course also suffers from the same lack of precision.

There are numerical methods for finding roots of equations that are much more practical than even the Ferro–Tartaglia–Cardano formula. The best known of these, the Newton–Raphson method, is usually taught in the first semester of calculus. It will nevertheless be reviewed here because we feel that this will bring about a better understanding of the difference between solvability of equations in general and solvability of equations by radicals.

Loosely speaking, the Newton–Raphson method, when applied to an equation of the form

$$f(x) = 0$$

says that if x_n is an estimate for a solution, then

$$x_{n+1} = x_n - \frac{f(x_n)}{f'(x_n)} \tag{7}$$

54 Solutions of Equations

where $f'(x)$ denotes the derivative of $f(x)$ with respect to x is a better estimate for the same solution. Consider equation (6). With $f(x) = x^5 - 6x + 3$, we find that $f(0) = 3$ and $f(1) = -2$. Thus this equation must have a solution between 0 and 1, and we begin with $x_1 = 0$ as our first estimate. Since $f'(x) = 5x^4 - 6$, the Newton–Raphson method yields the following successive estimates (correct to four decimal places) for the solution:

$$x_2 = 0 - \frac{0^5 - 6 \cdot 0 + 3}{5 \cdot 0^4 - 6} = .5$$

$$x_3 = .5 - \frac{(.5)^5 - 6 \cdot 5 + 3}{5 \cdot (.5)^4 - 6} = .5055$$

$$x_4 = .5055 - \frac{(.5055)^5 - 6 \cdot 5055 + 3}{5 \cdot (.5055)^4 - 6} = .5055$$

Once two successive estimates are equal to each other, there will of course be no further improvement in the estimates unless the number of decimal places is increased. It is easily verified that

$$(.5055)^5 - 6(.5055) + 3 = .000006981\ldots$$

and that the solution, correct to six decimal places, is .550501.

This is not the book in which to discuss the subtleties or attempt a rigorous proof of this marvelous technique. The underlying idea, though, is surprisingly simple. Suppose that we have arrived at the estimate of x_n for the

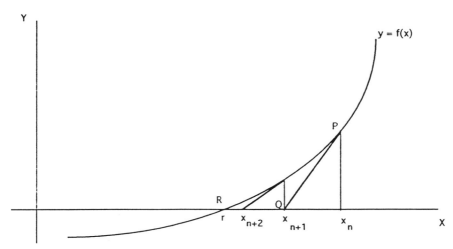

Figure 3.1. The Newton–Raphson method.

3.3 Other Types of Solutions

solution r of the equation $f(x) = 0$ (Figure 3.1). If $P = (x_n, f(x_n))$ is the point of the graph of $y = f(x)$ that lies directly above the x-axis point $(x_n, 0)$, then the equation of the line through P and tangent to this graph is given by the point-slope formula as

$$y - f(x_n) = f'(x_n)(x - x_n) \qquad (8)$$

Note that the diagram suggests that the x-intercept Q of the tangent line at P lies closer to our goal of $R = (r, 0)$ than P. We therefore choose the abscissa of this intersection as the next estimate x_{n+1}. Since Q is the x-intercept of the tangent line at P, the value of x_{n+1} is easily obtained by substituting $(x_{n+1}, 0)$ for (x, y) in (8). This yields

$$0 - f(x_n) = f'(x_n)(x_{n+1} - x_n)$$

When this equation is solved for x_{n+1}, we obtain the Newton–Raphson recursion of (7).

The Newton–Raphson method we presented here is not perfect. There are cases wherein this method will miss some rather obvious answers. Nevertheless, when this method does locate a root, the root will be correct.

EXERCISES 3.3

Using the Newton–Raphson method, find at least one real root of each of the equations in Exercises 1–8 (use four decimal places).

1. $x^3 - 3x^2 + 18x + 12 = 0$
2. $2x^3 - 5x^2 + x - 10 = 0$
3. $x^4 - 100x - 85 = 0$
4. $x^4 + x^3 - x^2 - 4x - 13 = 0$
5. $x^5 - 30x + 17 = 0$
6. $x^7 + x^5 - 3 = 0$
7. $\sin x = x + 1$ [*Hint:* Set $f(x) = \sin x - x - 1$, and apply the Newton–Raphson method to it.]
8. $\ln x = \sin x$

Using the Newton–Raphson method (with four decimal places), estimate the real roots in Exercises 9–12.

9. $\sqrt[3]{10}$ [*Hint:* Solve the equation $x^3 - 10 = 0$.]
10. $\sqrt[5]{100}$ 11. $\sqrt[7]{10000}$ 12. $\sqrt[8]{54321}$

For each of the equations in Exercises 13–18, explain why it does or does not have real solutions.

13. $x^7 + x^5 - 3 = 0$
14. $x^9 + 101x^8 + 93x^6 + 41x^4 + 1 = 0$

15. $x^8 - 2x^4 + 1 = 0$
16. $x^{16} - 4x^8 + 5 = 0$
17. $\ln x = \tan x$
18. $x^3 = \ln x$
19. Let a, b, c be real numbers. Prove that if $a^2 \leqslant 3b$, then the equation $x^3 + ax^2 + bx + c = 0$ has exactly one real root.

CHAPTER SUMMARY

We began by showing that every cubic equation is solvable by radicals. While the issue of solvability of equations by radicals eventually led to the creation of modern algebra, it was pointed out in this chapter that there are other, no less significant, aspects to the solvability of equations. The question of the existence of roots can be treated without regard to explicit derivations. Thus it is known that every equation with either complex or real coefficients has at least one, possibly complex, solution. *Ad hoc* arguments can be given for the existence of such roots that provide no information about its value. Finally, the Newton–Raphson method was informally discussed to show how roots can be found without the use of radicals.

Chapter Review Exercises

Mark the following true or false.

1. Every cubic equation has three distinct roots.
2. The equation $x^3 + ax^2 + bx + c = 0$ has at least one real root.
3. Every cubic equation has at least one imaginary solution.
4. Every equation can be solved by the Newton–Raphson method.
5. Every root of the equation $x^{23} - 1 = 0$ has an algebraic expression in the integers.

New Terms

Cyclotomic equation	51
Fifth-degree equation	51
Fourth-degree equation	51
Fundamental theorem of algebra	53
Newton–Raphson method	53

Supplementary Exercises

1. Implement the cubic formula on a computer.
2. Implement the Newton–Raphson method on a calculator or a computer.

CHAPTER 4

Modular Arithmetic

This chapter introduces some new number systems that are suggested by the properties of the exponents of the complex roots of unity. These number systems bear striking similarities to the more traditional rational and real numbers but also differ from them in crucial ways.

4.1 MODULAR ADDITION, SUBTRACTION, AND MULTIPLICATION

One of the steps in Gauss's proof of the ruler and compass constructibility of the regular 17-sided polygon called for the verification of the identity

$$(\zeta + \zeta^{16})(\zeta^{13} + \zeta^4) = \zeta^{14} + \zeta^5 + \zeta^{29} + \zeta^{20}$$
$$= \zeta^{14} + \zeta^5 + \zeta^{12} + \zeta^3$$

In his writings Gauss used an abbreviation that replaced each ζ^k with the symbol $[k]$. Since

$$\zeta^{k+17} = \zeta^k$$

it follows that, in Gauss's notation,

$$[k + 17] = [k]$$

for each integer k. This is an example of *modular arithmetic*. For any positive integer n, the two integers a and b are said to be *congruent modulo n*, and we write

$$a \equiv b \pmod{n}$$

58   Modular Arithmetic

whenever

$$n \text{ is a divisor of } a - b$$

This, by Corollary 2.15, is tantamount to saying that

$$\zeta^a = \zeta^b$$

where ζ is any primitive nth root of 1. Thus $7 \equiv 3 \pmod{4}$, $2 \equiv 14 \pmod{6}$, and $-3 \equiv 35 \pmod{19}$. Note that if $a \equiv a' \pmod{n}$ and $b \equiv b' \pmod{n}$, and if ζ is as above, then

$$\zeta^{a+b} = \zeta^a \zeta^b = \zeta^{a'} \zeta^{b'} = \zeta^{a'+b'}$$

and

$$\zeta^{ab} = (\zeta^a)^b = (\zeta^{a'})^b = (\zeta^b)^{a'} = (\zeta^{b'})^{a'} = \zeta^{a'b'}$$

Hence

$$a + b \equiv a' + b' \text{ and } ab \equiv a'b' \pmod{n}$$

It follows from this that when performing arithmetic modulo n, it suffices to consider the applications of the arithmetic operations to the integers $0, 1, 2, \ldots, n - 1$ alone. Tables 4.1 to 4.4 contain such abbreviated addition and multiplication tables for $n = 4, 5, 6, 7$. Motivated by the standard denotation of the set of integers by $\mathbb{Z}$, arithmetic modulo n, when restricted to the set $\{0, 1, \ldots, n - 1\}$ is denoted by $\mathbb{Z}_n$. An alternate, and more formal, definition of modular arithmetic is offered in Section 10.1 following Corollary 10.7.

The commutativity, associativity, and distributivity of modular addition and multiplication are consequences of the following identities:

$$\zeta^{a+b} = \zeta^{b+a}, \quad \zeta^{(a+b)+c} = \zeta^{a+(b+c)}$$

$$\zeta^{ab} = \zeta^{ba}, \quad \zeta^{(ab)c} = \zeta^{a(bc)}$$

TABLE 4.1. Arithmetic modulo 4

+	0	1	2	3
0	0	1	2	3
1	1	2	3	0
2	2	3	0	1
3	3	0	1	2

$(\mathbb{Z}_4, +)$

·	0	1	2	3
0	0	0	0	0
1	0	1	2	3
2	0	2	0	2
3	0	3	2	1

$(\mathbb{Z}_4, \cdot)$

TABLE 4.2. Arithmetic modulo 5

+	0	1	2	3	4
0	0	1	2	3	4
1	1	2	3	4	0
2	2	3	4	0	1
3	3	4	0	1	2
4	4	0	1	2	3

$(\mathbb{Z}_5, +)$

·	0	1	2	3	4
0	0	0	0	0	0
1	0	1	2	3	4
2	0	2	4	1	3
3	0	3	1	4	2
4	0	4	3	2	1

$(\mathbb{Z}_5, \cdot)$

TABLE 4.3. Arithmetic modulo 6

+	0	1	2	3	4	5
0	0	1	2	3	4	5
1	1	2	3	4	5	0
2	2	3	4	5	0	1
3	3	4	5	0	1	2
4	4	5	0	1	2	3
5	5	0	1	2	3	4

$(\mathbb{Z}_6, +)$

·	0	1	2	3	4	5
0	0	0	0	0	0	0
1	0	1	2	3	4	5
2	0	2	4	0	2	4
3	0	3	0	3	0	3
4	0	4	2	0	4	2
5	0	5	4	3	2	1

$(\mathbb{Z}_6, \cdot)$

and
$$\zeta^{a(b+c)} = \zeta^{ab+ac}$$

It is also clear that
$$a \cdot 0 \equiv 0 \cdot a \equiv 0 \quad \text{and} \quad a \cdot 1 \equiv 1 \cdot a \equiv a \pmod{n}$$

since
$$\zeta^{a \cdot 0} = \zeta^{0 \cdot a} = \zeta^0 \quad \text{and} \quad \zeta^{a \cdot 1} = \zeta^{1 \cdot a} = \zeta^a$$

TABLE 4.4. Arithmetic modulo 7

+	0	1	2	3	4	5	6
0	0	1	2	3	4	6	6
1	1	2	3	4	5	6	0
2	2	3	4	5	6	0	1
3	3	4	5	6	0	1	2
4	4	5	6	0	1	2	3
5	5	6	0	1	2	3	4
6	6	0	1	2	3	4	5

$(\mathbb{Z}_7, +)$

·	0	1	2	3	4	5	6
0	0	0	0	0	0	0	0
1	0	1	2	3	4	5	6
2	0	2	4	6	1	3	5
3	0	3	6	2	5	1	4
4	0	4	1	5	2	6	3
5	0	5	3	1	6	4	2
6	0	6	5	4	3	2	1

$(\mathbb{Z}_7, \cdot)$

Thus the operations of addition and multiplication possess the same desirable properties relative to modular equivalence that they have when applied to the integers in the context of conventional equality.

If a is any nonzero element of $\mathbb{Z}_n$, then $n - a$ is also in $\mathbb{Z}_n$ and

$$a + (n - a) = n \equiv 0 \pmod{n}$$

and so $n - a$ is the *additive inverse* $-a$ of a in $\mathbb{Z}_n$. Thus 3 and 4 are each other's additive inverses in $\mathbb{Z}_7$, and 3 and 5 are each other's additive inverses in $\mathbb{Z}_8$. The additive inverse of 0 is, by definition, itself. The guaranteed existence of these additive inverses makes it possible to define the difference $a - b$ of two elements of $\mathbb{Z}_n$ as the element $a + (-b)$. Thus, in $\mathbb{Z}_{13}$,

$$5 - 7 \equiv 5 + 6 \equiv 11 \quad \text{and} \quad 5 - 2 \equiv 5 + 11 \equiv 3$$

A moment's reflection will lead to the conclusion that whenever a and b are integers such that $0 \leq b \leq a \leq n - 1$, then $a - b$ is unambiguous, regardless of whether a and b are considered as conventional integers or as elements of $\mathbb{Z}_n$.

The issues of the existence of multiplicative inverses and the possibility of division in $\mathbb{Z}_n$ are more subtle. There is no integer x such that

$$2x \equiv 1 \pmod{4}$$

since in the $(\mathbb{Z}_4, \cdot)$ multiplication table (Table 4.1) the row corresponding to 2 does not contain the entry 1. On the other hand, in the $(\mathbb{Z}_5, \cdot)$ multiplication

4.1 Modular Addition, Subtraction, and Multiplication

table (Table 4.2) all the rows but the first contain a 1, implying that every nonzero element of $\mathbb{Z}_5$ has a multiplicative inverse in $\mathbb{Z}_5$. Thus

$$2 \cdot 3 \equiv 1 \quad \text{and} \quad 4 \cdot 4 \equiv 1 \quad (\text{mod } 5)$$

meaning that 2 and 3 are each other's multiplicative inverses in $\mathbb{Z}_5$, whereas 4 is its own multiplicative inverse, a not so surprising fact, since $4 \equiv -1 \,(\text{mod } 5)$.

A glance at the multiplication tables for $\mathbb{Z}_6$ and $\mathbb{Z}_7$ makes it clear that the situation here is entirely analogous to the above. The integer 2 has no multiplicative inverse in $\mathbb{Z}_6$, whereas every nonzero element of $\mathbb{Z}_7$ does have such an inverse. It will be seen in the next section that $\mathbb{Z}_n$ possesses all the requisite multiplicative inverses if and only if n is a prime number.

A word of caution is in order here. While it is true that $7 \equiv 2 \,(\text{mod } 5)$, it does not follow from this that

$$x^7 \equiv x^2 \quad (\text{mod } 5)$$

since

$$2^7 = 128 \not\equiv 4 = 2^2 \quad (\text{mod } 5)$$

Even very complicated looking equations can be easily solved in modular arithmetic by the naïve method of substitution. In $\mathbb{Z}_5$ the solution set of the equation

$$x^6 + 3x^4 + 2x + 4 = 0$$

is $\{1, 2\}$ because these are the only elements of $\mathbb{Z}_5$ whose substitution into the polynomial $x^6 + 3x^4 + 2x + 4$ yields $0 \,(\text{mod } 5)$.

EXERCISES 4.1

Solve the equations in Exercises 1–5 in $\mathbb{Z}_2$.

1. $x + 1 \equiv 0$
2. $x^2 + 1 \equiv 0$
3. $x^3 + x \equiv 0$
4. $x^5 + x^2 + x + 1 \equiv 0$
5. $x^{17} + x^5 + x^3 \equiv 0$
6. Solve the equations of Exercises 1–5 in $\mathbb{Z}_3$.
7. Solve the equations of Exercises 1–5 in $\mathbb{Z}_5$.

Solve the equations in Exercises 8–15 in $\mathbb{Z}_7$.

8. $3x + 2 \equiv 1$
9. $5x^2 + 2x - 1 \equiv 0$
10. $x^3 + x^2 + 4x + 1 \equiv 0$
11. $4x - 3y \equiv 1$ and $2x + 4y \equiv 3$
12. $4x + 3y \equiv 1$ and $x + 6y \equiv 3$

13. $x + 2y + 3z \equiv 1$, $3x + y + 2z \equiv 2$ and $2x + 3y + z \equiv 3$
14. $x^{123,456} + x + 5 \equiv 0$
15. $x^{54,321} + 2x^{5,432} + 3x + 1 \equiv 0$
16. Solve the equations in Exercises 11–13 in $\mathbb{Z}_5$.
17. Solve the equations in Exercises 11–12 in $\mathbb{Z}_{13}$.
18. Solve the equations in Exercises 14, 15 in $\mathbb{Z}_5$.
19. Solve the equations in Exercises 14, 15 in $\mathbb{Z}_{13}$.
20. Evaluate the (modulo n) sum of all the elements of $\mathbb{Z}_n$ for each positive integer n.
21. Evaluate $1 + 3 + 5 + \cdots + 1001$ in $\mathbb{Z}_{1002}$.
22. Evaluate $1 + 4 + 7 + \cdots + 1234$ in $\mathbb{Z}_{432}$.
23. Evaluate $1 + 2 + 4 + \cdots + 2^{63}$ in $\mathbb{Z}_{11}$.
24. Evaluate $1 + 3 + 9 + \cdots + 3^{99}$ in $\mathbb{Z}_7$.
25. Prove that an integer is divisible by 3 if and only if the sum of its digits in base 10 notation is divisible by 3.
26. Prove that an integer is divisible by 9 if and only if the sum of its digits in base 10 notation is divisible by 9.
27. Let ζ be the first nth root of unity, and let $\eta = \zeta^k$ be any other nth root of unity. Prove that there exists an integer m such that $\zeta = \eta^m$ if and only if k has a multiplicative inverse in $\mathbb{Z}_n$.

4.2 THE EUCLIDEAN ALGORITHM AND MODULAR INVERSES

It turns out that the best way to deal with the question of invertible elements in $\mathbb{Z}_n$ involves the Euclidean algorithm for finding the greatest common divisor of two integers. This is a problem that was considered by many of the earliest mathematicians, including Euclid. An integer that is a divisor of both the integers m and n is said to be their *common divisor*. Thus 2, 3, -6, 6, 12 are all common divisors of 24 and 36. The greatest of all the common divisors of the integers m and n is called their *greatest common divisor* and is denoted by (m, n). Thus

$$12 = (24, 36) = (-24, 36) = (-24, -36)$$

Note that since every integer divides 0, it follows that $(0, 0)$ does not exist. In Propositions 1 and 2 of Book VII of *The Elements*, Euclid suggests the following method for finding the greatest common divisor of the two positive integers $m \geqslant n$. Suppose first that n is a divisor of m. Then it is clear that $(m, n) = n$. If n is not a divisor of m, then n and $m - n$ are positive integers such that

$$(m - n, n) = (m, n)$$

4.2 The Euclidean Algorithm and Modular Inverses

The reason for this is that every common divisor of m and n is clearly also a common divisor of $m - n$ and n, and vice versa, every common divisor of $m - n$ and n is also a divisor of $m = (m - n) + n$ and n. In other words, the set of common divisors of the pair $\{m - n, n\}$ is identical with the set of common divisors of the pair $\{m, n\}$. Consequently the two pairs also have the same greatest common divisor.

This leads to the following derivation of the greatest common divisor of 481 and 74:

$$(481, 74) = (407, 74) = (333, 74) = (259, 74) = (185, 74)$$
$$= (111, 74) = (37, 74) = 37$$

It is clear that this procedure will always yield the greatest common divisor in a finite number of steps, and hence it does deserve to be called an algorithm. It is also clear that the repeated subtractions that lead from $(481, 74)$ to $(37, 74)$ could be abbreviated by observing that 37 is the remainder left by 481 when divided by 74. Thus, if p is greater than q and if p leaves remainder r when divided by q, we have $(p, q) = (r, q)$. This leads to a much faster algorithm for finding greatest common divisors. To find the greatest common divisor of 2227 and 12707, we note that the remainder left by 12707 upon division by 2227 is 1572, and hence $(12707, 2227) = (2227, 1572)$. Several applications of this reduction yield

$$(12707, 2227) = (1572, 2227) = (1572, 655) = (262, 655) = (262, 131) = 131$$

This highly efficient procedure for determining the greatest common divisor of two integers, commonly known as the *Euclidean algorithm*, has a very surprising range of nonobvious applications. Here, of course, we are interested in the question of which integers possess multiplicative inverses in modulo n arithmetic. If two numbers are said to be *relatively prime* whenever their greatest common divisor is 1, then it is exactly those integers that are relatively prime to n that turn out to possess multiplicative inverses modulo n. The following proposition paves the way to proving this fact:

■ **PROPOSITION 4.1.** *If g is the greatest common divisor of the two integers m and n, then there exist integers A and B such that*

$$Am + Bn = g$$

Proof. It suffices to prove this proposition when m and n are both positive (Exercise 25). Thus we assume that $m \geq n > 0$ and proceed by induction on the number of steps (divisions) in the Euclidean algorithm.

If the pair (m, n) is such that exactly one division is required by the Euclidean algorithm, that must be because m is a multiple of n, say, $m = dn$ for some integer d. In that case $g = n$ and

$$g = 0 \cdot m + 1 \cdot n$$

is the required expression.

Assume next that $m \geq n$ are a fixed pair of positive integers the derivation of whose greatest common divisor requires $k > 0$ divisions and that the proposition holds for all positive integers x, y, the derivation of whose greatest common divisor requires $k - 1$ divisions. Let $q \geq 0$ and $r < n$ be the respective quotient and remainder of m when divided by n so that

$$m = qn + r$$

By the abbreviated Euclidean algorithm, $(m, n) = (r, n)$. However, the derivation of (r, n) clearly requires one division less than the derivation of (m, n). In other words, the derivation of (r, n) requires only $k - 1$ divisions, so, by the induction hypothesis, there exist integers A' and B' such that

$$g = (m, n) = (n, r) = A'n + B'r$$

However, by the definition of q and r, $m = qn + r$ so that

$$g = A'n + B'r = A'n + B'(m - qn) = B'm + (A' - B'q)n \tag{1}$$

Hence

$$A = B' \quad \text{and} \quad B = A' - B'q \tag{2}$$

are the required integers. ∎

■ **COROLLARY 4.2.** *If m is relatively prime to n, then m has a multiplicative inverse in $\mathbb{Z}_n$.*

Proof. Let m and n be relatively prime so that $(m, n) = 1$. By Proposition 4.1 there exist integers A and B such that

$$Am + Bn = 1$$

Since $Bn \equiv 0 \pmod{n}$, it follows that

$$Am \equiv 1 \pmod{n}$$

Thus A is the multiplicative inverse of m in $\mathbb{Z}_n$. ∎

Since
$$1 \cdot 1 \equiv 3 \cdot 3 \equiv 5 \cdot 5 \equiv 7 \cdot 7 \equiv 1 \pmod{8}$$
it follows that 1, 3, 5 and 7, which are all relatively prime to 8, are their own multiplicative inverses in $\mathbb{Z}_8$. Similarly, since
$$1 \cdot 1 \equiv 2 \cdot 5 \equiv 4 \cdot 7 \equiv 8 \cdot 8 \equiv 1 \pmod{9}$$
it follows that the multiplicative inverses of 1, 2, 4, 5, 7 and 8 in $\mathbb{Z}_9$ are 1, 5, 7, 2, 4 and 8 respectively.

The converse of Corollary 4.2 also holds (Exercise 26). Whenever n is a prime number, each positive integer less than n is necessarily relatively prime to n, so each of those has a multiplicative inverse in $\mathbb{Z}_n$. This fact is crucial to the subsequent development of this book, and we now state it explicitly.

■ **PROPOSITION 4.3.** *If p is a prime number and m is an integer, $0 < m < p$, then m has a multiplicative inverse in $\mathbb{Z}_p$.*

The list below indicates the multiplicative inverses of all the nonzero elements of $\mathbb{Z}_2$, $\mathbb{Z}_3$, $\mathbb{Z}_5$ and $\mathbb{Z}_7$:

$$1 \cdot 1 \equiv 1 \pmod{2}$$
$$1 \cdot 1 \equiv 2 \cdot 2 \equiv 1 \pmod{3}$$
$$1 \cdot 1 \equiv 2 \cdot 3 \equiv 4 \cdot 4 \equiv 1 \pmod{5}$$
$$1 \cdot 1 \equiv 2 \cdot 4 \equiv 3 \cdot 5 \equiv 6 \cdot 6 \equiv 1 \pmod{7}$$

Let p be a fixed prime number, and let m be any integer that is not divisible by p. Denote the multiplicative inverse of m by m^{-1}. We can then define
$$\frac{a}{b} \equiv ab^{-1} \pmod{p}$$
whenever $b \not\equiv 0 \pmod{p}$. Accordingly
$$\tfrac{3}{4} \equiv 3 \cdot 2 \equiv 6 \pmod{7}$$

Thus, from the point of view of the four arithmetical operations, modulo p arithmetic (when p is a prime) is just as well behaved as the arithmetic of the rational numbers, or the real numbers, or the complex numbers. This observation will be formalized in Section 6.1.

The proof of Proposition 4.1 provides an effective method for finding a^{-1} whenever it exists in $\mathbb{Z}_n$. Thus, to find the multiplicative inverse of 37 in $\mathbb{Z}_{201}$, we begin to answer this question by applying the Euclidean algorithm to 201 and 37. This gives

$$(201, 37) = (37, 16) = (16, 5) = (5, 1) = 1$$

with

$$201 = 5 \cdot 37 + 16, \quad 37 = 2 \cdot 16 + 5, \quad 16 = 3 \cdot 5 + 1, \quad 5 = 5 \cdot 1$$

Now

$$1 = 0 \cdot 5 + 1 \cdot 1$$

and when $A' = 0$, $B' = 1$, $m = 16$, $n = 5$, $q = 3$ are substituted in equations (1) and (2) above, we get $A = 1$ and $B = -3$ so that

$$1 = 1 \cdot 16 + (-3) \cdot 5$$

Again, when $A' = 1$, $B' = -3$, $m = 37$, $n = 16$, $q = 2$ are substituted in equations (1, 2), we get $A = -3$ and $B = 7$ so that

$$1 = (-3) \cdot 37 + 7 \cdot 16$$

Finally, when $A' = -3$, $B' = 7$, $m = 201$, $n = 37$, $q = 5$ are substituted in equations (1, 2), we get $A = 7$ and $B = -38$ so that

$$1 = 7 \cdot 201 + (-38) \cdot 37$$

This means that $-38 \equiv 163$ is the multiplicative inverse of 37 in $\mathbb{Z}_{201}$.

This section's final propositions state some basic number theoretic facts that may seem rather obvious. It is greatly to Euclid's credit that he realized that these facts actually called for proofs. The proofs we offer boil down to a clever application of the Euclidean algorithm and are essentially the same as those that appear in Euclid's *The Elements*. The content of these propositions will be assumed by several subsequent proofs.

■ **LEMMA 4.4.** *Let k, m, n be integers such that k is a divisor of the product mn. If k is relatively prime to m, then k is a divisor of n.*

Proof. Suppose that k is relatively prime to m. By Proposition 4.1 there exist integers A and B such that

$$1 = Am + Bk$$

or

$$n = Amn + Bkn$$

Since k is a divisor of both Amn and Bkn, it follows that k divides their sum n. ■

■ **COROLLARY 4.5.** *Let m and n be relatively prime integers. If m and n are divisors of k, then so is mn a divisor of k.*

Proof. Suppose that $k = k_1 m$. Since n is a divisor of $k = k_1 m$ and n is relatively prime to m, it follows from Lemma 4.4 that n is a divisor of k_1. If $k_1 = k_2 n$, then

$$k = k_1 m = k_2 nm$$

and so mn is a divisor of k. ■

EXERCISES 4.2

Find the greatest common divisor of the pairs of integers in Exercises 1–6.

1. 0 and 365
2. 1 and 3600
3. 36 and 48
4. 3367 and 4277
5. 123,456 and 862,091
6. 14,540,165 and 85,050,243

Find the multiplicative inverses of the elements of $\mathbb{Z}_{67}$ in Exercises 7–11.

7. 25
8. 66
9. 41
10. 37
11. 2

Find the multiplicative inverses of the elements of $\mathbb{Z}_{73}$ in Exercises 12–16.

12. 25
13. 72
14. 41
15. 33
16. 2

17. Find the multiplicative inverse of 4096 in $\mathbb{Z}_{65,537}$.
18. Find the multiplicative inverse of 1000 in $\mathbb{Z}_{65,537}$.
19. Find the multiplicative inverse of each of the invertible elements of $\mathbb{Z}_{12}$.
20. Find the multiplicative inverse of each of the invertible elements of $\mathbb{Z}_{18}$.
21. Does the equation $399x + 703y = 114$ have an integer solution in x and y? If so, find one; if not, explain why not. [*Hint:* Use Proposition 4.1.]
22. Does the equation $399x + 703y = 115$ have an integer solution in x and y? If so, find one; if not, explain why not. [*Hint:* Use Proposition 4.1.]
23. Suppose that m and n are integers, characterize those integers k for which the equation $mx + ny = k$ has integer solutions in x and y. Prove your answer.
24. If $g = (m, n)$, where m and n are integers, show that g is the smallest positive integer that can be expressed in the form $Am + Bn$ where A and B vary over all integers.
25. Explain why it suffices to prove Proposition 4.1 for positive integers.
26. Prove that m has a multiplicative inverse in $\mathbb{Z}_n$ if and only if it is relatively prime to n.
27. Prove that if m and n are two integers, then (m, n) is divisible by every common divisor of m and n.

28. Let m and n be relatively prime positive integers. Prove that there exists an integer k_0 such that for any integer $k > k_0$

$$k = Am + Bn$$

for some *positive* integers A and B.

29. Prove that if p is a prime and a, b are any integers such that $a^2 \equiv b^2 \pmod{p}$, then $a \equiv \pm b \pmod{p}$. Is this also true when p is not a prime?

30. Prove that $1 \cdot 2 \cdot 3 \cdots (p-1) \equiv -1 \pmod{p}$ whenever p is a prime.

31. The smallest positive multiple of both the integers k and m is called their *least common multiple*.
    (a) Prove that every common multiple of two integers is divisible by their least common multiple.
    (b) Prove that the least common multiple of any two positive integers k and m is $km/(k, m)$.

32. Let $m_1, m_2, \ldots, m_r$ denote r positive integers that are pairwise relatively prime, and let $a_1, a_2, \ldots, a_r$ denote any r integers. Then the congruences $x \equiv a_i \pmod{m_i}$, $i = 1, 2, \ldots, r$ have common solutions. Any two solutions are congruent modulo $m_1 m_2 \ldots m_r$. (This is known as the Chinese Remainder Theorem.)

33. Suppose that a/b is a rational zero of the equation

$$a_0 x^n + a_1 x^{n-1} + \cdots + a_{n-1} x + a_n = 0$$

where a and b are relatively prime integers and $a_0, a_1, \ldots, a_n$ are arbitrary integers. Prove that a is a divisor of a_n and that b is a divisor of a_0.

34. Prove that the equation $3x^3 - 3x^2 + 17x - 4 = 0$ has no rational roots.

35. Let a, n be any positive integers such that a is not an nth power of any integer. Prove that $\sqrt[n]{a}$ is not a rational number.

36. Let a and b be nonzero integers such that $g = (a, b)$. Prove that $(a/g, b/g) = 1$.

37. Let a, b, c be any integers such that $g = (a, b)$ is a divisor of c. Prove that if x_0, y_0 are any integers such that $ax_0 + by_0 = (a, b)$, then the complete solution set of the equation $ax + by = c$ is

$$\left\{ (x, y) \mid x = \left(\frac{c}{g}\right) x_0 + \left(\frac{b}{g}\right) t, \ y = \left(\frac{c}{g}\right) y_0 - \left(\frac{a}{g}\right) t, \ t \in \mathbb{Z} \right\}$$

38. Prove that if p is a prime and $\alpha, \beta \in \sqrt[p]{1}$ and $\alpha \neq 1$, then there exists an integer m such that $\alpha^m = \beta$.

4.3 RADICALS IN MODULAR ARITHMETIC*

It might be of interest to consider briefly the issue of radicals in the context of modular arithmetic. In analogy with the more conventional number systems, if a is in $\mathbb{Z}_n$, then $\sqrt{a}$ is the set of all the elements x of $\mathbb{Z}_n$ such that

*Optional.

$x^2 \equiv a \pmod{n}$. Thus, in $\mathbb{Z}_5$,

$$\sqrt{0} = \{0\}, \quad \sqrt{1} = \{1, 4\}, \quad \text{and} \quad \sqrt{4} = \{2, 3\}$$

whereas 2 and 3 have no square roots in $\mathbb{Z}_5$. Similarly, in $\mathbb{Z}_7$,

$$\sqrt{1} = \{1, 6\}, \quad \sqrt{2} = \{3, 4\} \quad \text{and} \quad \sqrt{4} = \{2, 5\}$$

whereas 3, 5, and 6 have no square roots in $\mathbb{Z}_7$. It is interesting to note that in $\mathbb{Z}_5$,

$$\sqrt{-1} \equiv \sqrt{4} = \{2, 3\} \pmod{5}$$

whereas

$$\sqrt{-1} \equiv \sqrt{6} \quad \text{does not exist} \pmod{7}$$

The answer to the question of just which elements of $\mathbb{Z}_p$, p prime, do possess square roots in $\mathbb{Z}_p$ is the subject of the *Law of Quadratic Reciprocity*. This theorem, conjectured by both Euler and Legendre and first proved by Gauss, is one of the most important theorems of number theory. Its discussion does not properly fall in this book's domain and can be found in many textbooks on number theory.

Higher-order radicals are defined in a manner similar to that in which the square roots were defined. The question of existence here is much less understood, though.

EXERCISES 4.3

Which of the elements of $\mathbb{Z}_n$ have square roots in $\mathbb{Z}_n$ for the values of n in Exercises 1–4?

1. 7   **2.** 10   **3.** 13   **4.** 17

Which of the elements of $\mathbb{Z}_n$ have cube roots in $\mathbb{Z}_n$ for the values of n in Exercises 5–8?

5. 7   **6.** 11   **7.** 13   **8.** 17

9. Let p be any prime integer, and let a, b be nonzero elements of $\mathbb{Z}_p$. Show that ab has a square root in $\mathbb{Z}_p$ if and only if either both a and b have such square roots or both a and b fail to have such square roots.

4.4 THE FUNDAMENTAL THEOREM OF ARITHMETIC*

The Euclidean algorithm of Section 4.2 provides us with the means for proving the Fundamental Theorem of Arithmetic, which states that every positive

*Optional.

integer can be factored into the product of prime numbers in an essentially unique way.

■ **THEOREM 4.6.** *Let n be a positive integer that is greater than 1. Then there exist prime numbers $p_1 < p_2 < \cdots < p_h$ and positive integers $r_1, r_2, \ldots, r_h$ such that*

$$n = p_1^{r_1} p_2^{r_2} \cdots p_h^{r_h}$$

Moreover, if $q_1 < q_2 < \cdots < q_k$ is another list of primes and $s_1, s_2, \ldots, s_k$ is another list of positive integers such that

$$n = q_1^{s_1} q_2^{s_2} \cdots q_k^{s_k}$$

then $h = k$, $p_i = q_i$, and $r_i = s_i$ for $i = 1, 2, \ldots, h$.

Proof. We first prove the existence of such a factorization into primes by induction on n. If $n = 2$, we can clearly use $h = 1$, and $p_1 = 2$. Let n be any positive integer such that every smaller integer that exceeds 1 has a factorization into primes. If n is a prime, then we can again set $h = 1$ and $p_1 = n$ to get a trivial factorization of n into primes. If n is not prime, say,

$$n = n_1 n_2, \qquad n_1, n_2 > 1$$

then, by the induction hypothesis, n_1 and n_2 both have prime factorizations, and the product of these two factorizations yields a factorization of n into primes.

The uniqueness of the prime factorizations is also demonstrated by induction. If n is prime, it cannot be expressed as the product of other primes, so

$$n = n$$

is the only prime factorization of n. Let n be a positive integer such that every smaller integer that exceeds 1 has a unique prime factorization. Suppose now that

$$n = p_1^{r_1} p_2^{r_2} \cdots p_h^{r_h} = q_1^{s_1} q_2^{s_2} \cdots q_k^{s_k}$$

A repeated application of Lemma 4.4 leads us to the conclusion that $p_1 = q_i \geq q_1$ for some $i = 1, 2, \ldots, k$. A symmetrical argument allows us to conclude that, in fact, $p_1 = q_1$. Consequently

$$\frac{n}{p_1} = p_1^{r_1 - 1} p_2^{r_2} \cdots p_h^{r_h} = q_1^{s_1 - 1} q_2^{s_2} \cdots q_k^{s_k}$$

The theorem now follows from an application of the induction hypothesis (of the uniqueness of factorization) to the smaller number n/p_1. ∎

EXERCISES 4.4

Find the prime factorization of the numbers in Exercises 1–5.

1. 1,000,000
2. $2^{10} + 1$
3. $2^{2^5} + 1$
4. 53,357
5. 1,048,576
6. Show that the sum, difference, and product of any two elements of the set

$$\mathbb{Z}[\sqrt{-5}] = \{a + b\sqrt{-5} \mid a \text{ and } b \text{ are real integers}\}$$

   is also in that set.

7. For any element $z = a + b\sqrt{-5}$ of the set $\mathbb{Z}[\sqrt{-5}]$ above, define

$$N(z) = a^2 + 5b^2$$

   (a) Prove that $N(z) = 1$ if and only if $z = \pm 1$.
   (b) Prove that $N(z) \neq 3$ for all $z \in \mathbb{Z}\sqrt{-5}$.
   (c) Find all the solutions of $N(z) = 9$ in $\mathbb{Z}[\sqrt{-5}]$.
   (d) Prove that $N(zw) = N(z)N(w)$.

8. A nonzero element p of the set $\mathbb{Z}[\sqrt{-5}]$ is said to be *prime* if it is not ± 1, and if whenever

$$p = zw \quad \text{for some } z, w \in \mathbb{Z}[\sqrt{-5}]$$

   we may conclude that either $z = \pm 1$ or $w = \pm 1$. Show that
   (a) $3, 2 + \sqrt{-5}$, and $2 - \sqrt{-5}$ are all prime elements of $\mathbb{Z}[\sqrt{-5}]$,
   (b) 9 can be factored into primes of $\mathbb{Z}[\sqrt{-5}]$ in two different ways.

9. Prove that there is an infinite number of prime integers.
10. Prove that every element of $\mathbb{Z}[\sqrt{-5}]$ is expressible as the product of a finite number of primes.
11. Prove that there is an infinite number of primes in $\mathbb{Z}[\sqrt{-5}]$.
12. A positive integer is said to be *perfect* if it equals the sum of its proper divisors. For example, 6 and 28 are perfect because $6 = 1 + 2 + 3$ and $28 = 1 + 2 + 4 + 7 + 14$. Prove that if n is an integer such that $2^n - 1$ is a prime integer, then $(2^n - 1)2^{n-1}$ is a perfect number. Use this to find at least two more perfect numbers. (As of this writing only 33 perfect numbers have been found, the largest having $n = 859,433$.)
13. A positive integer is said to be *multiplicatively perfect* if it equals the product of all of its proper divisors. For example, 6 and 10 are multiplicatively perfect since $6 = 1 \cdot 2 \cdot 3$ and $10 = 1 \cdot 2 \cdot 5$. Find a simple characterization of all the multiplicatively perfect integers.

Find the number of distinct positive divisors of the numbers in Exercises 14–16.

14. $3^5 5^4$
15. 12^{12}

16. $p_1^{r_1} p_2^{r_2} \cdots p_h^{r_h}$, where $p_1, p_2, \ldots, p_h$ are distinct primes
17. Express the greatest common divisor and least common multiple of any two integers in terms of their factorizations into prime powers, and then redo Exercise 4.2.31.

CHAPTER SUMMARY

The modular number systems $\mathbb{Z}_p$ (p prime) share many of the properties of the rational, real, and complex number systems. They are closed with respect to the four arithmetic operations, and their elements may or may not possess square roots, cube roots, and the like. In particular, the question of the existence of multiplicative inverses in modular arithmetic is resolved by an application of the well-known concept of the greater common divisor of integers. We went ahead and applied this tool to prove the unique factorization of integers.

Chapter Review Exercises

Mark the following statements true or false.

1. $2^8 \equiv 8^2 \pmod{2^6}$.
2. 3 is a root of the congruence $x^5 + 2x + 1 \equiv 0 \pmod{50}$.
3. There exist integers x and y such that $25x + 137y = 1$.
4. 62 has a multiplicative inverse in $\mathbb{Z}_{91}$.
5. The multiplicative inverse of 63 in $\mathbb{Z}_{100}$ is 25.
6. The prime factorization of 2730 is $2 \cdot 3 \cdot 5 \cdot 91$.
7. $(-12, -18) = 6$.

New Terms

Congruence	57
Euclidean algorithm	63
Fundamental Theorem of Arithmetic	70
Greatest common divisor of integers	62
Modular addition	58
Modular inverses	60
Modular multiplication	58
Modular radicals	68
Relatively prime integers	63

Supplementary Exercises

1. Write a computer script that will find the greatest common divisor of any two integers.

2. Write a computer script that will find the multiplicative inverse of any element in $\mathbb{Z}_p$ for any prime p.
3. Write a computer script that will solve any equation $f(x) = 0$ in $\mathbb{Z}_p$ for any prime p.
4. Write a computer script that will solve any pair of simultaneous linear equations in two unknowns in $\mathbb{Z}_p$ for any prime p.
5. Write a computer script that will factor any integer into primes.
6. Write a computer script that will list all the primes up to some given integer n.
7. Write a computer script that will list all the primes in $\mathbb{Z}[\sqrt{-5}]$.
8. If n is any integer, let

$$\mathbb{Z}[\sqrt{n}] = \{a + b\sqrt{n} \mid a, b \in \mathbb{Z}\}$$

Investigate the question of unique factorization in $\mathbb{Z}[\sqrt{n}]$ for various specific values of n.
9. Compute the number of digits in the largest known perfect number $2^{859,432}(2^{859,433} - 1)$.

CHAPTER 5

The Binomial Theorem and Modular Powers

The well-known Binomial Theorem is proved in this chapter. In addition to its intrinsic interest, this theorem also leads to a simple proof of Fermat's Theorem which, in turn, is useful for evaluating powers in arithmetic modulo a prime.

5.1 THE BINOMIAL THEOREM

The formula
$$(a + b)^2 = a^2 + 2ab + b^2$$
is of course standard fare in high school algebra. We are concerned here with its generalization to higher exponents. The search for this generalization begins with the successive expansions of the third and fourth powers of the binomial $(a + b)$. These are easily obtained recursively as follows:

$$\begin{aligned}(a + b)^3 &= (a + b)^2(a + b) = (a^2 + 2ab + b^2)(a + b) \\ &= a^3 + 2a^2b + ab^2 + a^2b + 2ab^2 + b^3 \\ &= a^3 + (2 + 1)a^2b + (1 + 2)ab^2 + b^3 \\ &= a^3 + 3a^2b + 3ab^2 + b^3\end{aligned}$$

and

$$\begin{aligned}(a + b)^4 &= (a + b)^3(a + b) = (a^3 + 3a^2b + 3ab^2 + b^3)(a + b) \\ &= a^4 + 3a^3b + 3a^2b^2 + ab^3 + a^3b + 3a^2b^2 + 3ab^3 + b^4 \\ &= a^4 + (3 + 1)a^3b + (3 + 3)a^2b^2 + (1 + 3)ab^3 + b^4 \\ &= a^4 + 4a^3b + 6a^2b^2 + 4ab^3 + b^4\end{aligned}$$

5.1 The Binomial Theorem

These low-order examples suggest that *the expansion of $(a + b)^n$ consists of the sum of terms of the form $c_{n,k}a^{n-k}b^k$ $n \geq k \geq 0$, where the coefficient $c_{n,k}$ is a positive integer. Moreover $c_{n,0} = c_{n,n} = 1$, and for each other k, $c_{n,k}$ is the sum of two coefficients of the expansion of $(a + b)^{n-1}$. More precisely $c_{n,k} = c_{n-1,k} + c_{n-1,k-1}$.*

The above observations motivate the following inductive definition as well as the subsequent theorem. For any integers $n \geq k \geq 0$ the *binomial coefficient* $\binom{n}{k}$, which is the traditional way of writing $c_{n,k}$, is defined by

$$\binom{n}{0} = \binom{n}{n} = 1 \tag{1}$$

$$\binom{n}{k} = \binom{n-1}{k} + \binom{n-1}{k-1} \quad \text{for } n > k > 0 \tag{2}$$

Thus, by (1),

$$\binom{0}{0} = \binom{1}{0} = \binom{1}{1} = \binom{2}{0} = \binom{2}{2} = 1$$

whereas by (2),

$$\binom{2}{1} = \binom{1}{1} + \binom{1}{0} = 1 + 1 = 2$$

Similarly

$$\binom{3}{0} = \binom{3}{3} = \binom{4}{0} = \binom{4}{4} = 1$$

and

$$\binom{3}{1} = \binom{2}{1} + \binom{2}{0} = 2 + 1 = 3$$

$$\binom{3}{2} = \binom{2}{2} + \binom{2}{1} = 1 + 2 = 3$$

$$\binom{4}{1} = \binom{3}{1} + \binom{3}{0} = 3 + 1 = 4$$

$$\binom{4}{2} = \binom{3}{2} + \binom{3}{1} = 3 + 3 = 6$$

$$\binom{4}{3} = \binom{3}{3} + \binom{3}{2} = 1 + 3 = 4$$

Since $(a + b)^0 = 1$ and $(a + b)^1 = (a + b)$, these numbers are now easily seen to agree with the coefficients of the expansions of $(a + b)^n$ for $n = 0, 1, 2, 3, 4$.

■ **THE BINOMIAL THEOREM 5.1.** *Let n be a nonnegative integer. Then*

$$(a + b)^n = \binom{n}{0} a^n + \binom{n}{1} a^{n-1}b + \cdots + \binom{n}{k} a^{n-k}b^k + \cdots + \binom{n}{n} b^n$$

Proof. We proceed by induction on n, the cases $n = 0, 1, 2, 3, 4$ having been verified above. Assuming the theorem for $n = m$, we expand

$$(a + b)^{m+1} = (a + b)^m (a + b)$$

$$= \left[\binom{m}{0} a^m + \binom{m}{1} a^{m+1}b + \cdots + \binom{m}{k} a^{m-k}b^k + \cdots + \binom{m}{m} b^m \right] (a + b)$$

$$= \binom{m}{0} a^{m+1} + \binom{m}{1} a^m b + \binom{m}{2} a^{m-1}b^2 + \cdots + \binom{m}{k} a^{m+1-k}b^k + \cdots$$

$$+ \binom{m}{m} ab^m + \binom{m}{0} a^m b + \binom{m}{1} a^{m-1}b^2 + \cdots$$

$$+ \binom{m}{k-1} a^{m-(k-1)}b^k + \cdots + \binom{m}{m} b^{m+1}$$

$$= \binom{m}{0} a^{m+1} + \left[\binom{m}{1} + \binom{m}{0} \right] a^m b + \left[\binom{m}{2} + \binom{m}{1} \right] a^{m-1}b^2 + \cdots$$

$$+ \left[\binom{m}{k} + \binom{m}{k-1} \right] a^{m+1-k}b^k + \cdots + \binom{m}{m} b^{m+1}$$

$$= \binom{m+1}{0} a^{m+1} + \binom{m+1}{1} a^m b + \binom{m+1}{2} a^{m-1}b^2 + \cdots$$

$$+ \binom{m+1}{k} a^{m+1-k}b^k + \cdots + \binom{m+1}{m+1} b^{m+1}$$

This completes the induction step and the proof. ■

Equation (2) is known as *Pascal's identity*. It was visualized by Pascal in the form of a triangular array (Figure 5.1) wherein each number is the sum of the two entries directly above it and to its left (whenever these entries exist). The entries on the $(n + 1)$th diagonal line (from bottom left to top right) are the coefficients that appear in the expansion of $(a + b)^n$. These triangular arrays did not originate with Pascal. Figure 5.2 shows a similar array that appeared over three centuries before Pascal's birth.

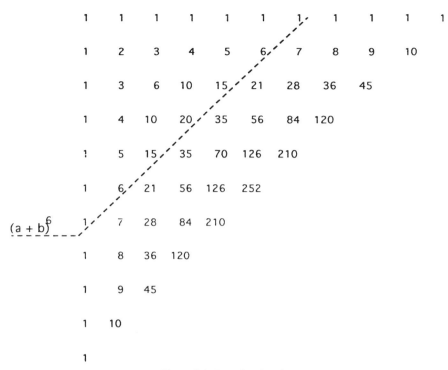

Figure 5.1. Pascal's triangle.

It should be noted that the Binomial Theorem is valid for all the number systems that we have studied so far. It is immaterial whether a and b are to be interpreted as real, complex, or modular numbers. Thus

$$(2 + i)^3 = 2^3 + 3 \cdot 2^2 \cdot i + 3 \cdot 2 \cdot i^2 + i^3 = 8 + 12i - 6 - i = 2 + 11i$$

Similarly, in $\mathbb{Z}_3$,

$$(a + 2)^3 = a^3 + 3 \cdot a^2 \cdot 2 + 3 \cdot a \cdot 2^2 + 2^3 \equiv a^3 + 0 + 0 + 2 = a^3 + 2$$

and, in $\mathbb{Z}_4$,

$$(a + b)^4 = a^4 + 4a^3b + 6a^2b^2 + 4ab^3 + b^4 \equiv a^4 + 2a^2b^2 + b^4$$

In the remainder of this section it may be assumed that the a and the b of the Binomial Theorem are real numbers.

Although Pascal's identity provides an effective procedure for computing the binomial coefficients, it would be clearly handy to have a more direct

78   The Binomial Theorem and Modular Powers

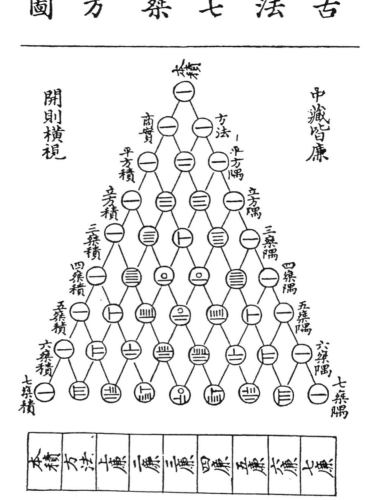

Figure 5.2. Pascal's triangle as depicted in a Chinese book in 1303. (Reproduced from J. Needham 1959, Vol. 3, p. 135. Reprinted with the kind permission of the Cambridge University Press.)

method. To obtain such a formula, the *a* of the Binomial Theorem is replaced by 1, and since we are about to differentiate, the *b* is replaced by *x*. The Binomial Theorem then assumes the form

$$(1+x)^n = 1 + \binom{n}{1}x + \binom{n}{2}x^2 + \cdots + \binom{n}{k}x^k + \cdots + \binom{n}{n}x^n \qquad (3)$$

5.1 The Binomial Theorem

When (3) is differentiated k times with respect to x, we obtain

$$n(n-1)(n-2)\cdots(n-k+1)(1+x)^{n-k}$$
$$= k(k-1)\cdots 2\cdot 1 \binom{n}{k} + d_1 x + d_2 x^2 + \cdots + d_{n-k} x^{n-k} \tag{4}$$

where $d_1, d_2, \ldots, d_{n-k}$ are some integers whose exact values turn out to be immaterial. The substitution of $x = 0$ in (4) yields

$$n(n-1)(n-2)\cdots(n-k+1) = k(k-1)\cdots 2\cdot 1 \binom{n}{k}$$

or

$$\binom{n}{k} = \frac{n(n-1)(n-2)\cdots(n-k+1)}{k(k-1)\cdots 2\cdot 1} \tag{5}$$

Accordingly, for $n = 6$,

$$\binom{6}{1} = \frac{6}{1} = 6, \quad \binom{6}{2} = \frac{6\cdot 5}{2\cdot 1} = 15, \quad \binom{6}{3} = \frac{6\cdot 5\cdot 4}{3\cdot 2\cdot 1} = 20$$
$$\binom{6}{4} = \frac{6\cdot 5\cdot 4\cdot 3}{4\cdot 3\cdot 2\cdot 1} = 15, \quad \binom{6}{5} = \frac{6\cdot 5\cdot 4\cdot 3\cdot 2}{5\cdot 4\cdot 3\cdot 2\cdot 1} = 6, \quad \binom{6}{6} = \frac{6\cdot 5\cdot 4\cdot 3\cdot 2\cdot 1}{6\cdot 5\cdot 4\cdot 3\cdot 2\cdot 1} = 1$$

so that

$$(a+b)^6 = a^6 + 6a^5 b + 15a^4 b^2 + 20a^3 b^3 + 15a^2 b^4 + 6ab^5 + b^6$$

While the formula for $\binom{n}{k}$ is given in (5) is convenient for numerical computations, it is quite frequently better to use the equivalent expression

$$\binom{n}{k} = \frac{n!}{k!(n-k)!} \tag{6}$$

where $k! = k(k-1)(k-2)\cdots 2\cdot 1$ for any positive integer k, and $0! = 1$ (Exercise 15).

The binomial coefficients $\binom{n}{k}$ are the subject of many surprising identities, and we will describe several methods for proving them. The most elementary approach uses formula (5). Such is the case in the proof of the identity

$$\binom{n}{k+1} = \frac{n-k}{k+1}\binom{n}{k} \tag{7}$$

For

$$\binom{n}{k+1} = \frac{n(n-1)\cdots[n-(k+1)+1]}{(k+1)k(k-1)\cdots 2\cdot 1} = \frac{n(n-1)\cdots(n-k)}{(k+1)k(k-1)\cdots 2\cdot 1}$$

$$= \frac{n-k}{k+1}\frac{n(n-1)(n-2)\cdots(n-k+1)}{k(k-1)\cdots 2\cdot 1} = \frac{n-k}{k+1}\binom{n}{k}$$

Equation (7) was put to very good use by Newton in his groundbreaking work on the extension of the Binomial Theorem to negative and fractional values of n.

As demonstrated by the proof of the Binomial Theorem, Pascal's identity (2) is itself very useful as it lays the groundwork for many inductive proofs of other identities. However, many of these identities are subject to shorter proofs by another method, commonly called the method of *generating functions*. Consider, for example, the identity

$$2^n = \binom{n}{0} + \binom{n}{1} + \cdots + \binom{n}{k} + \cdots + \binom{n}{n-1} + \binom{n}{n} \quad \text{for } n \geq 0 \quad (8)$$

While this identity can be proved directly, though somewhat laboriously, by mathematical induction, it can also be verified by substituting $a = b = 1$ in the Binomial Theorem. Thus this method calls for the recognition of some functional identity which, upon the replacement of the variable(s) by some cleverly chosen values, yields the required identity. In fact the above derivation of formula (5) is also an instance of the method of generating functions, as is the following. If both sides of (3) are differentiated with respect to x, we obtain

$$n(1+x)^{n-1} = \binom{n}{1} + 2\binom{n}{2}x + \cdots + k\binom{n}{k}x^{k-1} + \cdots + n\binom{n}{n}x^{n-1}$$

The substitution of $x = 1$ now yields the identity

$$n2^{n-1} = \binom{n}{1} + 2\binom{n}{2} + \cdots + k\binom{n}{k} + \cdots + n\binom{n}{n}$$

Finally, there is another interpretation of the binomial coefficients that allows for a completely different approach to the whole topic. For this we need to reexamine the Binomial Theorem from another point of view. Each summand in the expansion of

$$(a_1 + a_2)(b_1 + b_2)(c_1 + c_2)(d_1 + d_2)\cdots$$

has the form $a_i b_j c_k d_l \ldots$ where each of the indices i, j, k, l assumes the values 1 or 2. When this observation is applied to the expression

$$(1+x)^n = (1+x)(1+x)(1+x)(1+x)\ldots$$

it is clear that each of the summands of this expansion has form $X_1 X_2 \cdots X_n$ where each X_i is either 1 or x. In particular, the general summand $X_1 X_2 \cdots X_n$ is of degree 2 in x if it has the form

$$1 \cdot 1 \cdot \cdots \cdot 1 \cdot x \cdot 1 \cdot \cdots \cdot 1 \cdot x \cdot 1 \cdot \cdots \cdot 1$$

The coefficient $\binom{n}{2}$ clearly equals the number of such summands that appear in the expansion of $(1 + x)^n$. Since the number of such summands equals the number of pairs of symbols X_i, X_j that can be designated for replacement by x, it follows that we now have obtained a formula for the number of ways a pair of distinct objects can be selected from a given set of n distinct objects, namely this number is $\binom{n}{2}$.

Similarly $\binom{n}{3}$ equals the number of summands in the expansion of $(1 + x)^n$ that have degree 3 in x; put differently, it equals the number of summands of the form

$$1 \cdot 1 \cdot \cdots \cdot 1 \cdot x \cdot 1 \cdot \cdots \cdot 1 \cdot x \cdot 1 \cdot \cdots \cdot 1 \cdot x \cdot 1 \cdot \cdots \cdot 1$$

This number equals the number of distinct triples X_i, X_j, X_k that can be selected from $X_1, X_2, \ldots, X_n$. In other words, $\binom{n}{3}$ equals the number of three-element sets that can be formed using the integers $1, 2, \ldots, n$. This generalizes to the following statement:

■ **PROPOSITION 5.2.** *For each pair of nonnegative integers $k \leqslant n$, the binomial coefficient $\binom{n}{k}$ equals the number of k-element sets that can be formed from the integers $1, 2, \ldots, n$.*

The number of 3-person committees that can be selected from a group of 25 people is

$$\binom{25}{3} = \frac{25 \cdot 24 \cdot 23}{3 \cdot 2 \cdot 1} = 2300$$

This point of view provides us with a new tool for proving some facts about the binomial coefficients. Let us again consider identity (8). Its right-hand side now has the obvious interpretation of denoting the number of all the subsets of $\{1, 2, \ldots, n\}$, classified by their cardinality. This number, however, also happens to be 2^n as can be seen by the following argument. Choosing a subset of S of $\{1, 2, \ldots, n\}$ is tantamount to deciding for each $k = 1, 2, \ldots, n$ whether or not k belongs to S. Since n such decisions are to be made, and since each such decision can be settled in one of two ways (to belong or not to belong), it follows that there are 2^n such subsets S.

EXERCISES 5.1

Expand the binomials in Exercises 1–6.

1. $(2x + 3y^2)^7$
2. $(3x^2 - yz^3)^5$
3. $(3x^2 - yz^3)^5$ in $\mathbb{Z}_5$
4. $(3x^2 - yz^3)^5$ in $\mathbb{Z}_6$
5. $(z^2 + 2)^6$ in $\mathbb{Z}_4$
6. $\left(1 - \dfrac{2}{x}\right)^5$

7. Find the term containing a^{26} in the expansion of $(a - 4b^2c^3)^{30}$.
8. Find the coefficient of x^{18} in $\left(x^2 + \dfrac{3}{x}\right)^{15}$.
9. Find the coefficient of x^{18} in $(2x^4 - 3x)^9$.
10. Find the coefficients of x^{-4} and x^{-5} in $\left(x^3 - \dfrac{2}{x^2}\right)^{10}$.
11. Prove that for $m = 4n, 4n - 3, 4n - 6, \ldots, -2n$, the coefficient of x^m in $\left(x^2 + \dfrac{1}{x}\right)^{2n}$ is

$$\dfrac{(2n)!}{\dfrac{4n-m}{3}! \, \dfrac{2n+m}{3}!}$$

12. Prove that for $1 \leq k \leq n$, $\dbinom{n}{k-1} \lessgtr \dbinom{n}{k}$ according as $k \lessgtr \dfrac{n+1}{2}$.
13. Show that the middle coefficient(s) of the expansion of $(1 + x)^n$ is (or are) the largest.
14. Which power of x has the largest coefficient in the expansion of $(2 + 3x)^{17}$?

Prove Exercises 15–19 for any two integers k and n such that $0 \leq k \leq n$.

15. $\dbinom{n}{k} = \dfrac{n!}{k!(n-k)!}$

16. $\dbinom{n}{k} = \dbinom{n}{n-k}$

17. $\dbinom{n}{r}\dbinom{r}{k} = \dbinom{n}{k}\dbinom{n-k}{r-k}$, $(n \geq r \geq k)$

18. $n\dbinom{n}{r} = (r+1)\dbinom{n}{r+1} + r\dbinom{n}{r} = r\dbinom{n+1}{r+1} + \dbinom{n}{r+1}$

19. $\dbinom{n}{2}\dbinom{n}{r} = \dbinom{r+2}{2}\dbinom{n}{r+2} + 2\dbinom{r+1}{2}\dbinom{n}{r+1} + \dbinom{r}{2}\dbinom{n}{r}$

$= \dbinom{r}{2}\dbinom{n+2}{r+2} + 2r\dbinom{n+1}{r+2} + \dbinom{n}{r+2}$

5.1 The Binomial Theorem

20. Prove that the coefficient of the middle term of $(1 + x)^{2n}$ equals the sum of the coefficients of the two middle terms of $(1 + x)^{2n-1}$.

21. Prove that $2^n < \binom{2n}{n} < 4^n$ for $n > 1$.

22. Prove that the product of any k consecutive positive integers is an integer multiple of $k!$.

23. Prove that the number of different 0–1 strings that may be formed with p 0's and q 1's in which no two 1's are consecutive is $\binom{p+1}{q}$. For example, if $p = q = 3$, these strings are 010101, 100101, 101001, 101010.

24. Prove that $\binom{2n}{n}$ is even for $n > 0$.

Prove the identities in Exercises 25–36.

25. $\binom{n}{0} - \binom{n}{1} + \binom{n}{2} - \cdots + (-1)^n \binom{n}{n} = 0 \ (n > 0)$

26. $\binom{n}{1} + \binom{n}{3} + \binom{n}{5} + \cdots = 2^{n-1} \ (n > 0)$

27. $\binom{n}{1} - 2\binom{n}{2} + 3\binom{n}{3} - \cdots + (-1)^{n-1} n \binom{n}{n} = 0 \ (n > 1)$

28. $\binom{n}{0} + \frac{1}{2}\binom{n}{1} + \frac{1}{3}\binom{n}{2} + \cdots + \frac{1}{n+1}\binom{n}{n} = \frac{2^{n+1} - 1}{n+1}$

29. $\binom{n}{0} - \frac{1}{2}\binom{n}{1} + \frac{1}{3}\binom{n}{2} - \cdots + \frac{(-1)^n}{n+1}\binom{n}{n} = \frac{1}{n+1}$

30. $\binom{n}{0}^2 + \binom{n}{1}^2 + \binom{n}{2}^2 + \cdots + \binom{n}{n}^2 = \binom{2n}{n}$

31. $\binom{n}{0} - \binom{n}{2} + \binom{n}{4} - \binom{n}{6} + \cdots = 2^{n/2} \cos\left(\frac{n\pi}{4}\right)$

32. $\binom{n}{0} + \binom{n+1}{1} + \binom{n+2}{2} + \cdots + \binom{n+k}{k} = \binom{n+k+1}{k}$

33. $\binom{n}{0}\binom{m}{k} + \binom{n}{1}\binom{m}{k-1} + \binom{n}{2}\binom{m}{k-2} + \cdots + \binom{n}{k}\binom{m}{0} = \binom{n+m}{k}$

34. $\binom{n}{0}\binom{m}{0} + \binom{n}{1}\binom{m}{1} + \binom{n}{2}\binom{m}{2} + \cdots + \binom{n}{m}\binom{m}{m} = \binom{n+m}{m}$

35. $\dfrac{\binom{n}{1}}{\binom{n}{0}} + \dfrac{2\binom{n}{2}}{\binom{n}{1}} + \dfrac{3\binom{n}{3}}{\binom{n}{2}} + \cdots + \dfrac{n\binom{n}{n}}{\binom{n}{n-1}} = \binom{n+1}{2}$

36. $\left[\binom{n}{0}+\binom{n}{1}\right]\left[\binom{n}{1}+\binom{n}{2}\right]\cdots\left[\binom{n}{n-1}+\binom{n}{n}\right]=\dfrac{\binom{n}{1}\binom{n}{2}\cdots\binom{n}{n}(n+1)^n}{n!}$

Simplify the expressions in Exercises 37–39.

37. $\binom{n}{1}+2^2\binom{n}{2}+3^2\binom{n}{3}+\cdots+n^2\binom{n}{n}$

38. $\binom{n}{1}+2^3\binom{n}{2}+3^3\binom{n}{3}+\cdots+n^3\binom{n}{n}$

39. $\binom{n}{0}\binom{n}{1}+\binom{n}{1}\binom{n}{2}+\binom{n}{2}\binom{n}{3}+\cdots+\binom{n}{n-1}\binom{n}{n}$

40. The Fibonacci numbers F_n are defined inductively as

$$F_1 = F_2 = 1 \quad \text{and} \quad F_{n+2} = F_{n+1} + F_n \quad \text{for } n \geq 1$$

Prove that $F_{n+1} = 1 + \binom{n-1}{1} + \binom{n-2}{2} + \cdots + \binom{n-m}{m}$ for $n \geq 0$, where $m = n/2$ if n is even and $m = (n-1)/2$ if n is odd.

41. Prove that if m is an odd integer then $\binom{m-1}{2} \equiv 1 \pmod{m}$.

42. Prove that if p is a prime and m is an integer such that $m \not\equiv 0 \pmod{p}$, then

$$\binom{p^k m}{p^k} \not\equiv 0 \pmod{p}$$

for all positive integers k.

43. Prove that if k is any positive integer and p is a prime, then $\binom{p^k - 1}{j} \equiv (-1)^j$ $\pmod{p}$ for $j = 0, 1, 2, \ldots, p^k - 1$.

44. The grid $G_{m,n}$ consists of the graphs of the lines $x = i$, $y = j$, $i = 0, 1, \ldots, m$, $j = 0, 1, \ldots, n$, $0 \leq x \leq m$, $0 \leq y \leq n$. (See Figure 5.3.) Prove that $G_{m,n}$ contains $\binom{m+1}{2}\binom{n+1}{2}$ rectangles.

45. A $(0, 0) - (m, n)$ path in the grid $G_{m,n}$ consists of a sequence of integer points (x_0, y_0), $(x_1, y_1), (x_2, y_2), \ldots, (x_{m+n}, y_{m+n})$ such that

$$(x_0, y_0) = (0, 0), \quad (x_{m+n}, y_{m+n}) = (m, n)$$

and for $k = 1, 2, \ldots, m+n$, either

$$(x_k, y_k) = (x_{k-1} + 1, y_{k-1}) \quad \text{or} \quad (x_k, y_k) = (x_{k-1}, y_{k-1} + 1)$$

Prove that the number of distinct $(0, 0) - (m, n)$ paths in $G_{m,n}$ is $\binom{m+n}{n}$.

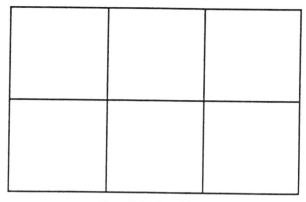

Figure 5.3. The grid $G_{3,2}$.

46. How many triangles are contained in Figure 5.4a?
47. How many triangles are contained in Figure 5.4b?
48. Find the error in Figure 5.2.

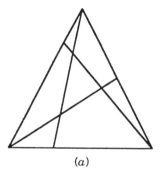   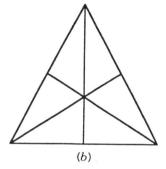

(a)   (b)

Figure 5.4. A triangle counting problem.

5.2 FERMAT'S THEOREM AND MODULAR POWERS

We now return to modular arithmetic and address the issue of exponentiation. Consider the infinite sequence

$$2, 2^2, 2^3, 2^4, 2^5, 2^6, 2^7, 2^8, 2^9, \ldots \tag{9}$$

in $\mathbb{Z}_5$. Since $\mathbb{Z}_5$ contains only five elements, it is clear that the actual values of these powers must display many repetitions. In fact, when these powers are evaluated, the sequence is transformed into

$$2, 4, 3, 1, 2, 4, 3, 1, 2, \ldots \tag{10}$$

and so sequence (9), and therefore also sequence (10), will cycle indefinitely through the values 2, 4, 3, 1. In general, if $a \in \mathbb{Z}_n$, the sequence

$$a, a^2, a^3, a^4, \ldots \tag{11}$$

is bound to eventually cycle, since its individual terms run through a finite number of values. Once $a^k \equiv a^m \pmod{n}$, we must clearly have $a^{k+s} \equiv a^{m+s} \pmod{n}$ for $s = 0, 1, 2, \ldots$. What is not obvious is that when n is a prime p, the cycling begins immediately and length of this cycle is either $p - 1$ or some divisor thereof. We now set out to prove these interesting facts.

As was observed in the previous section, the Binomial Theorem also holds in $\mathbb{Z}_m$. Under certain circumstances, this modular Binomial Theorem assumes a very simple form. If p is a prime number and $0 < k < p$, then the binomial coefficient $\binom{p}{k}$ is an integer that equals

$$\frac{p(p-1)\cdots(p-k+1)}{k(k-1)\cdots 1}$$

Since p is a prime greater than k, p is relatively prime to the denominator of this fraction, and since this fraction is known to cancel out to an integer, p must be a prime factor of this integer. In other words,

$$\binom{p}{k} \equiv 0 \pmod{p} \text{ if } 0 < k < p$$

When this observation is applied to the expansion of $(a + b)^p$, we obtain the following curious fact:

■ **PROPOSITION 5.3.** *If p is a prime integer and $a, b \in \mathbb{Z}_p$, then*

$$(a + b)^p \equiv a^p + b^p \pmod{p}$$

This proposition, in turn, has as its consequence one of the fundamental elementary and nonobvious theorems of number theory. It was first pointed out by Fermat as a tool in the search for *perfect numbers* (Exercise 4.4.12), but has since completely transcended that narrow context.

■ **FERMAT'S THEOREM 5.4.** *If p is any prime integer and a is any integer, then $a^p \equiv a \pmod{p}$.*

5.2. Fermat's Theorem and Modular Powers

Proof. We proceed by induction on a, the theorem being clearly true for $a = 0$. Assuming that the theorem holds for $a = k$, we note that

$$(k+1)^p \equiv k^p + 1^p \equiv k + 1 \pmod{p}$$

Hence, by induction, the theorem holds for all the nonnegative values of a. Since every negative integer is congruent to some positive integer modulo p, the theorem holds for all values of a. ∎

Fermat's Theorem guarantees that sequence (11) starts cycling immediately, a phenomenon that need not happen in $\mathbb{Z}_n$ when n is composite (Exercise 27). It still remains to demonstrate that the length of the cycle is a divisor of $p - 1$, and this will be done shortly.

An interesting and typical consequence of this theorem is the fact that if n is relatively prime to 7, then $n^6 - 1$ is divisible by 7. By Fermat's Theorem

$$n^7 \equiv n \pmod{7}$$

Since n is relatively prime to 7, it follows that $n \not\equiv 0 \pmod{7}$, and so division by n yields

$$n^6 \equiv 1 \pmod{7}$$

which is tantamount to saying that 7 divides $n^6 - 1$. It is clear that the same holds for every other prime.

■ **COROLLARY 5.5.** *If p is any prime and $a \not\equiv 0 \pmod{p}$, then $a^{p-1} \equiv 1 \pmod{p}$.*

Corollary 5.5 can be interpreted as saying that every nonzero element of $\mathbb{Z}_p$ (p prime) is a $(p-1)$th root of unity (mod p). Such roots of unity can exist in $\mathbb{Z}_n$ for composite n as well. Thus

$$3^4 \equiv 5^4 \equiv 7^2 \equiv 1 \pmod{16}$$

We will henceforth understand the term *root of unity* to include both the complex ones and the appropriate elements of $\mathbb{Z}_n$ for all integers n. When the need arises, we will refer to the latter as *modular roots of unity*.

TABLE 5.1. Orders of elements of $\mathbb{Z}_7$.

m	1	2	3	4	5	6
Order of m	1	3	6	3	6	2

For any modular root of unity r in $\mathbb{Z}_n$, we define $o(r)$, the *order* of r in $\mathbb{Z}_n$ to be the least positive integer k such that

$$r^k \equiv 1 \pmod{n}$$

Corollary 5.5 guarantees that every nonzero element of $\mathbb{Z}_p$, p prime, has a finite order which is in fact no greater than $p - 1$. Table 5.1 lists the orders of the nonzero elements of $\mathbb{Z}_7$.

The fact that all the orders of Table 5.1 are divisors of $6 = 7 - 1$ is no coincidence. Modular order enjoys the same properties as does the order of the complex roots of unity, and we restate these properties in this new context without proof. It is easily verified that the proofs of Proposition 2.14 and Corollary 2.15 carry over to this new context without any modifications whatsoever.

■ **PROPOSITION 5.6.** *If r is any modular root of unity in $\mathbb{Z}_n$ and k is any integer then $r^k \equiv 1 \pmod{n}$ if and only if k is a multiple of $o(r)$.*

■ **COROLLARY 5.7.** *Suppose r is a root of unity and a and b are any two integers. Then*

1. *$r^a = r^b$ if and only if $o(r)$ is a divisor of $a - b$,*
2. *$1, r, r^2, \ldots, r^{o(r)-1}$ are all distinct.*

Returning to (11), it follows from Proposition 5.6 that $o(a)$ is a divisor of $p - 1$. As the length of the repeating segment of (11) equals $o(a)$, we may conclude that this length is a divisor of $p - 1$.

An element of $\mathbb{Z}_p$ is said to be a *primitive root (mod p)* if its order in $\mathbb{Z}_p$ is $p - 1$. Thus, according to Table 5.1, 3 and 5 are the only primitive roots (mod 7). We will later see (Theorem 7.8) that for every prime number p there exist primitive roots (mod p). The solutions of Exercises 14–16 indicate that this fact is far from obvious.

We now go on to prove some more facts about orders of roots of unity. If ζ is any root of unity and k is any integer, then it stands to reason that the order of ζ^k should depend on k and the order of ζ. Thus, if $o(\zeta) = 12$, then the orders of $\zeta^2, \zeta^3, \ldots, \zeta^{11}$ are easily seen to be 6, 4, 3, 12, 2, 12, 3, 4, 6, 12, respectively. A little experimentation leads to the statement, if not the proof, of the following proposition:

■ **PROPOSITION 5.8.** *If ζ is a root of unity of order n, then*

$$o(\zeta^k) = \frac{n}{(k, n)} \quad \text{for all integers } k$$

Proof. Let g denote the greatest common divisor of k and n, and let n' and

k' be integers such that

$$n = gn' \quad \text{and} \quad k = gk'$$

Since $g = (k, n)$, it follows that k' and n' are relatively prime. If m is any integer, then the following statements are all equivalent:

$(\zeta^k)^m = 1$

n divides km

gn' divides $gk'm$

n' divides $k'm$

n' divides m

Hence $o(\zeta^k)$, the least positive integer m for which $(\zeta^k)^m = 1$, is also the least positive integer that is divisible by n', namely n' itself. Consequently

$$o(\zeta^k) = n' = \frac{n}{g} = \frac{n}{(k, n)} \qquad \blacksquare$$

The relevance of common divisors to the orders of roots of unity is reinforced by the next proposition for which we will eventually find several useful applications.

■ **PROPOSITION 5.9.** *If r and s are roots of unity (both complex or both modular) and if the orders of r and s are relatively prime, then*

$$o(rs) = o(r)o(s)$$

Proof. Let $R = o(r)$, $S = o(s)$, and $T = o(rs)$. Since

$$(rs)^{RS} = (r^R)^S (s^S)^R = 1$$

it follows that T is a divisor of RS. Conversely

$$1 = (rs)^{TS} = r^{TS} s^{ST} = r^{TS} \cdot 1 = r^{TS}$$

It therefore follows from Proposition 5.6 that R is a divisor of TS. Since R and S are relatively prime, it follows from Lemma 4.4 that R is a divisor of T. A similar argument permits us to conclude that S is also a divisor of T. Since R and S are relatively prime, it now follows from Corollary 4.5 that RS is a divisor of T. Thus we have shown that T and RS are each other's divisors and the proposition follows. ■

Thus, if

$$\zeta = \cos\frac{2\pi}{12} + i\sin\frac{2\pi}{12}$$

then the elements ζ^8 and ζ^9 of $\sqrt[12]{1}$ have orders 3 and 4, respectively, and the element $\zeta^5 = \zeta^8\zeta^9$ has order $12 = 3 \cdot 4$. Similarly, in $\mathbb{Z}_{19}$, $o(18) = 2$, $o(7) = 3$, and $o(18 \cdot 7) = o(12) = 6$.

Exercise 22 implies that the requirement of relative primeness in the above proposition is indeed necessary.

EXERCISES 5.2

1. Evaluate x^{1000} for each element x of $\mathbb{Z}_7$.
2. Evaluate x^{2000} for each element x of $\mathbb{Z}_{11}$.

Solve the equations in Exercises 3–5 in $\mathbb{Z}_7$.

3. $x^{7^{7777}} + x + 5 \equiv 0$
4. $x^{6^{7777}} + x + 5 \equiv 0$
5. $x^{5^{7777}} + x + 5 \equiv 0$
6. Solve Exercise 3 in $\mathbb{Z}_{11}$.
7. Solve Exercise 4 in $\mathbb{Z}_{13}$.
8. Solve Exercise 5 in $\mathbb{Z}_{17}$.
9. Prove that $n^7 - n$ is divisible by 42 for any integer n.
10. Prove that $n^{13} - n$ is divisible by 2730 for every integer n.
11. Prove that $n^5 - n$ is divisible by 30 for all integers n and by 240 for all odd integers n.
12. Prove that $n^{561} \equiv n \pmod{561}$. Note that this example disproves the converse of Fermat's Theorem. Composite numbers with this property are called *Carmichael numbers*; 561 is the smallest Carmichael number, and it was only recently proved that there are infinitely many such numbers.
13. For any prime p, if $a^p \equiv b^p \pmod{p}$, show that $a^p \equiv b^p \pmod{p^2}$.
14. Find the units digit of 2^{400}.
15. Find all the primitive roots (mod 11).
16. Find all the primitive roots (mod 19).
17. For each prime $p < 20$ find a number that is a primitive root (mod p).
18. Let p be a fixed prime, and let $o(a)$ denote the order of a in $\mathbb{Z}_p$. Prove that if $a \neq 1$, then

$$1 + a + a^2 + \cdots + a^{o(a)-1} \equiv 0 \pmod{p}$$

19. Let p be a fixed prime, and let $o(a)$ denote the order of a in $\mathbb{Z}_p$. Prove that

$$a \cdot a^2 \cdots \cdot a^{o(a)-1} \equiv (-1)^{o(a)-1} \pmod{p}$$

20. Let p be any prime number. Prove that $(p-1)! \equiv -1 \pmod{p}$.
21. Let p be any prime number. Prove that if there are primitive roots $\pmod{p}$, then the product of all of them is either 1 or 2 modulo p.
22. Prove that for any odd prime p there exist nonzero elements a and b of $\mathbb{Z}_p$ such that $o(ab) \neq o(a)o(b)$.
23. The Fibonacci number F_n is defined recursively as $F_1 = F_2 = 1$, and $F_n = F_{n-1} + F_{n-2}$ for $n > 2$.
    (a) Prove that $F_n = \dfrac{1}{\sqrt{5}} \left[\left(\dfrac{1+\sqrt{5}}{2} \right)^n - \left(\dfrac{1-\sqrt{5}}{2} \right)^n \right]$ for $n = 1, 2, \ldots$. (This is known as Binet's formula.)
    (b) Prove that if p is a prime distinct from 5, then $F_p \equiv \pm 1 \pmod{p}$.
24. Show that the equation $x^n + y^n \equiv z^n \pmod{3}$ has nonzero solutions in $\mathbb{Z}_3$ if and only if n is odd.
25. Show that the equation $x^n + y^n \equiv z^n \pmod{5}$ has nonzero solutions in $\mathbb{Z}_5$ if and only if n is odd.
26. For which positive integers n does the equation $x^n + y^n \equiv z^n \pmod{7}$ have nonzero solutions in $\mathbb{Z}_7$?
27. Find integers a, n such that $a \in \mathbb{Z}_n$ but the sequence (11) does not cycle immediately.

5.3 THE MULTINOMIAL THEOREM*

Having obtained the Binomial Theorem that describes the expansion of $(a+b)^n$, it is natural to ask for analogous expressions for $(a+b+c)^n$, $(a+b+c+d)^n$, and so on. It turns out that these expressions are harder to describe than to derive. We begin by changing the variables to subscripted x's and observe that for any v such variables and for any positive integer n, the expansion of

$$(x_1 + x_2 + \cdots + x_v)^n$$

consists of the sum of terms each of which has the form

$$c x_1^{k_1} x_2^{k_2} \cdots x_v^{k_v}$$

where $k_1 + k_2 + \cdots + k_v = n$ and c is some positive integer that depends on $n, k_1, k_2, \ldots, k_v$.

■ **THE MULTINOMIAL THEOREM 5.10.** *If n and v are any positive integers, then*

$$(x_1 + x_2 + \cdots + x_v)^n = \sum_K \frac{n!}{k_1! k_2! \cdots k_v!} x_1^{k_1} x_2^{k_2} \cdots x_v^{k_v}$$

*Optional.

where K varies over all the v-tuples $K = (k_1, k_2, \ldots, k_v)$ of nonnegative integers $k_1, k_2, \ldots, k_v$ such that $k_1 + k_2 + \cdots + k_v = n$.

Proof. Let n and v be fixed positive integers. As noted above, for each v-tuple $K = (k_1, k_2, \ldots, k_v)$ of the above format there exists an integer c_K such that

$$(x_1 + x_2 + \cdots + x_v)^n = \sum_K c_K x_1^{k_1} x_2^{k_2} \cdots x_v^{k_v} \qquad (12)$$

Fix some such $K = (k_1, k_2, \ldots, k_v)$ and differentiate k_i times both of the sides of (12) with respect to x_i for each $i = 1, 2, \ldots, v$. Since $k_1 + k_2 + \cdots + k_v = n$,

$$\frac{\partial^n}{\partial^{k_1} x_1 \, \partial^{k_2} x_2 \cdots \partial^{k_v} x_v} (x_1 + x_2 + \cdots + x_v)^n = n!$$

Moreover, for any v-tuple $(m_1, m_2, \ldots, m_v)$ that is different from $(k_1, k_2, \ldots, k_v)$ and for which $m_1 + m_2 + \cdots + m_v = n$, we must have $m_i < k_i$ for some i so that

$$\frac{\partial^n}{\partial^{k_1} x_1 \, \partial^{k_2} x_2 \cdots \partial^{k_v} x_v} (x_1^{m_1} x_2^{m_2} \cdots x_v^{m_v}) = \begin{cases} k_1! k_2! \cdots k_v! & \text{if } m_i = k_i \text{ for all } i \\ 0 & \text{otherwise} \end{cases}$$

Hence

$$n! = c_K k_1! k_2! \cdots k_v!$$

or

$$c_K = \frac{n!}{k_1! k_2! \cdots k_v!} \qquad \blacksquare$$

Accordingly the coefficient of $x^4 y^3 z$ in the expansion of $(x + y + z)^8$ is

$$\frac{8!}{4!3!1!} = 280$$

EXERCISES 5.3

Expand $(x + y + z)^n$ *for the values of n specified in Exercises 1–3.*

1. 2   2. 3   3. 4

Expand the multinomials in Exercises 4–6.

4. $(x^2 + y^3 + z)^3$
5. $(x^2 + y^3 + xy)^3$
6. $(xy + yz + zx)^3$
7. Determine the coefficient of $x^4 y^3 z^5$ in the expansion of $(x + y + 2z)^{12}$.

8. Find the coefficient of x^3 in the expansion of $(1 + 2x - 3x^2)^5$.
9. Find the coefficient of x^4 in the expansion of $(1 - 3x + 2x^2)^6$.
10. Find the coefficient of x^3 in the expansion of $(1 - 2x + 3x^2 - 4x^3)^7$.

Suppose that $(1 + x + x^2 + \cdots + x^k)^n = a_0 + a_1 x + a_2 x^2 + \cdots + a_{kn} x^{kn}$. *Solve Exercises 11–15.*

11. Show that $a_0 + a_1 + a_2 + \cdots + a_{kn} = (k + 1)^n$.
12. Evaluate $a_1 + 2a_2 + 3a_3 + \cdots + kna_{kn}$.
13. Show that when $k = 2$,

$$a_0^2 - a_1^2 + a_2^2 - a_3^2 + \cdots + (-1)^{n-1} a_{n-1}^2 = \frac{a_n}{2}[1 - (-1)^n a_n]$$

14. Show that when $k = 2$,

$$a_0 + a_3 + a_6 + \cdots = a_1 + a_4 + a_7 + \cdots = a_2 + a_5 + a_8 + \cdots = 3^{n-1}$$

15. Show that when $k = 3$, $a_m = \sum_{j=0}^{n} \binom{n}{j}\binom{n}{m-2j}$.
16. Prove that the number of terms in the expansion of $(x_1 + x_2 + \cdots + x_v)^n$ is $\binom{n+v-1}{n}$.

5.4 THE EULER ϕ-FUNCTION*

Proposition 4.2 guarantees that whenever m is relatively prime to n, it has a multiplicative inverse in $\mathbb{Z}_n$, and according to Exercise 4.2.26 this is actually the complete answer. Namely, if m is not relatively prime to n, then it does not possess a multiplicative inverse in $\mathbb{Z}_n$. Thus 1 and 5 are the only elements of $\mathbb{Z}_6$ that have multiplicative inverses, whereas 1, 3, 5, 7 are the only elements of $\mathbb{Z}_8$ that have multiplicative inverses. This raises the interesting question of just how many elements of $\mathbb{Z}_n$ do, in general, have such inverses. Since the resulting formula has some bearing on the complex roots of unity and on other subsequent issues, it will be derived here. For any positive integer n, let $\phi(n)$ denote the number of positive integers not greater than n that are relatively prime to n. This is known as the *Euler ϕ-function*. As noted above, $\phi(6) = 2$ and $\phi(8) = 4$. It is clear that if p is a prime, then $\phi(p) = p - 1$, since every positive number less than p is relatively prime to p. In fact, if p is a prime and m is a positive integer, then

$$\phi(p^m) = p^m - p^{m-1}$$

since the only numbers between 1 and p^m that are not relatively prime to p^m are $\{p, 2p, 3p, 4p, \ldots, (p^{m-1})p\}$, and there are clearly exactly p^{m-1} of those. As

*Optional.

every number n can be factored into the form

$$p_1^{m_1} p_2^{m_2} \cdots p_k^{m_k}$$

where $p_1, p_2, \ldots, p_k$ are distinct primes, it is now clear that the following lemma will eventually provide the complete answer:

■ **LEMMA 5.11.** *If m and n are relatively prime positive integers, then*

$$\phi(mn) = \phi(m)\phi(n)$$

Proof. Let

$$\zeta = \cos \frac{2\pi}{m} + i \sin \frac{2\pi}{m} \quad \text{and} \quad \eta = \cos \frac{2\pi}{n} + i \sin \frac{2\pi}{n}$$

It follows from Proposition 5.8 that $(k, m) = 1$ if and only if ζ^k is a primitive mth root of unity. Hence we can interpret $\phi(m)$ as the number of primitive mth roots of unity. The proposition will be proved by demonstrating that all of the primitive mnth roots of unity are obtained, without repetition, when an arbitrary primitive mth root of unity is multiplied by an arbitrary primitive nth root of unity.

We first dispose of the possible redundancies in this process. Thus suppose that ζ^a and ζ^x are two primitive mth roots of unity with $1 \leq a, x < m$, that η^b and η^y are two primitive nth roots of unity with $1 \leq b, y < n$, and that

$$\zeta^a \eta^b = \zeta^x \eta^y$$

Then

$$\zeta^{a-x} = \eta^{y-b}$$

and consequently

$$\zeta^{(a-x)n} = (\zeta^{a-x})^n = (\eta^{y-b})^n = (\eta^n)^{y-b} = (1)^{y-b} = 1$$

Since ζ is a primitive mth root of unity, it follows that m is a divisor of $(a-x)n$. However, m and n are relatively prime, and so we may conclude that m is a divisor of $(a-x)$ alone. Since both a and x are between 0 and m, it follows that $a = x$. A similar argument allows us to conclude that $b = y$. Thus it has been demonstrated that the set of all products of primitive mth roots by primitive nth roots contains exactly $\phi(m)\phi(n)$ distinct elements.

It follows from Proposition 5.9 that each of the products $\zeta^a \eta^b$, with ζ, η, a, b as above, has order mn and is therefore a primitive mnth root of unity. It therefore only remains to show that every primitive mnth root is accounted for by this process.

Let α be any primitive mnth root of unity, and let A and B be two integers such that

$$Am + Bn = 1 \tag{13}$$

Then clearly

$$\alpha = \alpha^{Am}\alpha^{Bn}$$

Now

$$(\alpha^{Am})^n = (\alpha^{mn})^A = 1^A = 1$$

and so α^{Am} is an nth root of unity. We know from (13) that $(A, n) = 1$. It therefore follows from Proposition 5.8 that α^{Am} is in fact primitive, since

$$o(\alpha^{Am}) = \frac{mn}{(Am, mn)} = \frac{mn}{m(A, n)} = n$$

A similar argument establishes that α^{Bn} is a primitive mth root of unity, and so products of the primitive mth roots of unity with the primitive nth roots do indeed cover all the primitive mnth roots, each exactly once. Thus $\phi(m)\phi(n) = \phi(mn)$. ∎

We are now ready to derive an explicit formula for the number of positive integers that are both less than n and relatively prime to it. By Proposition 5.8, this is also equal to the number of primitive nth roots of unity.

■ **THEOREM 5.12.** *If n is any number with prime factorization*

$$p_1^{m_1} p_2^{m_2} \cdots p_k^{m_k}$$

then

$$\phi(n) = \prod_{i=1}^{k} (p_i^{m_i} - p_i^{m_i - 1})$$

Proof. By the Lemma 5.11, it suffices to prove this theorem for the case where n is the power of a single prime, that is, where there exist a prime number p and an integer m such that $n = p^m$. However, as was noted just prior to Lemma 5.11, $\phi(p^m) = p^m - p^{m-1}$, and so the theorem now follows immediately. ∎

Since $100 = 2^2 5^2$, it follows that

$$\phi(100) = (2^2 - 2)(5^2 - 5) = 40$$

EXERCISES 5.4

1. Compute $\phi(24)$, $\phi(144)$, and $\phi(1000)$.
2. Prove that $\phi(n)$ is even for $n > 2$.
3. For what values of n is $\phi(n)$ a prime number? Justify your answer.
4. For what values of n is $\phi(n)$ the power of a single prime number? Justify your answer.
5. True or false: There is an infinite number of integers n such that $\phi(n) < 100$? Justify your answer.
6. Prove that if n is any positive integer, then

$$\sum_{d|n} \phi(d) = n$$

7. Let m and $n > 1$ be positive integers such that $\phi(mn) = \phi(m)$. Prove that $n = 2$ and m is odd.
8. Prove that if $g = (m, n)$, then

$$\phi(mn) = \frac{g\phi(m)\phi(n)}{\phi(g)}$$

9. Prove that if $d|n$, then $\phi(d)|\phi(n)$.

CHAPTER SUMMARY

Having proved the Binomial Theorem, we used it to derive Fermat's Theorem for exponents in arithmetic modulo p, which effectively states that the nonzero elements of $\mathbb{Z}_p$ are all $(p-1)$-th roots of unity. This allows us to extend the notion of order to modular arithmetic, and we derived some new theorems regarding the orders of roots of unity that apply to both the complex and the modular ones.

Chapter Review Exercises

Mark the following true or false.

1. $\binom{n}{2} = \frac{n(n-1)}{2}$.
2. $\binom{8}{3} = 56$.
3. The number of pairs that can be formed by selecting two elements from $\{a, b, c, d\}$ is 8.
4. $12^7 \equiv 1 \pmod{17}$.
5. $(3 + 8)^{11} \equiv 3^{11} + 8^{11} \pmod{11}$.
6. $13^{90} \equiv 1 \pmod{31}$.
7. If $o(\zeta) = 144$, then $o(\zeta^{120}) = 102$.

8. $(x + y + z)^5 = x^x + y^5 + z^5 + 5x^4y + 5xy^4 + 5x^4z + 5xz^4 + 5y^4z + 5yz^4 + 10x^2y^3 + 10x^3y^2 + 10x^2z^3 + 10x^3z^2 + 10y^2z^3 + 10y^3z^2$.
9. $\phi(n) < n$ for all integers $n > 2$.
10. $\phi(15) = 8$.

New Terms

Binomial coefficient	75
Generating function	80
Modular roots of unity	87
Order (mod p)	88
Pascal's identity	76
Pascal's triangle	77, 78
Primitive root (mod p)	88

Supplementary Exercises

1. Write a computer script that computes the order of any element modulo p.
2. Write a computer script that evaluates $\binom{m}{n}$ for any two positive integers $m \geq n$.
3. Find $\lim_{n \to \infty} \phi(n)$.
4. Let F_n be the Fibonacci number of Exercise 2.22. Prove that

$$(F_m, F_n) = F_{(m,n)}$$

   and investigate the question of which Fibonacci numbers are prime.
5. For which positive integers n does $\mathbb{Z}_n$ have an element of order $\phi(n)$?
6. For each positive integer n and $a \in \mathbb{Z}_n$ investigate the length of the (eventually) repeating segment of $a, a^2, a^3, a^4, \ldots$.
7. Find some more Carmichael numbers.

CHAPTER 6

Polynomials over a Field

The focus now shifts to the topic of polynomials. Since polynomials have numerical coefficients, and since we have by now encountered a great variety of disparate number systems, the polynomials are studied in the more general setting of abstract fields. We are mainly concerned here with the factorization of polynomials in one variable, but some attention is also given to the symmetric polynomials in several variables and their utility in solving the general quartic equation.

6.1 FIELDS AND THEIR POLYNOMIALS

We have by now encountered a host of algebraic structures within which the four arithmetical operations hold sway. These are the real numbers, the rational numbers, the complex numbers, and arithmetic modulo p where p is a prime. Another collection of such structures will be studied in detail in Chapter 7, and mathematicians have constructed many others that will not be mentioned here. It stands to reason that algebraic structures with such strong similarities will share yet other properties, and these are this chapter's concern. The notion of a *field* is used to extract the properties that are common to all these similar structures.

A *field* is a set F with two binary operations, usually denoted by $+$ and $\cdot$, for which the following hold: For any elements a, b, c of F,

$a + b \in F$	$a \cdot b \in F$	(*closure*)
$(a + b) + c = a + (b + c)$	$(a \cdot b) \cdot c = a \cdot (b \cdot c)$	(*associativity*)
$a + b = b + a$	$a \cdot b = b \cdot a$	(*commutativity*)
	$a \cdot (b + c) = a \cdot b + a \cdot c$	(*distributivity*)

there exist distinct elements $0, 1 \in F$ such that

$$a + 0 = a \qquad a \cdot 1 = 1 \qquad (identities)$$

there exists an element $-a \in F$ such that

$$a + (-a) = 0 \qquad (additive\ inverse)$$

if $a \neq 0$, then there exists an element $a^{-1} \in F$ such that

$$a \cdot a^{-1} = 1 \qquad (multiplicative\ inverse)$$

It will prove useful to label the most familiar fields with some symbols. Accordingly $\mathbb{Q}, \mathbb{R}$, and $\mathbb{C}$ will denote the rational, real, and complex number systems, respectively. The fact that all these number systems are indeed fields will not be belabored here. It should be noted, however, that not all algebraic structures are necessarily fields. If n is a composite number, then $\mathbb{Z}_n$ is not a field, since, as was noted in Chapter 4, no divisor of n except 1 has a multiplicative inverse in $\mathbb{Z}_n$. The set of polynomials with real coefficients is another example of an algebraic structure that is not a field. It is easy to convince oneself that the normal addition and multiplication of such polynomials have all the properties required of a field, except for the last one. The multiplicative inverses of polynomials are not polynomials. For example, there is no polynomial whose product with $x + 1$ is 1 (Exercise 4).

Since the above-listed properties are shared by all fields, it is clear that any proposition whose justification relies only on these common properties is necessarily valid in all fields. In particular, it will hold for $\mathbb{Q}, \mathbb{R}, \mathbb{C}$, for all $\mathbb{Z}_p$ with p prime, and for the new fields to be introduced in the next chapter. The following is an example of such a proposition:

■ **PROPOSITION 6.1.** *If a and b are two elements of the field F, then $a \cdot b = 0$ if and only if either a or b is zero.*

Proof. Set $x = a \cdot 0$. By the distributivity of addition and multiplication,

$$x = a \cdot 0 = a \cdot (0 + 0) = a \cdot 0 + a \cdot 0 = x + x$$

Consequently

$$a \cdot 0 = x = x + (x + (-x)) = (x + x) + (-x) = x + (-x) = 0$$

Conversely, suppose that $a \cdot b = 0$ but $a \neq 0$. Then a has a multiplicative inverse a^{-1}, and so

$$0 = a^{-1} \cdot 0 = a^{-1} \cdot (a \cdot b) = (a^{-1} \cdot a) \cdot b = 1 \cdot b = b$$

Hence either a or b is zero. ∎

It is important to note that Proposition 6.1 does not hold in all algebraic structures. Thus, in $\mathbb{Z}_6$, $2 \cdot 3 \equiv 0$, even though neither 2 nor 3 is zero. Another, more substantial example of a proposition that *does* hold for all fields is the Binomial Theorem. It was already stated in Section 5.1 that the proof of this theorem holds regardless of whether the numbers in question are complex or modular. In fact, the proof of Theorem 5.1 carries over verbatim to arbitrary fields once the meaning of the terms is clarified. For any positive integer m and any element a of some field F, let a^m denote the product of m a's, and let ma denote the sum of m a's. Accordingly

$$a^3 = a \cdot a \cdot a \quad \text{and} \quad 3a = a + a + a$$

If m is a negative integer, we define

$$a^m = (a^{-1})^{-m} \quad \text{and} \quad ma = -(-m)a$$

Finally, we set $a^0 = 1$ and $0a = 0$. It is easily seen that such identities as

$$a^m \cdot a^n = a^{m+n} \quad \text{and} \quad ma + na = (m+n)a$$

hold in this generalized context just as they do for the complex and modular fields (Exercises 24, 25).

■ **THE BINOMIAL THEOREM 6.2.** *Let F be a field, $a, b \in F$, and n a nonnegative integer. Then*

$$(a+b)^n = \binom{n}{0} a^n + \binom{n}{1} a^{n-1}b + \cdots + \binom{n}{k} a^{n-k}b^k + \cdots + \binom{n}{n} b^n$$

We now go on to show that many of the properties of polynomials with real coefficients are also valid when the coefficients are allowed to be the members of an arbitrary field. This will be accomplished by proving that the validity of these properties follows from the defining properties of fields listed above *alone*. For the sake of simplifying the notation, we adopt here the usual convention that the product $a \cdot b$ is abbreviated to ab.

Given a field F, the *variables* $x, y, z, \ldots$ are symbols, or place holders, that can be replaced by elements of F. A *polynomial in x over F* is an expression that is obtained by applying the operations of addition and multiplication to the variable x and/or some of the elements of F. Thus both

$$17x^3 - \frac{31}{2}x \quad \text{and} \quad 5x^2 + 6x + (-6)x^3 + 1 + \frac{2x}{9}$$

are polynomials in x over $\mathbb{Q}$. Actually they can also be interpreted as polynomials over $\mathbb{R}$, over $\mathbb{C}$, and over $\mathbb{Z}_7$ or any $\mathbb{Z}_p$ (for $p \neq 2, 3$) for that matter. On the other hand,

$$x^3 + ix^2 + 1 - 3i \tag{1}$$

is a polynomial over the field of complex numbers $\mathbb{C}$ but not over either the reals or the rational numbers. Moreover, since in $\mathbb{Z}_5$,

$$2^2 \equiv 3^2 \equiv 4 \equiv -1 \pmod{5}$$

it follows that in $\mathbb{Z}_5$ the quantity $i = \sqrt{-1}$ can be interpreted as either 2 or 3, and so the polynomial (1) can be interpreted as a polynomial over $\mathbb{Z}_5$. On the other hand, since $\mathbb{Z}_3$ contains no number a such that $a^2 \equiv 2 \equiv -1 \pmod{3}$, the polynomial (1) cannot be regarded as a polynomial over $\mathbb{Z}_3$. The set of polynomials in x over F is denoted by $F[x]$, and its members will generally be denoted by $P(x), Q(x), \ldots$. If $P(x) \in F[x]$, then we also say that F is the *ground field* of $P(x)$. The two polynomials

$$P(x) = a_0 x^n + a_1 x^{n-1} + a_2 x^{n-2} + \cdots + a_{n-1} x + a_n$$

and

$$Q(x) = b_0 x^m + b_1 x^{m-1} + b_2 x^{m-2} + \cdots + b_{m-1} x + b_m$$

are said to be equal if and only if $m = n$ and $a_i = b_i$ for $i = 0, 1, 2, \ldots, m = n$. In particular, the polynomials x^5 and x are considered to be distinct in $\mathbb{Z}_5[x]$, even though $n^5 \equiv n$ for all $n \in \mathbb{Z}_5$. The reason for this fine distinction will become clear in the next chapter.

Polynomials over an arbitrary field F can be added, subtracted, and multiplied, and these operations possess the usual properties of commutativity, associativity, and distributivity. Thus the following two polynomials over $\mathbb{Z}_5$ are added and multiplied as

$$(2x^3 + 3x + 1) + (4x^3 + 2) = 6x^3 + 3x + 3 = x^3 + 3x + 3$$

and

$$(2x^3 + 3x + 1)(4x^3 + 2) = 8x^6 + 4x^3 + 12x^4 + 6x + 4x^3 + 2$$
$$= 8x^6 + 12x^4 + 8x^3 + 6x + 2$$
$$= 3x^6 + 2x^4 + 3x^3 + x + 2$$

The details of these examples indicate that it is always necessary to keep the ground field in mind, since the final result clearly depends on which field the coefficients belong to. Many significant properties of a polynomial also depend on the ground field. The polynomial $x^2 + 1$ factors over the complex numbers, since

$$x^2 + 1 = (x + i)(x - i)$$

but it is well known that this polynomial cannot be factored over either the rationals or the real numbers. The same polynomial factors over $\mathbb{Z}_5$ as

$$x^2 + 1 = (x + 2)(x + 3)$$

but does not factor over $\mathbb{Z}_7$ (see Proposition 6.5 below).

The polynomial 0 is called the *zero polynomial*. If $P(x)$ is any nonzero polynomial, then it can clearly be written in the standard form

$$a_0 x^n + a_1 x^{n-1} + a_2 x^{n-2} + \cdots + a_{n-1} x + a_n, \qquad a_0 \neq 0$$

If $a_0 = 1$, the polynomial is said to be *monic*. The exponent n is the *degree* of $P(x)$. No degree is assigned to the zero polynomial. A polynomial of degree 0 is said to be a *constant polynomial*. The zero polynomial is also considered to be a *constant polynomial*. The proof of the following proposition is straightforward and is relegated to Exercises 22, 23:

■ **PROPOSITION 6.3.** *Let $P(x)$ and $Q(x)$ be polynomials of degrees m and n. Then*

1. *if $P(x) + Q(x)$ is nonzero, then degree of $P(x) + Q(x) \leq \max\{m, n\}$,*
2. *degree of $P(x)Q(x) = m + n$.*

We now address the issue of division of polynomials in $F[x]$. Much like the integers, polynomials are subject to a process of long division. Because of the fundamental significance of this process, we will prove its validity for polynomials over arbitrary ground fields. For the sake of completeness we next offer a proposition that justifies the process of long division in the general context of fields. The examples that follow the proof should clarify it and may actually obviate the need for such a proof. Given any two elements a and $b \neq 0$ of a field F, we use the symbol $\frac{a}{b}$ to denote $a \cdot b^{-1}$ in the usual way.

■ **PROPOSITION 6.4.** *If $P(x)$ and $D(x) \neq 0$ are two polynomials over F, then there exist polynomials $Q(x)$, $R(x) \in F[x]$ such that*

$$P(x) = D(x)Q(x) + R(x) \tag{2}$$

and if $R(x)$ is not the zero polynomial, then

$$\text{degree of } R(x) < \text{degree of } D(x)$$

Proof. Suppose that

$$P(x) = a_0 x^m + a_1 x^{m-1} + \cdots + a_m, \qquad a_0 \neq 0$$

and

$$D(x) = b_0 x^d + b_1 x^{d-1} + \cdots + b_d, \qquad b_0 \neq 0$$

If $m < d$, then we can clearly choose $Q(x) = 0$ and $R(x) = P(x)$. Hence we may assume that $m \geq d$. We now proceed by induction on m and assume that the theorem holds for all pairs of polynomials $P(x)$, $D(x)$ with degrees less than some fixed integer $m \geq d$. Let $P(x)$ and $D(x)$ be as given, and define the new polynomial

$$P_1(x) = \frac{a_0}{b_0} x^{m-d} D(x) = \frac{a_0}{b_0} x^{m-d} (b_0 x^d + b_1 x^{d-1} + \cdots + b_d)$$

$$= a_0 x^m + \frac{a_0 b_1}{b_0} x^{m-1} + \cdots + \frac{a_0 b_d}{b_0} x^{m-d}$$

Then, since $P(x)$ and $P_1(x)$ have the same degree and the same first coefficient, it follows that $P(x) - P_1(x)$ is either 0 or else it is a polynomial of degree less than their common degree m. In the first case $P(x) = P_1(x)$, and we can use

$$Q(x) = \frac{a_0}{b_0} x^{m-d} \quad \text{and} \quad R(x) = 0$$

to obtain (2). In the second case we use the induction hypothesis on the degrees. Accordingly there exist polynomials $Q_1(x)$ and $R(x)$, with $R(x)$ either 0 or else of degree less than d, such that

$$P(x) - P_1(x) = Q_1(x) D(x) + R(x)$$

But then

$$P(x) = P_1(x) + Q_1(x) D(x) + R(x)$$

$$= \frac{a_0}{b_0} x^{m-d} D(x) + Q_1(x) D(x) + R(x)$$

$$= \left[\frac{a_0}{b_0} x^{m-d} + Q_1(x) \right] D(x) + R(x)$$

Then, with

$$Q(x) = \frac{a_0}{b_0} x^{m-d} + Q_1(x)$$

the proof is concluded. ∎

Like most inductive proofs, the proof of the above proposition is constructive and contains a method for finding the quotient $Q(x)$ and the remainder $R(x)$ of any long division of polynomials. We demonstrate this by dividing the polynomial $x^5 + x^4 + x^2 + 1$ by the polynomial $x^3 + x + 1$.

$$\begin{array}{r}
x^2 + x - 1 \\
x^3 + x + 1 \overline{\smash{\big)} x^5 + x^4 + x^2 + 1} \\
\underline{x^5 + x^3 + x^2 } \\
x^4 - x^3 \\
\underline{x^4 + x^2 + x } \\
-x^3 - x^2 - x + 1 \\
\underline{-x^3 - x - 1} \\
-x^2 + 2
\end{array}$$

In this case, where the coefficients are real, $Q(x) = x^2 + x - 1$ and $R(x) = -x^2 + 2$.

On the other hand, if the coefficients are taken as elements of $\mathbb{Z}_2$, we get

$$\begin{array}{r}
x^2 + x + 1 \\
x^3 + x + 1 \overline{\smash{\big)} x^5 + x^4 + x^2 + 1} \\
\underline{x^5 + x^3 + x^2 } \\
x^4 + x^3 \\
\underline{x^4 + x^2 + x } \\
x^3 + x^2 + x + 1 \\
\underline{x^3 + x + 1} \\
x^2
\end{array}$$

so here $Q(x) = x^2 + x + 1$ and $R(x) = x^2$.

Finally, if the division is undertaken over $\mathbb{Z}_3$, we get

$$
\begin{array}{r}
x^2 + x + 2 \\
x^3 + x + 1 \overline{\smash{\big)}\, x^5 + x^4 + x^2 + 1}\\
\underline{x^5 + x^3 + x^2 }\\
x^4 + 2x^3 \\
\underline{x^4 + x^2 + x }\\
2x^3 + 2x^2 + 2x + 1\\
\underline{2x^3 + 2x + 2}\\
2x^2 + 2
\end{array}
$$

so in this case $Q(x) = x^2 + x + 2$ and $R(x) = 2(x^2 + 1)$.

EXERCISES 6.1

1. Find the quotient and remainder when $x^7 + x^4 + x + 1$ is divided by $x^3 + x^2 + 1$ over $\mathbb{Z}_2$.
2. Repeat Exercise 1 over $\mathbb{Z}_3$.
3. Repeat Exercise 1 over $\mathbb{Z}_5$.
4. Prove that there is no polynomial $P(x)$ with real coefficients such that

$$P(x)(x + 1) = 1$$

5. Prove that if n is a composite number, then $\mathbb{Z}_n$ has nonzero elements a and b such that $ab = 0$.

Prove that the identities in Exercises 6–9 hold in every field.

6. $(a + b)(a - b) = a^2 - b^2$
7. $(a + b)(a^2 - ab + b^2) = a^3 + b^3$
8. $(a - b)(a^2 + ab + b^2) = a^3 - b^3$
9. $1 + a + a^2 + \cdots + a^n = \dfrac{a^{n+1} - 1}{a - 1}$ if $a \neq 1$
10. Expand $(x + 1)^6$ over $\mathbb{Z}_2$.
11. Expand $(x + 1)^6$ over $\mathbb{Z}_3$.
12. Expand $(2x + 3)^6$ over $\mathbb{Z}_5$.
13. Expand $(2x + 3)^6$ over $\mathbb{Z}_7$.

Suppose that a, b, c, d, e, f are nonzero elements of field F such that

$$\frac{a}{b} = \frac{c}{d} = \frac{e}{f}$$

Prove the identities in Exercises 14–17 whenever the denominators in question are nonzero.

14. $\dfrac{a+b}{a-b} = \dfrac{c+d}{c-d}$

15. $\dfrac{a}{b} = \dfrac{a+c+e}{b+d+f}$

16. $\dfrac{a}{b} = \dfrac{a+2c-3e}{b+2d-3f}$

17. $\dfrac{a}{b} = \sqrt[3]{\dfrac{a^3-15c^3+9e^3}{b^3-15d^3+9f^3}}$

Let F be any field and let a, x be elements of F. Prove the statements in Exercises 18–21.

18. If $a + x = a$, then $x = 0$.
19. If $ax = a$ and $a \neq 0$, then $x = 1$.
20. If $a + x = 0$, then $x = -a$.
21. If $ax = 1$, then $x = a^{-1}$.
22. Prove part 1 of Proposition 6.3, and explain why equality need not hold.
23. Prove part 2 of Proposition 6.3.
24. Prove that if m and n are positive integers and $a \in F$, then $a^m \cdot a^n = a^{m+n}$ and $(a^m)^n = a^{mn}$.
25. Prove that if m and n are integers and $a \in F$, then $ma + na = (m+n)a$ and $m(na) = (mn)a$.
26. Prove Exercise 24 when m and n are arbitrary integers.

6.2 THE FACTORIZATION OF POLYNOMIALS

If $P(x)$ is a polynomial in $F[x]$, and if a is an element of the field F such that $P(a) = 0$, then we say that a is a *zero* of $P(x)$. Thus

-3 is a zero of $2x + 6$ if $F = \mathbb{Q}$

5 is a zero of $x^3 + 3x$ if $F = \mathbb{Z}_7$

$\sqrt{7}$ is a zero of $x^2 - 7$ if $F = \mathbb{R}$

$1 + i$ is a zero of $x^4 + 4$ if $F = \mathbb{C}$

Every precalculus algebra text contains a proposition that relates the zeros of a polynomial to its factorization. The same relationship holds for polynomials over arbitrary fields.

6.2 The Factorization of Polynomials

■ **PROPOSITION 6.5.** *Let $P(x)$ be a polynomial over the field F. Then $a \in F$ is a zero of $P(x)$ if and only if there exists a polynomial $Q(x)$ such that*

$$P(x) = Q(x)(x - a)$$

Proof. If such a polynomial $Q(x)$ does exist, then clearly

$$P(a) = Q(a)(a - a) = 0$$

and so a is indeed a zero of $P(x)$. Conversely, suppose that a is a zero of $P(x)$. Set $D(x) = x - a$, and let $Q(x)$ and $R(x)$ be the polynomials whose existence is guaranteed by Proposition 6.4. Then

$$P(x) = Q(x)(x - a) + R(x)$$

If $R(x)$ is the zero polynomial, we are done. Otherwise, the degree of $R(x)$ is less than that of $D(x) = x - a$ which is 1. Hence $R(x)$ is a polynomial of degree zero, and for some $r \in F$,

$$P(x) = Q(x)(x - a) + r$$

If we now substitute $x = a$, we obtain

$$0 = P(a) = Q(a)(a - a) + r = 0 + r = r$$

so

$$P(x) = Q(x)(x - a) \qquad ■$$

This proposition has a corollary whose straightforward inductive proof is relegated to Exercise 22.

■ **COROLLARY 6.6.** *If $P(x)$ is a polynomial over the field F and if $a_1, a_2, \ldots, a_k \in F$ are distinct zeros of $P(x)$, then there exists a polynomial $Q(x)$ such that*

$$P(x) = Q(x)(x - a_1)(x - a_2) \cdots (x - a_k)$$

If $P(x)$, $Q(x)$, and $R(x)$ are polynomials over F such that

$$P(x) = Q(x)R(x)$$

then we say that $Q(x)R(x)$ is a *factorization* of $P(x)$ over F and that $Q(x)$ and $R(x)$ are *factors* or *divisors* of $P(x)$. If $P(x)$ is a nonconstant polynomial such that in every factorization $P(x) = Q(x)R(x)$ either $Q(x)$ or $R(x)$ has degree 0 (i.e., is a nonzero constant), then $P(x)$ is said to be *irreducible over F*.

The above proposition greatly facilitates the task of factoring polynomials. The polynomial $P_1(x) = x^2 + x + 3$ has the two zeros 1 and 3 over $\mathbb{Z}_5$, since

$$P_1(1) = 5 \equiv 0 \quad \text{and} \quad P_1(3) = 15 \equiv 0 \pmod{5}$$

Hence it follows from Proposition 6.5 that $(x - 1) = (x + 4)$ and $(x - 3) = (x + 2)$ are factors of $x^2 + x + 3$ (over $\mathbb{Z}_5$). Since $x^2 + x + 3$ has degree 2, it can have at most two factors. Consequently, by Corollary 6.6,

$$x^2 + x + 3 = (x + 2)(x + 4) \quad \text{over } \mathbb{Z}_5$$

Similarly the polynomial $P_2(x) = 2x^2 + 2x + 3$ has the two zeros 1 and 5 (mod 7). Thus $(x - 1)(x - 5) = (x + 6)(x + 2)$ is a divisor of $P_2(x)$. Since the leading coefficient of $P_2(x)$ is 2, it follows that

$$2x^2 + 2x + 3 = 2(x + 2)(x + 6) \quad \text{over } \mathbb{Z}_7$$

The polynomial $P_3(x) = x^2 + x + 1$ has no zeros in $\mathbb{Z}_2$, since

$$P_3(0) = 1 \neq 0 \quad \text{and} \quad P_3(1) = 3 \neq 0 \pmod{2}$$

Since every nonconstant factor of $P_3(x)$ would have form $x - a$, $a \in \{0, 1\}$, and since neither 0 nor 1 are zeros of $P_3(x)$, it now follows from Proposition 6.5 that this polynomial is irreducible over $\mathbb{Z}_2$. On the other hand, since $P_3(1) \equiv 0$ (mod 3), it follows that $P_3(x)$ does factor over $\mathbb{Z}_3$ in a nontrivial way.

It is easy to read too much into Corollary 6.6. The polynomial $x^4 + x^2 + 1$ has no zeros in $\mathbb{Z}_2$ for the same reasons that $P_3(x)$ does not. Nevertheless,

$$(x^2 + x + 1)^2 = (x^2 + x + 1)(x^2 + x + 1)$$
$$= x^4 + x^3 + x^2 + x^3 + x^2 + x + x^2 + x + 1 = x^4 + x^2 + 1$$

Thus *Corollary 6.6 only supplies information about first degree factors. It may fail to detect the existence of factors of a higher degree.*

The question of which polynomials of degree greater than 3 are irreducible is quite difficult and cannot be discussed here in its full generality. In the case where the ground field is $\mathbb{Z}_p$, however, the finiteness of p can be used to make some headway. Thus the irreducible polynomials of degree 1, 2, or 3 over $\mathbb{Z}_2$ are (Exercise 1)

$$x, \quad x + 1, \quad x^2 + x + 1, \quad x^3 + x + 1, \quad x^3 + x^2 + 1$$

Hence the list of reducible fourth-degree polynomials over $\mathbb{Z}_2$ is

x^4
$x^2(x+1)^2 = x^4 + x^2$
$(x+1)^4 = x^4 + 1$
$x(x+1)(x^2+x+1) = x^4 + x$
$(x^2+x+1)^2 = x^4 + x^2 + 1$
$x(x^3+x^2+1) = x^4 + x^3 + x$
$(x+1)(x^3+x^2+1) = x^4 + x^2 + x + 1$

$x^3(x+1) = x^4 + x^3$
$x(x+1)^3 = x^4 + x^3 + x^2 + x$
$x^2(x^2+x+1) = x^4 + x^3 + x^2$
$(x+1)^2(x^2+x+1) = x^4 + x^3 + x + 1$
$x(x^3+x+1) = x^4 + x^2 + x$
$(x+1)(x^3+x+1) = x^4 + x^3 + x^2 + 1$

The remaining 3 fourth-degree polynomials over $\mathbb{Z}_2$, namely

$$x^4 + x^3 + 1, \quad x^4 + x + 1, \quad \text{and} \quad x^4 + x^3 + x^2 + x + 1$$

must therefore be irreducible.

It is important to realize that the same polynomial may be factorable over one field and irreducible over another. Thus we saw above that $x^2 + x + 1$ is irreducible over $\mathbb{Z}_2$, whereas it is easily verified that

$$x^2 + x + 1 = (x + 2)^2 \quad \text{over } \mathbb{Z}_3$$

Similarly the polynomial $x^2 + 1$ is irreducible over $\mathbb{R}$ (it has degree 2 and no zeros in $\mathbb{R}$), whereas it factors into $(x + i)(x - i)$ over $\mathbb{C}$.

We conclude this section by extending to arbitrary fields yet another fact that is well known for polynomials over the real numbers.

■ **PROPOSITION 6.7.** *If $P(x)$ is a polynomial of degree n over any field F, then the equation $P(x) = 0$ has at most n distinct solutions.*

Proof. This is proved by induction on n. When $n = 0$, $P(x)$ must be a constant polynomial c for some $c \neq 0$. In this case the polynomial equation $P(x) = 0$ has the form

$$c = 0$$

which has no (i.e., 0) solutions. Hence the induction process has been anchored at $n = 0$.

Let $P(x)$ be a polynomial of degree $n > 0$, and suppose that the theorem has been proved for all polynomials of degree less than n. If $P(x)$ has no zeros, then we are done. Suppose therefore that $r \in F$ is a zero of $P(x)$. Proposition 6.5 implies the existence of a polynomial $Q(x)$ over F such that

$$P(x) = Q(x)(x - r)$$

It is clear that $Q(x)$ has degree $n - 1$. If s is any zero of $P(x)$ that is distinct

from r, then
$$0 = P(s) = Q(s)(s - r)$$

Since s is distinct from r, it follows that $s - r \neq 0$, and hence, by Proposition 6.1, we may conclude that $Q(s) = 0$. Thus all the zeros of $P(x)$ that are distinct from r are zeros of $Q(x)$. Since $Q(x)$ has degree $n - 1$, there are, by the induction hypothesis, at most $n - 1$ zeros of $P(x)$ that are distinct from r. In other words, $P(x)$ has at most n distinct zeros. ∎

The actual number of distinct solutions of a polynomial equation will vary with the polynomial. It is easily verified by direct substitution that the equation

$$x^3 + x^2 + x + 1 \equiv 0 \pmod{5}$$

has the solutions $x = 2, 3, 4$ in $\mathbb{Z}_5$, whereas the equation

$$x^3 + 3x + 4 \equiv 0 \pmod{5}$$

has only two distinct solutions, namely $x = 3, 4$. However, since

$$x^3 + 3x + 4 = (x - 3)^2(x - 4) \qquad \text{over } \mathbb{Z}_5$$

we say that 3 is a double zero of the polynomial $x^3 + 3x + 4$, and consequently, counting multiplicities, this polynomial has three zeros. In general, if r is any zero of the polynomial $P(x)$, it is said to have *multiplicity* m if m is the largest integer such that $(x - r)^m$ divides $P(x)$. Thus, as was seen above, 3 and 4 are zeros of multiplicities 2 and 1 respectively, of the polynomial $x^3 + 3x + 4$ over $\mathbb{Z}_5$. A slight modification of the proof of Proposition 6.7, together with the Fundamental Theorem of Algebra that was stated without proof in Section 3.3, yields the following fact whose proof is relegated to Exercise 20.

∎ **PROPOSITION 6.8.** *Counting multiplicities, every polynomial of degree n over the complex numbers has exactly n complex zeros.*

This proposition is in some sense also true even when $P(x)$ has its coefficients in other fields (Exercise 10.2.23).

EXERCISES 6.2

1. List and completely factor all the polynomials of degree $d \leq 4$ over $\mathbb{Z}_2$.
2. List and completely factor all the polynomials of the form $x^5 + ax^2 + bx + c$ over $\mathbb{Z}_2$.
3. List and completely factor all the monic quadratic polynomials over $\mathbb{Z}_3$.
4. List and completely factor all the cubic polynomials of the form $x^3 + x^2 + ax + b$ over $\mathbb{Z}_3$.
5. Completely factor all the polynomials of the form $x^3 + ax + 1$ over $\mathbb{Z}_5$.
6. Prove that the number of irreducible monic quadratic polynomials over $\mathbb{Z}_p$ is $\binom{p}{2}$.

6.3 The Euclidean Algorithm for Polynomials

7. Find a formula for the number of irreducible monic cubic polynomials over $\mathbb{Z}_p$.

8. Prove that if F is a finite field then there is a quadratic polynomial in $F[x]$ that is irreducible over F.

9. Suppose that the polynomial $a_0 x^n + a_1 x^{n-1} + \cdots + a_{n-1} x + a_n$ is irreducible over a field F, with $a_0, a_n \neq 0$. Prove that the polynomial $a_n x^n + a_{n-1} x^{n-1} + \cdots + a_1 x + a_0$ is also irreducible over F.

10. Prove that if the polynomial $P(x) \in F[x]$ is divided by $x - a$, then the remainder is $P(a)$.

11. Let a be any element of the field F, and let $P(x)$ be a polynomial over F. Prove that $P(x)$ is irreducible over F if and only if $P(x + a)$ is irreducible over F.

Find the remainder when $P(x) = x^{25} - 2x^4 + 3x^3 - 4x^2 + 5x - 1 \in \mathbb{C}[x]$ is divided by the polynomials in Exercises 12–15.

12. $x - 1$ 13. $x^2 - x$

14. $x^2 - 1$ 15. $x^2 + 1$

16. For which values of a and b will $x^2 + 2$ be a factor of $x^{17} + ax + b$ over $\mathbb{Z}_3$?

17. Repeat Exercise 16 over $\mathbb{Z}_{11}$.

18. Let F be an arbitrary field. Do there exist $a, b \in F$ such that $x^2 + 2$ is a divisor of $x^{17} + ax + b$ over F?

19. Repeat Exercise 16 over $\mathbb{R}$.

20. Repeat Exercise 16 over $\mathbb{C}$.

21. Prove Proposition 6.8 assuming the Fundamental Theorem of Algebra (Section 3.3).

22. Prove Corollary 6.6.

23. Prove that the polynomial $x^4 + x^3 + x^2 + x + 1$ is irreducible over $\mathbb{Q}$.

6.3 THE EUCLIDEAN ALGORITHM FOR POLYNOMIALS

The process of long division was used once before in Chapter 4 in connection with the Euclidean algorithm for finding the greatest common divisor of two integers. It proves equally handy in finding the greatest common divisor of two polynomials. We define a *greatest common divisor* of the polynomials $M(x)$ and $N(x)$ as a common divisor of maximum degree. The reason for the use of the indefinite article in this definition is that if the polynomial $P(x)$ is a divisor of $Q(x)$ and c is any nonzero number in the ground field, then so is $cP(x)$ a divisor of $Q(x)$. Indeed, if

$$Q(x) = P(x)R(x)$$

then also

$$Q(x) = [cP(x)]\left[\frac{1}{c}R(x)\right]$$

Thus, if $G(x)$ is any greatest common divisor of two polynomials, then so is $cG(x)$ whenever $c \neq 0$. Exercise 8 calls for a proof that any two greatest common divisors of $M(x)$ and $N(x)$ are indeed such multiples of each other, but until then some caution must be exercised.

To find a greatest common divisor of two given polynomials it suffices to mimic the Euclidean algorithm for integers. We begin with a lemma that implies that the long division process of Proposition 6.4 can be used to reduce the degrees of the polynomials in question.

■ **LEMMA 6.9.** *Suppose that $P(x)$, $Q(x)$, $D(x)$, and $R(x)$ are polynomials over the field F such that*

$$P(x) = D(x)Q(x) + R(x)$$

then every greatest common divisor of $P(x)$ and $D(x)$ is also a greatest common divisor of $D(x)$ and $R(x)$, and vice versa.

Proof. Every common divisor of $D(x)$ and $R(x)$ is also a divisor of $D(x)P(x) + R(x) = P(x)$, and hence it is also a common divisor of $D(x)$ and $P(x)$. Conversely, every common divisor of $P(x)$ and $D(x)$ is also a divisor of $P(x) - D(x)Q(x) = R(x)$, and hence it is also a common divisor of $D(x)$ and $R(x)$. The complete statement of the lemma now follows immediately ■

Let $M(x)$ and $N(x)$ be two polynomials. Set $M_1(x) = M(x)$ and $N_1(x) = N(x)$, and let $Q_1(x)$ and $R_1(X)$ be the appropriate quotient and remainder so that $M_1(x) = Q_1(x)N_1(x) + R_1(x)$. For $i = 1, 2, 3, \ldots$ we set

$$M_{i+1}(x) = N_i(x) \quad \text{and} \quad N_{i+1}(x) = R_i(x) \qquad (4)$$

with $Q_{i+1}(x)$ and $R_{i+1}(x)$ being the respective quotient and remainder when $M_{i+1}(x)$ is divided by $N_{i+1}(x)$. Figure 6.1 should be helpful.

Note that if $R_i(x)$ is not the zero polynomial, then either $R_{i+1}(x)$ is the zero polynomial or else

$$\text{degree of } R_{i+1}(x) < \text{degree of } N_{i+1}(x) = \text{degree of } R_i(x)$$

Hence this procedure is bound to eventually produce a remainder $R_k(x)$ that is the zero polynomial, at which point the algorithm stops. We claim that $R_{k-1}(x)$ is a greatest common divisor of $M(x)$ and $N(X)$. To see this, first note

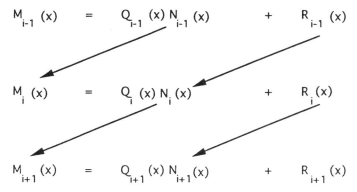

Figure 6.1. The Euclidean algorithm for polynomials.

that the bottom line of Figure 6.1 is equivalent to

$$R_{i-1}(x) = Q_{i+1}(x)R_i(x) + R_{i+1}(x), \quad i = 2, 3, \ldots, k-1$$

Hence, by the above lemma, for $i = 2, 3, \ldots, k-1$ any greatest common divisor of $R_{i-1}(x)$ and $R_i(x)$ is also a greatest common divisor of $R_i(x)$ and $R_{i+1}(x)$, and vice versa.

A similar argument allows us to conclude that any greatest common divisor of $R_2(x)$ and $R_1(x)$ is also a greatest common divisor of $M(x) = M_1(x)$ and $N(x) = N_1(x)$ (Exercise 20). Hence, since $R_{k-1}(x)$ is a greatest common divisor of $R_{k-1}(x)$ and $R_k(x)$, it is also a greatest common divisor of $M(x)$ and $N(x)$.

If the ground field is $\mathbb{Z}_2$ and

$$M(x) = M_1(x) = x^8 + x^7 + x^6 + x^4 + x^3 + x + 1$$

and

$$N(x) = N_1(x) = x^5 + x^4 + x^3 + x^2 + x + 1$$

then two long divisions yield

$$M_2(x) = N_1(x) = x^5 + x^4 + x^3 + x^2 + x + 1$$
$$N_2(x) = R_1(x) = x^4 + x^3 + x^2$$
$$M_3(x) = N_2(X) = x^4 + x^3 + x^2$$
$$N_3(x) = R_2(x) = x^2 + x + 1$$

and

$$R_3(x) = 0$$

Thus $R_2(x) = x^2 + x + 1$ is the required greatest common divisor of $M(x)$ and $N(x)$.

It will subsequently prove useful to have a polynomial version of Proposition 4.1 available.

■ **PROPOSITION 6.10.** *If $G(x)$ is a greatest common divisor of the polynomials $M(x)$ and $N(x)$ over the field F, then there exist polynomials $A(x)$ and $B(x)$ over F such that*

$$A(x)M(x) + B(x)N(x) = G(x)$$

Proof. The proof we give applies only to the greatest common divisor obtained by the Euclidean algorithm, which we will call the *Euclidean greatest common divisor*. The proposition's validity for all greatest common divisors then follows from Exercises 7 and 8.

We mimic the inductive proof of Proposition 4.1 and use the notation employed above in the description of the Euclidean algorithm for polynomials. Let k be the number of divisions in the application of the Euclidean algorithm to $M(x)$ and $N(x)$. If $k = 1$, this means that $R_1(x)$ is the zero polynomial so that $N(x)$ is a divisor of $M(x)$ and the Euclidean greatest common divisor $G(x)$ of $M(x)$ and $N(x)$ is $N(x)$ itself. Thus, choosing $A(x) = 0$ and $B(x) = 1$, we get

$$0 \cdot M(x) + 1 \cdot N(x) = G(x)$$

Assume that the theorem holds for all pairs of polynomials for which the Euclidean algorithm requires $k - 1$ divisions. If $M(x)$ and $N(x)$ are a pair that require k divisions to arrive at their Euclidean greatest common divisor $G(x)$, then the pair $M_2(x)$ and $N_2(x)$ given by (4) requires only $k - 1$ divisions to arrive at their Euclidean greatest common divisor which is also $G(x)$. By the induction hypothesis, there exist polynomials $A'(x)$, $B'(x)$ such that

$$A'(x)M_2(x) + B'(x)N_2(x) = G(x)$$

However, if $Q_1(x)$ and $R_1(x)$ are the quotient and remainder obtained when $M(x)$ is divided by $N(x)$, then

$$M_2(x) = N_1(x) \quad \text{and} \quad N_2(x) = R_1(x) = M_1(x) - Q_1(x)N_1(x)$$

We can conclude that

$$A'(x)N_1(x) + B'(x)[M_1(x) - Q_1(x)N_1(x)] = G(x)$$

or
$$B'(x)M(x) + [A'(x) - B'(x)Q_1(x)]N(x) = G(x)$$

so $A(x) = B'(x)$ and $B(x) = A'(x) - B'(x)Q_1(x)$ are the required polynomials. ∎

Let $x^4 + x^3 + x + 1$ and $x^5 + x^2 + x + 1$ be two polynomials in $\mathbb{Z}_2[x]$. Then

$$x^5 + x^2 + x + 1 = (x + 1)(x^4 + x^3 + x + 1) + (x^3 + x)$$
$$x^4 + x^3 + x + 1 = (x + 1)(x^3 + x) + (x^2 + 1)$$
$$x^3 + x = x(x^2 + 1)$$

Hence $x^2 + 1$ is a greatest common divisor of the two given polynomials, and

$$x^2 + 1 = (x^4 + x^3 + x + 1) + (x + 1)(x^3 + x)$$
$$= (x^4 + x^3 + x + 1) + (x + 1)[(x^5 + x^2 + x + 1) + (x + 1)(x^4 + x^3 + x + 1)]$$
$$= [1 + (x + 1)^2](x^4 + x^3 + x + 1) + (x + 1)(x^5 + x^2 + x + 1)$$
$$= x^2(x^4 + x^3 + x + 1) + (x + 1)(x^5 + x^2 + x + 1)$$

Irreducible polynomials are the analogs of prime numbers, and just like the integers, polynomials have a unique factorization theorem. The formal statement of this theorem and its proof are relegated to Exercises 15, 16.

Two polynomials are said to be *relatively prime* if their only common divisors are constants (i.e., elements of the ground field F). It is clear that such relatively prime polynomials have 1 as their greatest common divisor. Consequently we have the following proposition, which will turn out to be very useful, in a nonobvious way, in Section 7.1.

■ **PROPOSITION 6.11.** *If $M(x)$ and $N(x)$ are relatively prime polynomials over a field F, then there exist polynomials $A(x)$, $B(x) \in F[x]$ such that*

$$A(x)M(x) + B(x)N(x) = 1$$

EXERCISES 6.3

1. Find a greatest common divisor of the polynomials $x^5 + x + 1$ and $x^6 + x^5 + x^4 + x^3 + 1$ over $\mathbb{Z}_2$.
2. Find a greatest common divisor of the polynomials $x^6 + x^4 + x + 1$ and $x^7 + x^4 + x^3 + 1$ over $\mathbb{Z}_2$.
3. Find a greatest common divisor of the polynomials $x^5 + x^4 + 2x^3 + x + 2$ and $x^6 + 2x^4 + x^2 + 2$ over $\mathbb{Z}_3$.

4. Find a greatest common divisor of the polynomials $x^6 + x^5 + x^4 + 2$ and $x^7 + x^6 + x^5 + 2x^3 + x^2 + 2x + 1$ over $\mathbb{Z}_3$.
5. Repeat Exercise 1 over $\mathbb{Z}_5$.
6. Repeat Exercise 3 over $\mathbb{Z}_5$.
7. Let $M(x)$ and $N(x)$ be any two polynomials over an arbitrary field F, and suppose that $D(x)$ is another such polynomial that divides both $M(x)$ and $N(x)$. Prove that $D(x)$ divides the Euclidean greatest common divisor of $M(x)$ and $N(x)$.
8. Let $G(x)$ be the Euclidean greatest common divisor of $M(x)$, $N(x) \in F[x]$, and let $H(x)$ be any polynomial over F. Prove that $H(x)$ is a greatest common divisor of $M(x)$ and $N(x)$ if and only if there exists a nonzero constant $c \in F$ such that $H(x) = cG(x)$.
9. Do there exist polynomials $A(x)$ and $B(x)$ over $\mathbb{R}$ such that

$$A(x)(x^2 + 3x + 2) + B(x)(x^2 - 1) = x + 2$$

10. Do there exist polynomials $A(x)$ and $B(x)$ over $\mathbb{R}$ such that

$$A(x)(x^2 + 3x + 2) + B(x)(x^2 - 1) = x^2 - x - 2$$

11. Do there exist polynomials $A(x)$ and $B(x)$ over $\mathbb{R}$ such that

$$A(x)(x^2 - 5x + 6) + B(x)(x^2 + x - 6) = x^2 + 1$$

12. Do there exist polynomials $A(x)$ and $B(x)$ over $\mathbb{R}$ such that

$$A(x)(x^2 - 4) + B(x)(x^2 + 2x - 8) = x^2 - 2x$$

13. Let $M(x)$ and $N(x)$ be any two polynomials over an arbitrary field F. Characterize all the polynomials K that can be expressed in the form

$$K(x) = A(x)M(x) + B(x)N(x)$$

for some polynomials $A(x)$ and $B(x)$ over F.
14. Let $M(x)$ and $N(x)$ be any two polynomials over an arbitrary field F. Of all the polynomials that can be expressed in the form

$$A(x)M(x) + B(x)N(x)$$

for some $A(x), B(x) \in F[x]$, let $H(x)$ be one that possesses minimum degree. Prove that $H(x)$ is a greatest common divisor of $M(x)$ and $N(x)$.
15. Let $P(x), M(x), N(x)$ be polynomials over F such that $P(x)$ is irreducible and $P(x)$ is a divisor of the product $M(x)N(x)$. Prove that $P(x)$ is a divisor of either $M(x)$ or $N(x)$.
16. Let $N(x)$ be any monic polynomial over the field F. Prove that there exist monic irreducible polynomials $P_1(x), P_2(x), \ldots, P_h(x)$ and positive integers $r_1, r_2, \ldots, r_h$

6.3 The Euclidean Algorithm for Polynomials

such that

$$N(x) = P_1^{r_1}(x) P_2^{r_2}(x) \cdots P_h^{r_h}(x)$$

Moreover, if $Q_1(x), Q_2(x), \ldots, Q_k(x)$ is another set of irreducible monic polynomials, and $s_1, s_2, \ldots, s_k$ is another set of positive integers such that

$$N(x) = Q_1^{s_1}(x) Q_2^{s_2}(x) \cdots Q_k^{s_k}(x)$$

then $h = k$, and the Q_i's can be reindexed so that $P_i = Q_i$ and $r_i = s_i$ for $i = 1, 2, \ldots, h$. (Hint: See Theorem 4.6.)

17. Suppose the polynomials $x^3 + 3px^2 + 3qx + r$ and $x^2 + 2px + q$ have a nonconstant greatest common divisor. Show that

$$4(p^2 - q)(q^2 - pr) - (pq - r)^2 = 0$$

18. Prove that the polynomial $P(x) \in \mathbb{C}[x]$ has a zero of multiplicity at least two if and only if $(P(x), P'(x))$ is not a constant, where $P'(x)$ is the derivative of $P(x)$.

19. Prove that if the polynomial $ax^3 + 3bx^2 + 3cx + d \in \mathbb{C}[x]$ has a zero of multiplicity 2, then this zero is

$$\frac{bc - ad}{2(ac - b^2)}$$

20. Supply the missing details in the verification of the Euclidean algorithm for polynomials by proving that any greatest common divisor of $R_2(x)$ and $R_1(x)$ is also a greatest common divisor of $M(x)$ and $N(x)$.

21. Let ζ be a complex primitive nth root of unity. Prove that

$$\prod_{k=1}^{n} (x - \zeta^k) = x^n - 1$$

22. Let a be any primitive root modulo p for some prime p. Prove that

$$\prod_{k=1}^{p-1} (x - a^k) = x^{p-1} - 1 \qquad \text{over } \mathbb{Z}_p$$

23. For each positive integer n, let $\zeta_{1,n}, \zeta_{2,n}, \ldots, \zeta_{m,n}$ be all the complex primitive nth roots of unity (m will depend on n). Prove that the polynomial

$$P_n(x) = \prod_{k=1}^{m} (x - \zeta_{k,n})$$

has real rational coefficients for each n.

24. Prove that if $P(x)$ and $Q(x)$ are polynomials over both the fields $F \subset F'$, then the greatest common divisor of $P(x)$ and $Q(x)$ over F is also their greatest common divisor over F'.

25. The polynomials $x + 1$ and 1 are greatest common divisors of $x + 1$ and $x^2 + 1$ over $\mathbb{Z}_2$ and $\mathbb{Z}_3$, respectively. Explain why this does not contradict Exercise 24.

26. Let $M(x) = x^5 + x + 1$ and $N(x) = x^6 + x^5 + x^4 + x^3 + 1$ be the polynomials of Exercise 1, and let $G(x)$ be their greatest common divisor over $\mathbb{Z}_2$. Find polynomials $A(x)$ and $B(x)$ such that $G(x) = A(x)M(x) + B(x)N(x)$ over $\mathbb{Z}_2$.

27. Let $M(x) = x^6 + x^4 + x + 1$ and $N(x) = x^7 + x^4 + x^3 + 1$ be the polynomials of Exercise 2, and let $G(x)$ be their greatest common divisor over $\mathbb{Z}_2$. Find polynomials $A(x)$ and $B(x)$ such that $G(x) = A(x)M(x) + B(x)N(x)$ over $\mathbb{Z}_2$.

6.4 ELEMENTARY SYMMETRIC POLYNOMIALS

It was already observed in Chapter 1 that there is a close relationship between the coefficients of a quadratic equation and its roots. Namely, if r and s are the roots of the quadratic equation

$$ax^2 + bx + c = 0$$

then

$$r + s = -\frac{b}{a} \quad \text{and} \quad rs = \frac{c}{a}$$

This will now be generalized to polynomial equations of arbitrary degrees and with coefficients in arbitrary fields. First, however, we simplify the statements of the subsequent theorems by restricting attention to monic polynomials. Thus, for the monic quadratic equation,

$$x^2 + bx + c = 0$$

we have, by Proposition 1.1,

$$r + s = -b \quad \text{and} \quad rs = c$$

In general, suppose that we have a monic polynomial

$$P(x) = x^n + a_1 x^{n-1} + a_2 x^{n-2} + \cdots + a_{n-1} x + a_n \qquad (5)$$

with coefficients in an arbitrary field F. Suppose further that $P(x)$ has been factored into linear factors so that

$$P(x) = (x - r_1)(x - r_2) \cdots (x - r_i) \cdots (x - r_n) \qquad (6)$$

Then, if all the $(x - r_i)$ of (6) are multiplied out, and if all like terms are added,

6.4 Elementary Symmetric Polynomials

the right-hand side of (5) should be obtained. Before these summands are added, there are 2^n of them, each having the form

$$(-1)^k A_1 A_2 \cdots A_i \cdots A_n \tag{7}$$

where each A_i is either x or r_i, and k is the number of the A_i's that equal r_i. For example, there are n summands that contain $n - 1$ x's, namely

$$(-1)^1 r_1 xxx \cdots xx = -r_1 x^{n-1}$$
$$(-1)^1 xr_2 xx \cdots xx = -r_2 x^{n-1}$$
$$(-1)^1 xxr_3 x \cdots xx = -r_3 x^{n-1}$$
$$\cdots$$
$$(-1)^1 xxxx \cdots xr_n = -r_n x^{n-1}$$

The sum of these terms must agree with the x^{n-1} term in the right-hand side of (5), and we conclude that

$$-r_1 x^{n-1} - r_2 x^{n-1} - r_3 x^{n-1} - \cdots - r_n x^{n-1} = a_1 x^{n-1}$$

or

$$r_1 + r_2 + r_3 + \cdots + r_n = -a_1$$

Similarly, each summand (7) that contains $n - 2$ x's has two of its A_i's equal to the corresponding r_i's, the rest of the A_i's being x, and the value of k is 2. Thus we conclude that

$$r_1 r_2 x^{n-2} + r_1 r_3 x^{n-2} + \cdots + r_1 r_n x^{n-2} + r_2 r_3 x^{n-2}$$
$$+ r_2 r_4 x^{n-2} + \cdots + r_{n-1} r_n x^{n-2} = a_2 x^{n-2}$$

or

$$r_1 r_2 + r_1 r_3 + \cdots + r_1 r_n + r_2 r_3 + r_2 r_4 + \cdots + r_{n-1} r_n = a_2$$

The pattern and its justification should now be clear, and we only need a definition before this discussion can be summarized in a theorem. Let $r_1, r_2, \ldots, r_n$ be a sequence of numbers (in any field), and let k be any positive integer $1 \leq k \leq n$. Then we denote by $\Sigma r_1 r_2 \cdots r_k$ the sum of all the products of the form

$$r_{i_1} r_{i_2} \cdots r_{i_k}$$

where $1 \leq i_1 < i_2 < \cdots < i_k \leq n$. Thus

$$\sum r_1 = r_1 + r_2 + \cdots + r_n$$
$$\sum r_1 r_2 = r_1 r_2 + r_1 r_3 + \cdots + r_1 r_n + r_2 r_3 + r_2 r_4 + \cdots + r_{n-1} r_n$$

and

$$\sum r_1 r_2 \cdots r_n = r_1 r_2 \cdots r_n$$

At this point it is convenient to extend the notion of a polynomial to several variables. If $x, y, z, \ldots$ are variables, then any function that is obtained by adding, subtracting, or multiplying these variables and/or elements of the ground field F is called a *polynomial over F*. Thus

$$197 x^6 y^7 z^{73} + \frac{33}{17} x^2 - \frac{xy + yz + xz}{6}$$

is a polynomial over any field F in which 17, 3, and 2 are all distinct from 0. The polynomials $\sum r_1, \sum r_1 r_2, \ldots, \sum r_1 r_2 \cdots r_n$ are called the *elementary symmetric polynomials*. The above considerations prove the following theorem:

■ **THEOREM 6.12.** *Suppose that the monic polynomial*

$$P(x) = x^n + a_1 x^{n-1} + a_2 x^{n-2} + \cdots + a_{n-1} x + a_n$$
$$= (x - r_1)(x - r_2) \cdots (x - r_n)$$

Then

$$\sum r_1 r_2 \cdots r_k = (-1)^k a_k, \qquad k = 1, 2, \ldots, n$$

This theorem can of course be used to conclude that the sum of the roots of the equation $x^3 + 6x + 5 = 0$ is 0 (the coefficient of x^2) and that their product is $(-1)^3 5 = -5$. It can, however, be brought to bear on other interesting expressions as well. Let $\sum r_i^\alpha r_j^\beta r_k^\gamma \ldots$ denote the sum of all the *distinct* monomials obtained by permuting the indices $i, j, k, \ldots$ in the monomial $r_i^\alpha r_j^\beta r_k^\gamma \ldots$ Then $\sum r_1^\alpha r_2^\beta r_3^\gamma \ldots$ can be expressed in terms of the elementary symmetric polynomials. This general fact, which is known as the *Fundamental Theorem of Symmetric Polynomials*, will not be proved here. Instead, some special cases will be considered. The expression $\sum r_i^2$, which denotes the sum of the squares of the zeros of the polynomial $P(x)$ of (5) can be evaluated, with the help of the Multinomial Theorem, as follows:

$$\sum r_1^2 = \left(\sum r_1 \right)^2 - 2 \sum r_1 r_2 = a_1^2 - 2(-1)^2 a_2 = a_1^2 - 2a_2$$

6.4 Elementary Symmetric Polynomials

In particular, the sum of the squares of the roots of the equation $x^3 + 6x + 5 = 0$ is $0^2 - 2 \cdot 6 = -12$. Similarly, if r_1, r_2, r_3, r_4 are the roots of the equation $x^4 + 5x^3 - 3x^2 + 7x + 10 = 0$, then

$$\sum \frac{1}{r_1} = \frac{\sum r_1 r_2 r_3}{r_1 r_2 r_3 r_4} = -\frac{7}{10}$$

The next proposition provides the general framework for verifying that any proposed set of roots does indeed constitute a complete solution set.

■ **PROPOSITION 6.13.** *If $r_1, r_2, \ldots, r_n$ are elements of the field F and*

$$P(x) = x^n + a_1 x^{n-1} + a_2 x^{n-2} + \cdots + a_{n-1} x + a_n$$

is a polynomial over F such that

$$\sum r_1 r_2 \cdots r_k = (-1)^k a_k, \qquad k = 1, 2, \ldots, n$$

then

$$P(x) = (x - r_1)(x - r_2) \cdots (x - r_n)$$

Proof. Set

$$Q(x) = (x - r_1)(x - r_2) \cdots (x - r_n)$$
$$= x^n + b_1 x^{n-1} + b_2 x^{n-2} + \cdots + b_{n-1} x + b_n$$

Then, by Theorem 6.12,

$$b_k = (-1)^k \sum r_1 r_2 \cdots r_k = (-1)^k (-1)^k a_k = a_k, \qquad k = 1, 2, \ldots, n$$

Hence $P(x)$ and $Q(x)$ are identical polynomials. ■

We are now in position to eliminate the rough edges in Chapter 3's solution of the cubic equation. That solution was incomplete in that it was not proved that the three roots x_1, x_2, x_3 selected in equations (5) of Chapter 3 constitute the complete solution set of the general cubic equation $x^3 + ax^2 + bx + c = 0$.

■ **COROLLARY 6.14.** *The values x_1, x_2, x_3 given by equations (5) of Chapter 3 are the complete solution set of the cubic equation*

$$x^3 + ax^2 + bx + c = 0$$

Proof. Since the reduced cubic equation $y^3 + py + q = 0$ was obtained by setting $x = y - a/3$, it follows from the same equations (5) of Chapter 3 that it suffices to show that y_1, y_2, y_3 are the complete solution set of this reduced cubic equation. By Proposition 6.13, it suffices to prove that

1. $y_1 + y_2 + y_3 = 0$,
2. $y_1 y_2 + y_2 y_3 + y_3 y_1 = p$,
3. $y_1 y_2 y_3 = -q$.

We demonstrate only the last of these equalities, leaving the other two to Exercise 31. It follows from equations (5) of Chapter 3 that

$$y_1 y_2 y_3 = \left(z_1 - \frac{p}{3z_1}\right)\left(\omega z_1 - \frac{p\omega^2}{3z_1}\right)\left(\omega^2 z_1 - \frac{p\omega}{3z_1}\right)$$

$$= \omega^3 z_1^3 - \frac{pz_1}{3}(\omega^3 + \omega^2 + \omega^4) + \frac{p^2}{9z_1}(\omega^3 + \omega^2 + \omega^4) - \frac{p^3 \omega^3}{27 z_1^3}$$

$$= z_1^3 - \frac{pz_1}{3}(1 + \omega^2 + \omega) + \frac{p^2}{9z_1}(1 + \omega^2 + \omega) - \frac{p^3}{27 z_1^3}$$

$$= z_1^3 - \frac{p^3}{27 z_1^3}$$

Now, by Equation (3) of Chapter 3, z_1^3 is one of the roots of the quadratic equation $u^2 + qu - p^3/27 = 0$, and hence the other one is $-p^3/27 z_1^3$. Since the sum of the roots of this quadratic is $-q$, it now follows that

$$z_1^3 - \frac{p^3}{27 z_1^3} = -q$$

and so

$$y_1 y_2 y_3 = -q \qquad \blacksquare$$

EXERCISES 6.4

Let r, s, t be the roots of the equation $x^3 + ax^2 + bx + c = 0$. Rewrite the expressions in Exercises 1–7 in terms of a, b, c. (Wherever necessary, you may assume that the denominators are not zero.)

1. $r^2 + s^2 + t^2$
2. $(r + s)^2 + (r + t)^2 + (s + t)^2$
3. $(r + s)(r + t)(s + t)$
4. $r^2 s^2 + r^2 t^2 + s^2 t^2$
5. $\dfrac{1}{r} + \dfrac{1}{s} + \dfrac{1}{t}$
6. $\dfrac{1}{r^2} + \dfrac{1}{s^2} + \dfrac{1}{t^2}$

7. $\dfrac{1}{r+s} + \dfrac{1}{r+t} + \dfrac{1}{s+t}$

Let r, s, t, be the zeros of the real polynomial $x^3 + 23x + 1$. Find the real cubic polynomial whose zeros are those that appear in Exercises 8–13.

8. r^2, s^2, t^2
9. $r+s, s+t, t+r$
10. rs, st, tr
11. $1+r, 1+s, 1+t$
12. kr, ks, kt
13. $\dfrac{k}{r}, \dfrac{k}{s}, \dfrac{k}{t}$

Let r_1, r_2, r_3, r_4 be the zeros of the polynomial $x^4 + 3x^3 - 5x + 1$. Evaluate the expressions in Exercises 14–16.

14. $r_1^2 + r_2^2 + r_3^2 + r_4^2$
15. $\dfrac{1}{r_1} + \dfrac{1}{r_2} + \dfrac{1}{r_3} + \dfrac{1}{r_4}$
16. $(r_1 r_2 - r_3 r_4)^2 + (r_1 r_3 - r_2 r_4)^2 + (r_1 r_4 - r_2 r_3)^2$
17. Solve the equation $x^5 - 4x^4 + 2x^3 - 8x^2 - 35x + 140 = 0$ whose solutions have the form $r, -r, s, -s, t$.
18. The equation $x^5 + 3x^4 - x^3 - 11x^2 - 17x - 5 = 0$ has two roots whose product is 1. Determine these two roots.
19. The equation $x^5 - 409x + 285 = 0$ has two roots whose sum equals 5. Determine these two roots.

Let $r_1, r_2, r_3, \ldots, r_n$ be the zeros of the polynomial

$$x^n + a_1 x^{n-1} + a_2 x^{n-2} + \cdots + a_{n-1} x + a_n$$

Prove the identities in Exercises 20–25.

20. $\sum r_1^2 r_2 = 3a_3 - a_1 a_2$
21. $\sum r_1^3 = 3a_1 a_2 - a_1^3 - 3a_3$
22. $\sum r_1^2 r_2 r_3 = a_1 a_3 - 4a_4$
23. $\sum r_1^2 r_2^2 = a_2^2 - 2a_1 a_3 + 6a_4$
24. $\sum r_1^3 r_2 = a_1^2 a_2 - 2a_2^2 - a_1 a_3 - 4a_4$
25. $\sum r_1^4 = a_1^4 - 4a_1^2 a_2 + 2a_2^2 + 4a_1 a_3 - 4a_4$
26. Let $r_1, r_2, \ldots, r_n$ be the zeros of the polynomial

$$x^n + a_1 x^{n-1} + a_2 x^{n-2} + \cdots + a_{n-1} x + a_n$$

where $a_n \neq 0$. Prove that the zeros of

$$x^n + \dfrac{a_{n-1}}{a_n} x^{n-1} + \dfrac{a_{n-2}}{a_n} x^{n-2} + \cdots + \dfrac{a_1}{a_n} x + \dfrac{1}{a_n}$$

are

$$\dfrac{1}{r_1}, \dfrac{1}{r_2}, \ldots, \dfrac{1}{r_n}$$

Let $r_1, r_2, r_3, \ldots, r_n$ be the zeros of the polynomial

$$x^n + a_1 x^{n-1} + a_2 x^{n-2} + \cdots + a_{n-1} x + a_n$$

where $a_n \neq 0$. Express the sums in Exercises 27–30 in terms of $a_1, a_2, \ldots, a_n$.

27. $\sum \dfrac{1}{r_1}$ 28. $\sum \dfrac{1}{r_1^2}$

29. $\sum \dfrac{1}{r_1 r_2}$ 30. $\sum \dfrac{1}{r_1^3}$

31. Complete the proof of Corollary 6.14.

6.5 LAGRANGE'S SOLUTION OF THE QUARTIC EQUATION*

In 1771 Lagrange wrote a lengthy treatise titled *Reflexions sur la Résolution Algébrique des Equations* in which he summarized what was known about solvability of equations by radicals. He also added some thoughts of his own and in fact proved several theorems that eventually did lead to the resolution of this issue by the next generation of mathematicians. It is with this contribution of Lagrange's as well as some of its subsequent developments that most of the rest of this book is concerned.

One of the methods that Lagrange offered for the solution of quartic equations began with the seemingly innocuous observation that when the roots r_1, r_2, r_3, r_4 are permuted (in other words substituted for each other), the expression

$$r_1 r_2 + r_3 r_4$$

assumes only three values, namely itself and

$$r_1 r_3 + r_2 r_4 \quad \text{and} \quad r_1 r_4 + r_2 r_3$$

For example, when the variables are interchanged by cycling them to the left, $r_1 r_2 + r_3 r_4$ becomes $r_2 r_3 + r_4 r_1 = r_1 r_4 + r_2 r_3$. If, on the other hand, only r_1 and r_4 are switched, the polynomial is transformed into $r_4 r_2 + r_3 r_1 = r_1 r_3 + r_2 r_4$.

This fact can be used to solve the quartic in the following manner. Let r_1, r_2, r_3, r_4 denote the four roots of the equation

$$x^4 + ax^3 + bx^2 + cx + d = 0$$

*Optional.

and set

$$A = r_1r_2 + r_3r_4$$
$$B = r_1r_3 + r_2r_4$$
$$C = r_1r_4 + r_2r_3$$

Clearly

$$A + B + C = \Sigma r_1r_2 = b.$$

Next

$$\begin{aligned}AB + AC + BC &= (r_1r_2 + r_3r_4)(r_1r_3 + r_2r_4) \\ &\quad + (r_1r_2 + r_3r_4)(r_1r_4 + r_2r_3) + (r_1r_3 + r_2r_4)(r_1r_4 + r_2r_3) \\ &= \Sigma r_1^2 r_2 r_3 = (\Sigma r_1)(\Sigma r_1 r_2 r_3) - 4(\Sigma r_1 r_2 r_3 r_4) \\ &= (-a)(-c) - 4d = ac - 4d\end{aligned}$$

Finally,

$$\begin{aligned}ABC &= (r_1r_2 + r_3r_4)(r_1r_3 + r_2r_4)(r_1r_4 + r_2r_3) \\ &= \Sigma r_1^3 r_2 r_3 r_4 + \Sigma r_1^2 r_2^2 r_3^2 \\ &= r_1 r_2 r_3 r_4 \Sigma r_1^2 + (\Sigma r_1 r_2 r_3)^2 - 2 \Sigma r_1^2 r_2^2 r_3 r_4 \\ &= d[(\Sigma r_1)^2 - 2 \Sigma r_1 r_2] + (-c)^2 - 2r_1 r_2 r_3 r_4 \Sigma r_1 r_2 \\ &= d(a^2 - 2b) + c^2 - 2db = a^2 d + c^2 - 4bd\end{aligned}$$

These computations lead to the observation that *if r_1, r_2, r_3, r_4 are the roots of the quartic equation*

$$x^4 + ax^3 + bx^2 + cx + d = 0$$

then $r_1r_2 + r_3r_4$, $r_1r_3 + r_2r_4$, and $r_1r_4 + r_2r_3$ are the roots of the cubic equation

$$y^3 - by^2 + (ac - 4d)y - (a^2 d + c^2 - 4bd) = 0$$

Since every cubic equation is already known to be solvable by radicals, it follows that A, B, and C have algebraic expressions in a, b, c, d. The actual values of r_1, r_2, r_3, r_4 are now extracted with relative ease. By Exercise 8 we may assume that $r_1 r_2 \neq r_3 r_4$. Since

$$r_1 r_2 + r_3 r_4 = A \quad \text{and} \quad (r_1 r_2)(r_3 r_4) = d$$

it follows that $\alpha = r_1 r_2$ and $\beta = r_3 r_4$ are the solutions of the quadratic

$$z^2 - Az + d = 0$$

and so they too have algebraic expressions in a, b, c, d. Moreover

$$(r_1 + r_2) + (r_3 + r_4) = -a$$

and

$$r_3 r_4 (r_1 + r_2) + r_1 r_2 (r_3 + r_4) = \sum r_1 r_2 r_3 = -c$$

When this system of simultaneous equations is solved for $(r_1 + r_2)$ and $(r_3 + r_4)$, we get

$$r_1 + r_2 = \frac{c - a r_1 r_2}{r_1 r_2 - r_3 r_4} = \frac{c - a\alpha}{\alpha - \beta} = \gamma$$

and

$$r_3 + r_4 = \frac{a r_3 r_4 - c}{r_1 r_2 - r_3 r_4} = \frac{a\beta - c}{\alpha - \beta} = \delta$$

Note that γ and δ both have algebraic expressions in a, b, c, d. Finally, since $r_1 + r_2 = \gamma$ and $r_1 r_2 = \alpha$, it follows that r_1 and r_2 are the solutions of the quadratic

$$u^2 - \gamma u + \alpha = 0$$

and similarly r_3 and r_4 are the solutions of the quadratic

$$v^2 - \delta v + \beta = 0$$

Thus we have proved the following theorem.

■ **THEOREM 6.15.** *Every fourth-degree equation is solvable by radicals.*

We illustrate Lagrange's method with an equation whose solutions can of course be found in a much shorter way. Consider the equation $x^4 - x^2 = 0$. Here the auxiliary cubic turns out to be $y^3 + y^2 = 0$, and we choose $A = -1$ (usually the choice is arbitrary, though in this case choosing $A = 0$ would lead to problems). This gives us the auxiliary quadratic $z^2 + z = 0$, and we set $\alpha = -1, \beta = 0$. This gives us $\gamma = \delta = 0$, and so the roots of the original quartic are those of the quadratics $u^2 - 1 = 0$ and $v^2 = 0$, namely $\pm 1, 0, 0$.

EXERCISES 6.5

Use Lagrange's method to solve the equations in Exercises 1–4.

1. $x^4 - 1 = 0$
2. $x^4 + 1 = 0$
3. $x^4 - x = 0$
4. $x^4 + x = 0$

Explain why the equations in Exercises 5–7 are resolvable by radicals.

5. $x^8 - 2x^7 + 3x^6 - 5x^5 + 11x^4 - 5x^3 + 3x^2 - 2x + 1 = 0$
6. $x^8 - 3x^6 + 5x^4 - 2x^2 + 17 = 0$
7. $x^{12} - x^9 + 5x^6 - 32 = 0$
8. Suppose that the roots of $x^4 + ax^3 + bx^2 + cx + d = 0$ are such that the product of any two equals the product of the other two. Prove that $x^4 + ax^3 + bx^2 + cx + d$ factors into either $(x + r)^4$ or $(x^2 - r^2)^2$ for some suitable complex number r.

CHAPTER SUMMARY

The notion of a field was isolated and identified as the underlying structure common to the rational, real, complex, and prime modulus number systems. Polynomials were studied in this abstract context, and the correspondence between their zeros and their linear factors was pointed out. The Euclidean algorithm was demonstrated to contain much useful information about polynomials. Some attention was paid to the relation between the coefficients of a polynomial and its zeros. It was also shown that information of this type can be used to yield a formula for the solution of the general quartic equation.

Chapter Review Exercises

Mark the following true or false

1. The remainder of $x^{17} + 1$ when divided by $x^2 + 1$ is $x^3 + 1$.
2. In $\mathbb{R}[x]$, the remainder of $x^{72} + 1$ when divided by $x + 2$ is 4321.
3. In $\mathbb{Z}_5[x]$, $x^4 - x$ is divisible by $x^2 - 3x + 2$.
4. $x^2 + 1$ is reducible over $\mathbb{Z}_2$.
5. $x^2 + 1$ is reducible over every field.
6. The equation $5x^4 + (1 + i)x^3 - ix + 17 = 0$ has four complex roots.
7. The greatest common divisor of $x^2 + 1$ and $x^2 - ix$ over $\mathbb{C}$ is $x - i$.
8. The product of all the complex roots of the equation $x^{17} - 3x + 5 = 0$ is -5.
9. The quartic equation is solvable by radicals.

New Terms

Degree of a polynomial	102
Division of polynomials	102
Elementary symmetric polynomial	120
Euclidean algorithm for polynomials	111
Factorization of polynomials	107
Field	98
Ground field	101
Irreducible polynomial	108
Monic polynomial	102
Multiplicity of zero	110
Relatively prime polynomials	115
Zero of polynomial	106

Supplementary Exercises

1. Write a computer script that finds the greatest common divisor of any two polynomials with real coefficients.
2. Write a computer script that finds the greatest common divisor of any two polynomials with coefficients in $\mathbb{Z}_2$.
3. Write a computer script that finds the greatest common divisor of any two polynomials with coefficients in $\mathbb{Z}_p$.
4. Write a computer script that implements Lagrange's solution of the general quartic with complex coefficients.
5. Write a computer script that lists the monic irreducible polynomials of degree d over $\mathbb{Z}_p$.
6. Find a formula for the number of irreducible monic polynomials of degree d over $\mathbb{Z}_p$.
7. Prove that every symmetric polynomial in the variables $x_1, x_2, \ldots, x_n$ is expressible as a polynomial in the elementary symmetric polynomials $\Sigma x_1, \Sigma x_1 x_2, \ldots, \Sigma x_1 x_2 \cdots x_n$.

CHAPTER 7

Galois Fields

In this chapter some new fields are introduced and studied in detail. These fields combine some of the features of both the complex and the modular numbers systems. The existence of primitive roots (modulo p) is proved in this new context.

7.1 GALOIS'S CONSTRUCTION OF HIS FIELDS

The following quotation consists of the opening paragraphs of the article *On the Theory of Numbers* by Évariste Galois, which appeared in the June 1830 issue of the *Bulletin des Sciences Mathematiques*. Some of the notation has been modernized for pedagogical reasons, and a more faithful translation appears in Appendix D.

> When it is agreed to consider as zero all the quantities which are the multiples of a given prime number p, and, subject to this convention, one looks for solutions to the polynomial equation $F(x) = 0$, i.e., the equations that Mr. Gauss denotes by $F(x) \equiv 0$, it is customary to consider only integer solutions to these sorts of questions. Having been led by some specific researches to consider their irrational solutions, I have arrived at some results that I believe to be new.
>
> Let there be given such an equation or congruence, $F(x) = 0$, and let p be the modulus. Suppose first that the congruence in question admits no rational factors, that is, there exist no three polynomials $\phi(x)$, $\psi(x)$, $\chi(x)$ such that
>
> $$\phi(x) \cdot \psi(x) = F(x) + p \cdot \chi(x).$$
>
> In that case the congruence has no integer roots, nor any irrational root of smaller degree. One should therefore regard the roots of this congruence as

some kind of imaginary symbols (since they do not satisfy the same questions as integers), symbols whose employment, in calculations, will often prove as useful as that of the imaginary $\sqrt{-1}$ in ordinary analysis.

We are concerned here with the classification of these imaginaries and their reduction to the smallest possible number.

Let i denote one of the roots of the congruence $F(x) = 0$, which can be supposed to have degree v.

Consider the general expression

$$a_0 + a_1 i + a_2 i^2 + \cdots + a_{v-1} i^{v-1} \tag{A}$$

where $a_0, a_1, a_2, \ldots, a_{v-1}$ represent integers. When these numbers are assigned all their possible values, expression (A) runs through p^v values which possess, as I shall demonstrate, the same properties as the natural numbers in the theory of residues of powers.

In the first paragraph Galois states that it is his intention to consider the solutions of polynomial equations with coefficients in $\mathbb{Z}_p$. He then goes on to explain what is meant by irreducibility of polynomials modulo p, a topic that was covered in Section 6.2. The closing sentence of the second paragraph is extraordinarily creative and imaginative. Just as the polynomial $x^2 + 1$, which is irreducible over the real numbers yields the imaginary but useful number $\sqrt{-1}$ so do these polynomials that are irreducible over $\mathbb{Z}_p$ yield a new species of imaginary symbols. Accordingly we will refer to these as *Galois imaginaries*.

Galois next proceeds to draw some further consequences from this analogy. It was seen in Section 2.1 that if $\sqrt{-1}$ is a zero of the irreducible quadratic $x^2 + 1$, that is, if

$$(\sqrt{-1})^2 + 1 = 0 \quad \text{or} \quad (\sqrt{-1})^2 = -1 \tag{1}$$

then any rational expression in $\sqrt{-1}$ can be reduced to the form $a + b\sqrt{-1}$. Galois now asserts that if i is the new imaginary number associated with an irreducible polynomial $F(x)$ of degree v over $\mathbb{Z}_p$, then every rational expression in i can be reduced to the form

$$a_0 + a_1 i + a_2 i^2 + \cdots + a_{v-1} i^{v-1} \tag{2}$$

Let us digress here into some concrete computations. Consider the irreducible polynomial $x^2 + x + 1$ over $\mathbb{Z}_2$, and suppose that it has α as a Galois imaginary so that, in analogy with (1), α satisfies the equation

$$\alpha^2 + \alpha + 1 = 0 \quad \text{over } \mathbb{Z}_2$$

or

$$\alpha^2 = 1 + \alpha$$

Consequently any second-degree polynomial in α can be reduced to a linear function of α. For example,

$$\alpha^2 + 1 = 1 + \alpha + 1 = 2 + \alpha = 0 + \alpha = \alpha$$

The same holds for cubic polynomials, since

$$\alpha^3 = \alpha^2 \alpha = (1 + \alpha)\alpha = \alpha + \alpha^2 = \alpha + 1 + \alpha$$
$$= 1 + 2\alpha = 1 + 0\alpha = 1$$

Similarly

$$\alpha^4 = \alpha^3 \alpha = 1\alpha = \alpha, \quad \alpha^5 = \alpha^4 \alpha = \alpha\alpha = \alpha^2 = 1 + \alpha$$

and so on. In other words, the successive powers of α cycle through the values 1, α, and $1 + \alpha$, and hence every polynomial function of α can be reduced to the form

$$a_0 + a_1 \alpha, \quad a_0\, a_1 \in \mathbb{Z}_2 \tag{3}$$

The preceding considerations make it clear that the sum and product of any two of the expressions 0, 1, α, $\alpha + 1$ is again an expression of the same form. Let us now examine the issue of division. Does each of the nonzero elements of the form (3) have an inverse? Since the coefficients a_0 and a_1 can assume only the values 0 or 1, there are only three nonzero elements to consider

$$1 = \alpha^0, \quad \alpha = \alpha^1, \quad \text{and} \quad 1 + \alpha = \alpha^2$$

As it is already known that $\alpha^3 = 1$, it follows that

$$1^{-1} = 1 \quad \text{(of course)}$$
$$\alpha^{-1} = \alpha^2$$
$$(\alpha^2)^{-1} = \alpha^1 = \alpha$$

Thus the elements 0, 1, α, $1+\alpha$ form a field provided that 0 and 1 are understood to be elements of $\mathbb{Z}_2$ and provided it is assumed that $\alpha^2 + \alpha + 1 = 0$. With the sole exception of the issue of the existence of multiplicative inverses, the validity of the field properties in this new context follows from their validity for polynomials, since α and $1 + \alpha$ are treated much the same as the polynomials x and $1 + x$.

Let us consider another example in detail before we comment on Galois's brainchild in general. The polynomial $x^3 + x^2 + 1$ is irreducible of degree 3 over $\mathbb{Z}_2$. Let β be a Galois imaginary associated with this polynomial. That is, β is a number such that

$$\beta^3 + \beta^2 + 1 = 0 \quad \text{or} \quad \beta^3 = 1 + \beta^2 \qquad \text{over } \mathbb{Z}_2$$

Then

$$\beta^4 = \beta^3 \beta = (1 + \beta^2)\beta = \beta + \beta^3 = \beta + 1 + \beta^2 = 1 + \beta + \beta^2$$
$$\beta^5 = \beta^4 \beta = (1 + \beta + \beta^2)\beta = \beta + \beta^2 + \beta^3 = \beta + \beta^2 + 1 + \beta^2$$
$$= 1 + \beta + 2\beta^2 = 1 + \beta$$
$$\beta^6 = \beta^5 \beta = (1 + \beta)\beta = \beta + \beta^2$$
$$\beta^7 = \beta^6 \beta = (\beta + \beta^2)\beta = \beta^2 + \beta^3 = \beta^2 + 1 + \beta^2 = 1 + 2\beta^2 = 1$$

It is again clear that every polynomial function of β is reducible to the form

$$a_0 + a_1 \beta + a_2 \beta^2$$

where each of the coefficients a_0, a_1, a_2 can assume the values 0 or 1. There are exactly $2^3 = 8$ such values, and as seen above, these eight values can also be listed as 0, 1, β, β^2, β^3, β^4, β^5, β^6. Since the fact that $\beta^7 = 1$ implies that

$$(\beta^k)^{-1} = \beta^{7-k}, \qquad k = 0, 1, 2, 3, 4, 5, 6$$

it follows that this set of eight Galois imaginaries constitutes a field.

It turns out that the set of elements of the form (2) generated by a Galois imaginary is always a field, a fact that will be proved shortly. That such a set is closed with respect to addition, subtraction, and multiplication is quite clear. The existence of multiplicative inverses, however, is another, less obvious, matter. The polynomial $x^4 + x^3 + x^2 + x + 1$ is irreducible over $\mathbb{Z}_2$ and so it has a Galois imaginary γ associated with it, such that

$$\gamma^4 = 1 + \gamma + \gamma^2 + \gamma^3 \qquad \text{over } \mathbb{Z}_2$$

Form (2) gives rise to $2^4 = 16$ associated elements. However, if we now proceed to list the powers of γ as was done earlier for α and β, we encounter a difficulty, for

$$\gamma^5 = \gamma^4 \gamma = (1 + \gamma + \gamma^2 + \gamma^3)\gamma = \gamma + \gamma^2 + \gamma^3 + \gamma^4$$
$$= \gamma + \gamma^2 + \gamma^3 + 1 + \gamma + \gamma^2 + \gamma^3 = 1$$

For the first time the successive powers of the Galois imaginary have failed to cycle through all the elements of the form (2). Consequently the device used in the previous examples to identify the inverse of each nonzero element is not available here. Nevertheless, the Galois imaginary γ does generate a corresponding field. To demonstrate this, it is only necessary to show that when the symbol i of (2) is replaced by γ or by any Galois imaginary, then each nonzero element of the form (2) does indeed have a multiplicative inverse of the same form. The same Euclidean algorithm that was used to prove the existence of inverse in $\mathbb{Z}_p$ works in this new context as well.

■ **LEMMA 7.1.** *Let $P(x)$ be an irreducible polynomial of degree v over $\mathbb{Z}_p$, and let δ be the associated Galois imaginary. For each element*

$$\zeta = a_0 + a_1\delta + a_2\delta^2 + \cdots + a_{v-1}\delta^{v-1}, \qquad a_i \in \mathbb{Z}_p, \ i = 0, 1, 2, 3, \ldots$$

if the coefficients $a_0, a_1, \ldots, a_{v-1}$ are not all zero, then there exists an element η of the same form such that

$$\eta\zeta = 1$$

Proof. With ζ as above, define

$$Q(x) = a_0 + a_1 x + a_2 x^2 + \cdots + a_{v-1} x^{v-1}$$

Since $P(x)$ is an irreducible polynomial of degree v and $Q(x)$ is a nonzero polynomial of degree less that v, it follows that $P(x)$ and $Q(x)$ have 1 as a greatest common divisor. Hence, by Proposition 6.11, there exist polynomials $A(x)$ and $B(x)$ such that

$$A(x)Q(x) + B(x)P(x) = 1 \qquad \text{over } \mathbb{Z}_p$$

Consequently

$$A(\delta)Q(\delta) + B(\delta)P(\delta) = 1 \qquad \text{over } \mathbb{Z}_p$$

However, by the definition of δ, $P(\delta) = 0$, and by the definition of $Q(x)$, $Q(\delta) = \zeta$. Hence

$$A(\delta)\zeta = 1 \qquad \text{over } \mathbb{Z}_p$$

and we can choose $\eta = A(\delta)$. ■

If $P(x)$ is any irreducible polynomial over $\mathbb{Z}_p$ and i is a corresponding Galois imaginary, then the set of all the elements of the form (2) is denoted by $GF(p, P(x))$ and is called a *Galois field*. The three Galois fields described above are GF(2, $x^2 + x + 1$), GF(2, $x^3 + x^2 + 1$), and GF(2, $x^4 + x^3 + x^2 + x + 1$). It follows from Lemma 7.1 that every Galois field is indeed a field in the sense of Section 6.1, and we state this explicitly.

■ **THEOREM 7.2.** *If $P(x)$ is an irreducible polynomial over $\mathbb{Z}_p$, then the Galois field GF(p, $P(x)$) is a field.*

Just as the polynomial $x^2 + 1$, which is irreducible over $\mathbb{R}$ has the *two* imaginaries i and $-i$ associated with it, so is it the case that every degree v polynomial $P(x)$ that is irreducible over $\mathbb{Z}_p$ has v distinct Galois imaginaries. However, it so happens that the Galois field generated by any one of these imaginaries of $P(x)$ contains all the other $v - 1$ imaginaries of $P(x)$, and thus all of the Galois imaginaries of $P(x)$ define the same Galois field GF(p, $P(x)$). A more detailed discussion of this phenomenon will be found in Section 7.4.

Galois does not prove Theorem 7.2 explicitly. Later in his paper he writes:

Next it can be proven, just as is done in the theory of numbers, that there exist primitive roots α... which... reproduce, by their powers, the complete sequence of all the other roots.

In other words, Galois claims that the finite field GF(p, $P(x)$) contains an element α whose powers

$$1, \alpha, \alpha^2, \alpha^3, \ldots$$

run through all the nonzero values of (2) above. Because of the similarity that this bears to the powers of the complex roots of unity, such an element is also called *primitive*. The examples preceding Lemma 7.1 show how such a primitive element can be used to demonstrate the existence of multiplicative inverses. A table that expresses all the powers of a primitive Galois imaginary in form (2) is called a *cyclic table*. Table 7.1 contains the cyclic table of the primitive Galois

TABLE 7.1. The cyclic table of GF(2, $x^3 + x^2 + 1$).

1
β
β^2
$\beta^3 = 1 + \beta^2$
$\beta^4 = 1 + \beta + \beta^2$
$\beta^5 = 1 + \beta$
$\beta^6 = \beta + \beta^2$
$\beta^7 = 1$

7.1 Galois's Construction of His Fields

imaginary β associated with GF(2, $x^3 + x^2 + 1$). The detailed calculations were displayed above.

The elements α and β of this chapter's first two examples are primitive elements of GF(2, $x^2 + x + 1$) and GF(2, $x^3 + x^2 + 1$), respectively. On the other hand, the element γ fails to be a primitive element of GF(2, $x^4 + x^3 + x^2 + x + 1$). The *order* $o(\zeta)$ of any element ζ of some Galois field is the least positive integer k such that

$$\zeta^k = 1$$

Thus, in the aforementioned Galois field, γ has order 5. On the other hand, in GF(2, $x^2 + x + 1$), the elements α and $\alpha + 1$ both have order 3. Similarly, with the exception of 0 and 1, each of the elements of GF(2, $x^3 + x^2 + 1$) has order 7. As it would be pedagogically useful to have at least one Galois field that is not associated with a polynomial over $\mathbb{Z}_2$, we compute the cyclic table of GF(3, $x^2 + x + 2$). That this polynomial is indeed irreducible over $\mathbb{Z}_3$ follows from the fact that

$$0^2 + 0 + 2 = 2 \neq 0, \quad 1^2 + 1 + 2 = 4 \neq 0, \quad 2^2 + 2 + 2 = 8 \neq 0 \pmod{3}$$

If σ is the associated Galois imaginary, then $\sigma^2 + \sigma + 2 = 0$ over $\mathbb{Z}_3$, and so

$$\sigma^2 = -2 - \sigma = 1 + 2\sigma$$

Consequently

$$\sigma^3 = \sigma^2\sigma = (1 + 2\sigma)\sigma = \sigma + 2\sigma^2 = \sigma + 2(1 + 2\sigma)$$
$$= 2 + 5\sigma = 2 + 2\sigma$$
$$\sigma^4 = \sigma^3\sigma = (2 + 2\sigma)\sigma = 2\sigma + 2\sigma^2 = 2\sigma + 2(1 + 2\sigma)$$
$$= 2 + 6\sigma = 2$$
$$\sigma^5 = \sigma^4\sigma = 2\sigma$$
$$\sigma^6 = \sigma^5\sigma = 2\sigma^2 = 2 + 4\sigma = 2 + \sigma$$
$$\sigma^7 = \sigma^6\sigma = (2 + \sigma)\sigma = 2\sigma + \sigma^2 = 2\sigma + 1 + 2\sigma$$
$$= 1 + \sigma$$
$$\sigma^8 = \sigma^7\sigma = (1 + \sigma)\sigma = \sigma + \sigma^2 = \sigma + 1 + 2\sigma = 1$$

It follows that σ is indeed a primitive element of GF(3, $x^2 + x + 2$).

The following proposition provides us with some general information about the number of elements of a Galois field and their orders:

Galois Fields

■ **PROPOSITION 7.3.** *Let F be a Galois field associated with the irreducible polynomial $P(x)$ of degree v over $\mathbb{Z}_p$. Then*

1. *F has exactly p^v elements;*
2. *the order of each nonzero element of F is finite.*

Proof. Let i be the Galois imaginary associated with $P(x)$. As was remarked above, every element of F is expressible in the form (2). Hence F contains at most p^v elements. To show that F contains exactly that number of elements, it suffices to show that no two distinct expressions of the form (2) can equal each other. Suppose that ζ_1 and ζ_2 are distinct elements of F. Then

$$\zeta_1 = a_0 + a_1 i + a_2 i^2 + \cdots + a_{v-1} i^{v-1}$$

and

$$\zeta_2 = b_0 + b_1 i + b_2 i^2 + \cdots + b_{v-1} i^{v-1}$$

where

$$a_k \neq b_k \quad \text{for some } k = 0, 1, \ldots, v-1$$

Then, when Lemma 7.1 is applied to the element

$$\zeta = \zeta_1 - \zeta_2$$
$$= (a_0 - b_0) + (a_1 - b_1)i + (a_2 - b_2)i^2 + \cdots + (a_{v-1} - b_{v-1})i^{v-1}$$

we conclude that there is an element η such that $\zeta\eta = 1$. By Proposition 6.1, ζ is not zero, and so

$$\zeta_1 \neq \zeta_2$$

This concludes the proof of part 1.

Let α be any nonzero element of F. Since all the terms of the infinite sequence

$$1, \alpha, \alpha^2, \alpha^3, \ldots$$

are elements of F, and since F contains only a finite number of elements, it follows that there exist two distinct exponents $m > n$ such that

$$\alpha^m = \alpha^n$$

But then

$$\alpha^{m-n} = 1$$

and so $o(\alpha) \leq m - n$. This concludes the proof of part 2. ■

EXERCISES 7.1

1. Write out the cyclic table of GF(2, $x^3 + x + 1$).
2. Write out the cyclic table of GF(2, $x^4 + x + 1$).

Verify that the Galois imaginary associated with each of the polynomials in Exercises 3, 4 is primitive over $\mathbb{Z}_2$, and write out the associated cyclic table.

3. $x^5 + x^2 + 1$ 4. $x^5 + x^3 + 1$

Verify that the Galois imaginary associated with each of the polynomials in Exercises 5–7 is primitive over $\mathbb{Z}_3$, and write out the associated cyclic table.

5. $x^2 + 2x + 2$ 6. $x^3 + 2x + 1$ 7. $x^3 + 2x^2 + 1$
8. Verify that the Galois imaginary associated with $x^2 + 4x + 2$ is primitive over $\mathbb{Z}_5$, and construct its cyclic table.
9. Verify that the Galois imaginary associated with $x^2 + 6x + 3$ is primitive over $\mathbb{Z}_7$, and construct its cyclic table.

Let β be the Galois imaginary associated with the irreducible polynomial $x^3 + x^2 + 1$ over $\mathbb{Z}_2$. Solve the (systems of simultaneous) equations in Exercises 10–13 in GF(2, $x^3 + x^2 + 1$).

10. $(1 + \beta)x + \beta = 1 + \beta^2$
11. $x + y = \beta$ and $x + \beta y = 1$
12. $x + (1 + \beta)y = \beta + \beta^2$, $(1 + \beta^2)x + y = 0$
13. $x + \beta y + \beta^2 z = 1 + \beta$, $(1 + \beta)x + (1 + \beta^2)y + z = 1$, $\beta y + z = \beta$
14. Solve Exercise 10 with GF(2, $x^3 + x^2 + 1$) replaced by GF(3, $x^2 + 2x + 2$).
15. Solve Exercise 11 with GF(2, $x^3 + x^2 + 1$) replaced by GF(3, $x^2 + 2x + 2$).
16. Solve Exercise 12 with GF(2, $x^3 + x^2 + 1$) replaced by GF(3, $x^2 + 2x + 2$).
17. Solve Exercise 13 with GF(2, $x^3 + x^2 + 1$) replaced by GF(3, $x^2 + 2x + 2$).
18. Explain why GF($p, x - 1$) = $\mathbb{Z}_p$.
19. Explain why the Binomial Theorem holds for elements of Galois fields.
20. Prove that $(a \pm b)^p = a^p \pm b^p$ for any $a, b \in$ GF($p, P(x)$).
21. Prove that GF($p, P(x)$) contains exactly one pth root of unity.
22. Prove that for any $a, b \in$ GF($p, P(x)$), $a^p = b^p$ if and only if $a = b$.
23. Prove that for any positive integer k and for any $a, b \in$ GF($p, P(x)$), $a^{p^k} = b^{p^k}$ if and only if $a = b$.
24. True or false: For any $a, b \in$ GF($p, P(x)$), $a^2 = b^2$ if and only if $a = b$? Justify your answer.
25. Show that for any $a \in$ GF($p, P(x)$), $a^p = a$ if and only if $a \in \mathbb{Z}_p$.
26. Let k be any fixed positive integer. Show that $\{a \in$ GF($p, P(x)$) $\mid a^{p^k} = a\}$ is also a field.
27. Prove that if $p \neq 2$, then the sum of all the elements of GF($p, P(x)$) is 0.
28. Prove that the product of all the nonzero elements of GF($p, P(x)$) is -1.
29. Let a be an element of the Galois field GF($p, P(x)$). Prove that
    (a) $1 + a + a^2 + \cdots + a^{o(a) - 1} = 0$
    (b) $1 \cdot a \cdot a^2 \cdots a^{o(a) - 1} = (-1)^{o(a) - 1}$

7.2 THE GALOIS POLYNOMIAL

The orders of the elements of Galois fields, defined in the previous section, possess the same properties as the orders of the complex and modular roots of unity, which are restated here for the sake of completeness. Since the proofs of Propositions 2.14, 2.15, 5.8, and 5.9 work in the new context verbatim, these properties are restated without proof.

■ **PROPOSITION 7.4.** *Let α and β be any roots of unity in some field F. Then*

1. $\alpha^n = 1$ *if and only if n is a multiple of* $o(\alpha)$;
2. $\alpha^a = \alpha^b$ *if and only if* $o(a)$ *is a divisor of* $a - b$, *and so* $1, \alpha, \alpha^2, \ldots, \alpha^{o(\alpha)-1}$ *are all distinct;*
3. *if* $o(\alpha) = n$, *then* $o(\alpha^k) = \dfrac{n}{(k, n)}$;
4. $o(\alpha\beta) = o(\alpha)o(\beta)$ *if* $o(\alpha)$ *and* $o(\beta)$ *are relatively prime.*

If α is any element of order k, it must clearly be a zero of the polynomial $x^m - 1$ whenever m is a multiple of k. Hence, by part 1 of the above proposition, if e is the least common multiple of the orders of all the nonzero elements of the Galois field GF($p, P(x)$), then these elements are all zeros of $x^e - 1$. This number e is of course of interest, and it will eventually be demonstrated (Theorem 7.8) that $e = p^v - 1$, where v is the degree of $P(x)$. We begin this process by picking up where the previous section's quotation from Galois's paper left off.

> Of the expressions (2) we shall only take the $p^v - 1$ values obtained when $a_0, a_1, a_2, \ldots, a_{v-1}$ are not all zero; let α be one of these expressions.
>
> If α is successively raised to the second, third, ... powers, a sequence of quantities all of which have the same form is obtained (since every function of i is reducible to the $(v - 1)$th degree). Hence it must be that $\alpha^n = 1$ for some n; let n be the smallest number such that $\alpha^n = 1$. Then the set of numbers
>
> $$1, \alpha, \alpha^2, \alpha^3, \ldots, \alpha^{n-1}$$
>
> are all distinct. Next, multiply these n numbers by another expression C of the same form. We then obtain another new group of quantities all different from the first group as well as from each other. If the quantities (2) have not been exhausted yet, the powers of α can be multiplied by a new expression γ, and so on. Consequently the number n necessarily divides the total number of quantities of form (2). Since this number is $p^v - 1$, we see that n divides $p^v - 1$. From this it also follows that
>
> $$\alpha^{p^v - 1} = 1, \quad \text{or} \quad \alpha^{p^v} = \alpha.$$

Two sentences later we find the statement:

> We note here the remarkable result that all the algebraic quantities that arise in this theory are roots of equations of the form
>
> $$x^{p^v} = x.$$

To illustrate the above procedure, consider the Galois imaginary γ associated with the polynomial $x^4 + x^3 + x^2 + x + 1$ which is irreducible over $\mathbb{Z}_2$. We saw that $\gamma^5 = 1$. Since $\gamma^4 = 1 + \gamma + \gamma^2 + \gamma^3$, the process begins with

$$1, \quad \gamma, \quad \gamma^2, \quad \gamma^3, \quad 1 + \gamma + \gamma^2 + \gamma^3 \tag{5}$$

According to Proposition 7.3.1, any two elements of the form (2) are distinct, and hence $1 + \gamma$ is different from all the elements listed in (5). Using this $1 + \gamma$ as C, the next set of elements produced by Galois's procedure is

$$(1+\gamma)1, \quad (1+\gamma)\gamma, \quad (1+\gamma)\gamma^2, \quad (1+\gamma)\gamma^3, \quad (1+\gamma)(1+\gamma+\gamma^2+\gamma^3)$$

or, upon simplification,

$$1+\gamma, \quad \gamma+\gamma^2, \quad \gamma^2+\gamma^3, \quad 1+\gamma+\gamma^2, \quad \gamma+\gamma^2+\gamma^3 \tag{6}$$

The element $1 + \gamma^2$ has not been listed yet, so the next list is

$$1+\gamma^2, \quad (1+\gamma^2)\gamma, \quad (1+\gamma^2)\gamma^2, \quad (1+\gamma^2)\gamma^3, \quad (1+\gamma^2)(1+\gamma+\gamma^2+\gamma^3)$$

or, upon simplification,

$$1+\gamma^2, \quad \gamma+\gamma^3, \quad 1+\gamma+\gamma^3, \quad 1+\gamma^3, \quad 1+\gamma^2+\gamma^3 \tag{7}$$

It is easily verified by inspection that, as Galois claims, the three sets listed in (5), (6), and (7) exhaust all the nonzero elements of $GF(2, x^4 + x^3 + x^2 + x + 1)$. We now state Galois's theorem and supply some of the missing details in his proof. Another, much more general and succinct proof will be provided later by Proposition 9.16 and Exercise 9.5.12.

■ **THEOREM 7.5 (Galois).** *Let $P(x)$ be an irreducible polynomial of degree v over $\mathbb{Z}_p$. Then all the elements of $GF(p, P(x))$ are zeros of the polynomial*

$$x^{p^v} - x$$

Proof. Let α be any nonzero element of $F = GF(p, P(x))$, and suppose that it has order n. By Proposition 7.4.2, the list

$$\{1, \alpha, \alpha^2, \ldots, \alpha^{n-1}\} \tag{8}$$

consists of n distinct elements. If this list does not exhaust all the nonzero elements of F, let β be a nonzero element that does not appear in this list, and consider the new list

$$\{\beta, \alpha\beta, \alpha^2\beta, \ldots, \alpha^{n-1}\beta\} \tag{9}$$

All the elements of list (9) are distinct from each other, since otherwise some two elements of (8) would also be nondistinct. Moreover lists (8) and (9) are also disjoint, since otherwise we would have for some integers k and m,

$$\alpha^k \beta = \alpha^m$$

implying that β is a power of α, which we know not to be the case. If lists (8) and (9) do not exhaust the field, we choose an element γ that is in neither list and repeat the process. Since the field F is known to be finite, this process must eventually terminate and leave the nonzero elements of the field partitioned into disjoint lists, each of which contains exactly n elements. It follows that n is a divisor of the total number of nonzero elements of F, which, by Proposition 7.3.1, is $p^\nu - 1$. Hence, by Proposition 7.4.1,

$$\alpha^{p^\nu - 1} = 1 \tag{10}$$

Thus α, an arbitrary nonzero element of F, is a zero of the polynomial

$$x^{p^\nu - 1} - 1$$

Consequently, each of the elements of F, 0 included, is a zero of the polynomial

$$x(x^{p^\nu - 1} - 1) = x^{p^\nu} - x \tag{11}$$

∎

As Galois notes, this is a remarkable fact, for it provides us with a single, very simple polynomial (11) that contains all the elements of $GF(p, P(x))$ as its zeros. We will refer to this polynomial as the *Galois polynomial of $GF(p, P(x))$*. The observation that the order of each nonzero element of F divides $p^\nu - 1$ implies that the least common multiple of all the orders, denoted by e in this section's opening paragraph, divides $p^\nu - 1$.

It should be of interest to examine the case $\nu = 1$. The irreducible polynomials over $\mathbb{Z}_p$ of degree $\nu = 1$ are of course the binomials $x - a$, where

$a \in \mathbb{Z}_p$. Now the Galois imaginary associated with $x - a$ is none other than the known quantity a, so

$$\text{GF}(p, x - a) = \mathbb{Z}_p$$

Here the Galois polynomial is simply $x^p - x$, and Theorem 7.5 reduces to Fermat's Theorem 5.4.

The correspondence between the linear factors of a polynomial and its zeros yields the following result.

■ **COROLLARY 7.6.** *If $P(x)$ is an irreducible polynomial of degree v over $\mathbb{Z}_p$, then*

$$x^{p^v} - x = \prod_{i=1}^{p^v} (x - \alpha_i)$$

where $\alpha_1, \alpha_2, \ldots, \alpha_{p^v}$ is any listing of the elements of $\text{GF}(p, P(x))$.

Proof. By the previous theorem the polynomial $x^{p^v} - x$ has the p^v distinct elements of $\text{GF}(p, P(x))$ as its zeros. By Proposition 6.7, the polynomial $x^{p^v} - x$ cannot have any more zeros. Thus $\alpha_1, \alpha_2, \ldots, \alpha_{p^v}$ constitute all the zeros of the Galois polynomial $x^{p^v} - x$. The statement of the corollary now follows from Proposition 6.6. ∎

If α is the Galois imaginary of the polynomial $x^2 + x + 1$ over $\mathbb{Z}_2$, then the elements of $\text{GF}(2, x^2 + x + 1)$ are $0, 1, \alpha, 1 + \alpha$ and

$$(x - 0)(x - 1)(x - \alpha)(x - (1 + \alpha)) = [x(x + 1)][(x + \alpha)(x + 1 + \alpha)]$$
$$= (x^2 + x)(x^2 + x + \alpha^2 + \alpha)$$
$$= (x^2 + x)(x^2 + x + 1)$$
$$= x^4 + x^3 + x^2 + x^3 + x^2 + x$$
$$= x^4 + x = x^{2^2} - x$$

EXERCISES 7.2

1. Find the orders of all the nonzero elements of $\text{GF}(2, x^3 + x^2 + 1)$.
2. Find the orders of all the nonzero elements of $\text{GF}(2, x^4 + x + 1)$.
3. Find the orders of all the nonzero elements of $\text{GF}(2, x^5 + x^2 + 1)$.
4. Find the orders of all the nonzero elements of $\text{GF}(3, x^2 + 2x + 2)$.
5. Find the orders of all the nonzero elements of $\text{GF}(5, x^2 + 4x + 2)$.
6. Find the orders of all the nonzero elements of $\text{GF}(7, x^2 + 6x + 3)$.
7. Verify Corollary 7.6 directly for $\text{GF}(2, x^3 + x^2 + 1)$.
8. Verify Corollary 7.6 directly for $\text{GF}(3, x^2 + 2x + 2)$.

142 Galois Fields

9. Show that if GF(p, $P(x)$) is a Galois field and α is any of its elements, then

$$\alpha^{1+p+p^2+\cdots+p^{v-1}} \in \mathbb{Z}_p$$

where v is the degree of $P(x)$.

Let F be the Galois field GF(p, $P(x)$).

10. Show that the sum of all the elements of F is either 0 or 1.
11. Use the Galois polynomial to prove that the product of all the nonzero elements of F is -1.
12. What is the sum of the squares of the elements of F?
13. Evaluate the sum of the reciprocals of all the nonzero elements of F.
14. Let p be a prime number, and let n be relatively prime to $p^v - 1$. Prove that there is exactly one nth root of unity in GF(p, $P(x)$).
15. Suppose that $a, b \in$ GF(2, $P(x)$) for some degree 9 polynomial $P(x)$ that is irreducible over $\mathbb{Z}_2$. Suppose further that $a^2 + ab + b^2 = 0$. Prove that $a = b$.

7.3 THE PRIMITIVE ELEMENT THEOREM

Toward the end of his paper, Galois lets on that his purpose in constructing these new number systems was to find new contexts within which primitive roots exist and to which Gauss's techniques, which proved so effective for the algebraic resolution of the cyclotomic equation (see Section 2.4), could be applied to produce new algebraically resolvable equations. Galois does not prove the existence of these primitive elements, contenting himself with a comment to the effect that Gauss's proof of the existence of primitive roots modulo p carries over intact to this new setting. We will not follow Gauss's proof here and give instead a more modern, and somewhat shorter, proof.

■ **LEMMA 7.7.** *If F is a Galois field with f elements, and if q^m is the largest power of the prime number q that divides $f - 1$, then F contains an element a of order q^m.*

Proof. The polynomial $x^{(f-1)/q} - 1$ has degree $(f-1)/q < f - 1$, and so it follows from Proposition 6.7 that there is a nonzero element $b \in F$ that is *not* a zero of this polynomial:

$$b^{(f-1)/q} \neq 1$$

Set $a = b^{(f-1)/q^m}$. Then

$$a^{q^{m-1}} = b^{(f-1)/q} \neq 1$$

while, by Proposition 7.3.1 and Theorem 7.5,

$$a^{q^m} = b^{f-1} = 1$$

Thus $o(a)$ divides q^m but not q^{m-1}, whence $o(a) = q^m$. ∎

We are ready for this chapter's main theorem.

■ **THE PRIMITIVE ELEMENT THEOREM 7.8 (Galois).** *Every Galois field has a primitive element.*

Proof. Let F be a Galois field, and suppose that it contains f elements. If the prime factorization of $f - 1$ is

$$f - 1 = p_1^{m_1} p_2^{m_2} \cdots p_k^{m_k}$$

then, by the lemma, there exist elements $a_1, a_2, \ldots, a_k \in F$ such that

$$o(a_i) = p_i^{m_i}, \quad i = 1, 2, \ldots, k$$

It follows from Proposition 7.4.4 that

$$o(a_1 a_2 \cdots a_k) = p_1^{m_1} p_2^{m_2} \cdots p_k^{m_k} = f - 1$$

Hence $a_1 a_2 \cdots a_k$ is the required primitive element of F. ∎

It was pointed out above that $\mathbb{Z}_p = \mathrm{GF}(p, x - 1)$, so $\mathbb{Z}_p$ is also a Galois field and hence Theorem 7.8 guarantees the existence of primitive elements in $\mathbb{Z}_p$. Thus 3 is such a primitive element of $\mathbb{Z}_{17}$, since its first 16 powers are

1, 3, 9, 10, 13, 5, 15, 11, 16, 14, 8, 7, 4, 12, 2, 6, 1

which are all the nonzero elements of $\mathbb{Z}_{17}$. This sequence is of course identical with the exponents in the sum

$$\zeta + \zeta^3 + \zeta^9 + \zeta^{10} + \zeta^{13} + \zeta^5 + \zeta^{15} + \zeta^{11} + \zeta^{16}$$
$$+ \zeta^{14} + \zeta^8 + \zeta^7 + \zeta^4 + \zeta^{12} + \zeta^2 + \zeta^6$$

that was used in Section 2.4 to prove the constructibility of the regular 17-sided polygon.

The primitive elements of $\mathbb{Z}_p$, whose existence is guaranteed by Theorem 7.8, are identical with the primitive roots (modulo p) that were defined in

Section 5.2. It was Euler who first proved the existence of these primitive roots (modulo p). Gauss expanded on this work of Euler's and also applied it to his analysis of the cyclotomic equation $x^p - 1 = 0$. Galois, in attempting to generalize Gauss's method to prove the resolvability of other equations, invented what came to be known as the Galois fields and observed that the same methods that were used to prove the existence of primitive roots (modulo p) could also be used to establish the existence of primitive elements in his fields (Theorem 7.8).

The identification of primitive elements is an issue that puzzled Euler. Both Gauss and Galois later introduced some methodology into this question. Since this would take us outside the scope of this text, we merely point out that trial and error can always be used to locate primitive elements in relatively small Galois fields. Thus the Galois imaginaries α and β of Section 7.1 are clearly primitive elements of $GF(2, x^2 + x + 1)$ and $GF(2, x^3 + x^2 + 1)$, respectively, whereas γ is not a primitive element of $GF(2, x^4 + x^3 + x^2 + x + 1)$, since $\gamma^5 = 1$. However, the element $1 + \gamma$ is a primitive element of this latter field. To see this, note that by Proposition 7.4.1 and Theorem 7.5, $o(1 + \gamma)$ is a divisor of 15. However, by the Binomial Theorem 6.2,

$$(1 + \gamma)^3 = 1 + 3\gamma + 3\gamma^2 + \gamma^3 = 1 + \gamma + \gamma^2 + \gamma^3 \neq 1$$

and

$$(1 + \gamma)^5 = 1 + 5\gamma + 10\gamma^2 + 10\gamma^3 + 5\gamma^4 + \gamma^5 = 1 + \gamma + \gamma^4 + \gamma^5$$
$$= 1 + \gamma + (1 + \gamma + \gamma^2 + \gamma^3) + 1 = 1 + \gamma^2 + \gamma^3 \neq 1$$

Hence $o(1 + \gamma) = 15$, and $1 + \gamma$ is indeed a primitive element of $GF(2, x^4 + x^3 + x^2 + x + 1)$. We also note in passing that once it is known that a certain element ζ is a primitive element of some Galois field, then it follows, by Proposition 7.4.3, that the other primitive elements of this field are the powers ζ^m where m is relatively prime to $p^v - 1$.

EXERCISES 7.3

List all the primitive elements of the fields in Exercises 1–9.

1. $GF(2, x^2 + x + 1)$
2. $GF(2, x^3 + x^2 + 1)$
3. $\mathbb{Z}_5$
4. $\mathbb{Z}_{17}$
5. $GF(2, x^4 + x + 1)$
6. $GF(2, x^4 + x^3 + x^2 + x + 1)$
7. $GF(3, x^2 + x + 2)$
8. $GF(5, x^2 + 4x + 2)$
9. $GF(7, x^2 + 6x + 3)$

10. For any element $a \in GF(p, P(x))$, let $r(a)$ denote the number of distinct elements b in $GF(p, P(x))$ such that $b^{p-1} = a$. Prove that if $a \neq 0$, then $r(a) = 0$ or $p - 1$.

11. Let GF(p, $P(x)$) be any Galois field, and let r be a positive integer such that $r \not\equiv 0$ (mod $p^v - 1$), where v is the degree of $P(x)$. Prove that the sum of the rth powers of the elements of GF(p, $P(x)$) is zero.
12. Prove that the product of all the primitive elements of GF(p, $P(x)$) is 1, unless GF(p, $P(x)$) is $\mathbb{Z}_3$, in which case this product is 2.
13. Prove that if ζ is a primitive element of GF(p, $P(x)$), where $P(x)$ has degree v, then

$$x^{p^v-1} - 1 = \prod_{i=1}^{p^v-1} (x - \zeta^i)$$

14. Prove that for every prime p, the polynomial $x^p - x - 1$ is irreducible over $\mathbb{Z}_p$.

7.4 ON THE VARIETY OF GALOIS FIELDS*

We now know that for every polynomial $P(x)$ that is irreducible over $\mathbb{Z}_p$ there is a corresponding Galois field GF(p, $P(x)$). This observation made it possible to construct a variety of new fields, each of which contains p^v elements, where p is some prime and v is some positive integer. It is appropriate at this point to address the issue of classifying these new structures. There are two questions that every experienced mathematician would ask in this context. Given any such p and v, does there exist a Galois field of order p^v? Given any two such Galois fields, when are they in fact one and the same? Unfortunately, the complete resolutions of these questions lie beyond the bounds of this book, and the following discussion provides only informal answers.

One way of proving that a field of order p^v exists is to display a polynomial of degree v that is irreducible over $\mathbb{Z}_p$. This task is not easy. In fact, it turns out to be easier to count the number of all such polynomials than to produce even one. Exercises 6.2.6–7 and 7.3.14 deal with some special cases of this issue.

Next we turn to the second question and reexamine the first Galois field constructed in this chapter, GF(2, $x^2 + x + 1$). This field was constructed by stipulating that α is a zero of the polynomial $x^2 + x + 1$ over $\mathbb{Z}_2$ and then extracting some arithmetical consequences. However, our experience with all the previously constructed fields, namely the rationals, reals, complex, and modulo p arithmetic, leads us to expect any quadratic to have *two* zeros. Should we therefore stipulate the existence of *another* zero α' of $x^2 + x + 1$ over $\mathbb{Z}_2$ and then proceed to create its Galois field? This is of course unnecessary, since α' gives rise to a field that behaves exactly like that associated with α, except that each occurrence of α in the latter will be replaced by an α'. It turns out that the redundancy goes even deeper. The element α' is already in GF(2, $x^2 + x + 1$). In fact $\alpha' = \alpha^2$, for

$$(\alpha^2)^2 + \alpha^2 + 1 = \alpha^4 + \alpha^2 + 1 = \alpha + \alpha^2 + 1 = 0$$

*Optional.

A similar phenomenon occurs in $GF(2, x^3 + x^2 + 1)$ whose Galois imaginary was denoted by β. Note that

$$(\beta^2)^3 + (\beta^2)^2 + 1 = \beta^6 + \beta^4 + 1$$
$$= (\beta + \beta^2) + (1 + \beta + \beta^2) + 1 = 0 \quad \text{over } \mathbb{Z}_2$$

implying that β^2 is another zero of $x^3 + x^2 + 1$ in this field. Cubics, however, can have up to three zeros, so we can expect yet another zero of this polynomial. The element β^3 fails to be such a zero, since

$$(\beta^3)^3 + (\beta^3)^2 + 1 = \beta^9 + \beta^6 + 1 = \beta^2 + (\beta + \beta^2) + 1$$
$$= 1 + \beta \ne 0 \quad \text{over } \mathbb{Z}_2$$

However, β^4 turns out to be the third zero of $x^3 + x^2 + 1$, since

$$(\beta^4)^3 + (\beta^4)^2 + 1 = \beta^{12} + \beta^8 + 1 = \beta^5 + \beta + 1$$
$$= (1 + \beta) + \beta + 1 = 0 \quad \text{over } \mathbb{Z}_2$$

With hindsight, we could have argued as follows: Assuming that β was a zero of $x^3 + x^2 + 1$ over $\mathbb{Z}_2$, it was shown that β^2 was also such a zero. Hence, beginning with the fact that β^2 is a zero of this polynomial, it follows that its square, namely $(\beta^2)^2 = \beta^4$, should also be such a zero, as is indeed the case. Note that the square of β^4 is $\beta^8 = \beta$, which is also a zero of $x^3 + x^2 + 1$ but not a new one.

Let us examine another example before announcing the general principle. Consider the polynomial $x^2 + x + 2$ which is irreducible over $\mathbb{Z}_3$, and whose cyclic table was constructed in Section 7.1 in terms of the Galois imaginary σ. Since

$$(\sigma^2)^2 + \sigma^2 + 2 = \sigma^4 + \sigma^2 + 2 = 2 + (1 + 2\sigma) + 2 = 2 + 2\sigma \ne 0$$

it follows that σ^2 is not another zero of $x^2 + x + 2$ over $\mathbb{Z}_3$. However,

$$(\sigma^3)^2 + \sigma^3 + 2 = \sigma^6 + \sigma^3 + 2 = (2 + \sigma) + (2 + 2\sigma) + 2$$
$$= 0$$

Hence σ^3 is the other zero of $x^2 + x + 2$ and so of course is $(\sigma^3)^3 = \sigma^9 = \sigma$. The pattern is clear and is formulated in Proposition 7.10 below.

■ **LEMMA 7.9.** *If $a, b, c, \ldots \in GF(p, P(x))$, then*

$$(a + b + c + \cdots)^p = a^p + b^p + c^p + \cdots$$

Proof. Inasmuch as the Binomial Theorem holds in arbitrary fields (Theorem 6.2), and since, as was argued in the proof of Proposition 5.3, $\binom{p}{k} \equiv 0$ (mod p) for $0 < k < p$, it is seen that

$$(a + b)^p = a^p + b^p \quad \text{for any } a, b \in \text{GF}(p, P(x))$$

The lemma now follows by an easy induction argument. ∎

■ **PROPOSITION 7.10.** *If α is a zero of the polynomial $P(x)$ over $\mathbb{Z}_p$, then so are*

$$\alpha, \alpha^p, \alpha^{p^2}, \alpha^{p^3}, \ldots$$

zeros of $P(x)$.

Proof. It clearly suffices to show that if α is a zero of $P(x)$, then so is α^p. Suppose that

$$P(x) = a_0 x^n + a_1 x^{n-1} + \cdots + a_{n-1} x + a_n, \quad a_0, \ldots, a_n \in \mathbb{Z}_p$$

If α is any zero of $P(x)$, then it follows from the lemma and Fermat's Theorem 5.4 that

$$\begin{aligned} P(\alpha^p) &= a_0(\alpha^p)^n + a_1(\alpha^p)^{n-1} + \cdots + a_{n-1}\alpha^p + a_n \\ &= a_0^p(\alpha^n)^p + a_1^p(\alpha^{n-1})^p + \cdots + a_{n-1}^p \alpha^p + a_n^p \\ &= (a_0 \alpha^n + a_1 \alpha^{n-1} + \cdots + a_{n-1} \alpha + a_n)^p = 0^p = 0 \end{aligned}$$

Thus, if α is any zero of $P(x)$, so is α^p. ∎

When $P(x)$ is not irreducible, the list given in the statement of the above proposition does not need to contain all of its zeros (Exercise 15). When $P(x)$ is irreducible, a stronger claim can be made. We state this without justification, and relegate the proof to Exercises 7–9:

■ **PROPOSITION 7.11.** *If α is a zero of the irreducible polynomial $P(x)$ of degree v over $\mathbb{Z}_p$, then*

$$\alpha, \alpha^p, \alpha^{p^2}, \alpha^{p^3}, \ldots, \alpha^{p^{v-1}}$$

are all of its zeros.

We have argued that Galois fields that arise from different imaginaries that correspond to the same irreducible polynomial are in fact one and the same. Surprisingly a similar phenomenon occurs when the Galois imaginaries correspond to different irreducible polynomials that have the same degree. Such is

TABLE 7.2. The cyclic table of $GF(2, x^3 + x + 1)$.

$$1$$
$$\tau$$
$$\tau^2$$
$$\tau^3 = 1 + \tau$$
$$\tau^4 = \tau + \tau^2$$
$$\tau^5 = 1 + \tau + \tau^2$$
$$\tau^6 = 1 + \tau^2$$
$$\tau^7 = 1$$

the case for the fields $GF(2, x^3 + x^2 + 1)$ and $GF(2, x^3 + x + 1)$, both of which have $2^3 = 8$ elements. The cyclic table of the first of these appears in Section 7.1, and that of the second is displayed in Table 7.2 (Exercise 7.1.1). These two tables are evidently quite different. Nevertheless, these two fields are really one and the same. To see this, observe that β^3 of $GF(2, x^3 + x^2 + 1)$ is also a zero of $x^3 + x + 1$, the same polynomial that gave rise to the Galois imaginary τ, for

$$(\beta^3)^3 + \beta^3 + 1 = \beta^9 + \beta^3 + 1 = \beta^2 + \beta^3 + 1 = 0$$

Thus, having constructed the field $GF(2, x^3 + x^2 + 1)$, there is no call for associating a new Galois imaginary τ to the polynomial $x^3 + x + 1$ which is irreducible over $\mathbb{Z}_2$. The element β^3 of $GF(2, x^3 + x^2 + 1)$ is already a zero of this polynomial, as are $(\beta^3)^2 = \beta^6$ and $(\beta^6)^2 = \beta^5$. In other words, there is an unanticipated redundancy in the process Galois used to create his fields. This is what Galois had in mind when he wrote in the above quoted passage:

> We are concerned here with the classification of these imaginaries and their reduction to the smallest possible number.

It is very tempting at this point to argue as follows. If $P(x)$ and $Q(x)$ are irreducible polynomials of degree v over $\mathbb{Z}_p$, then, by Corollary 7.6, both $GF(p, P(x))$ and $GF(p, Q(x))$ consist of all the zeros of $x^{p^v} - x$, and consequently these two fields should be one and the same. The flaw in this argument is exposed by the application of an analogous argument to the polynomial $x^2 + 1$. This polynomial has zeros $\{2, 3\}$ in $\mathbb{Z}_5$ and $\{4, 13\}$ in $\mathbb{Z}_{17}$, yet we cannot conclude that $\{2, 3\} = \{4, 13\}$ in any sense. Similarly the zeros that $x^{p^v} - x$ has in $GF(p, P(x))$ have, in principle, nothing to do with its zeros in $GF(p, Q(x))$. The faultiness of this argument notwithstanding, its conclusion happens to be valid. We state it informally below, since the careful formulation and the proof of this theorem lie beyond the scope of this text.

7.4 On the Variety of Galois Fields

■ **THEOREM 7.12.** *For any prime number p and any positive integer v, there is exactly one Galois field containing p^v elements.*

The field whose existence is guaranteed by this theorem is denoted $GF(p^v)$. Accordingly

$$GF(2, x^3 + x^2 + 1) = GF(2, x^3 + x + 1) = GF(2^3)$$

and

$$GF(3, x^2 + x + 2) = GF(3^2)$$

We conclude this section with another aspect of primitivity. A polynomial $P(x)$ of degree v is said to be *primitive* over $\mathbb{Z}_p$ if it is irreducible over $\mathbb{Z}_p$ and its associated Galois imaginary ξ has order $p^v - 1$. In other words, $P(x)$ is primitive over $\mathbb{Z}_p$ when it is irreducible over $\mathbb{Z}_p$, and its Galois imaginary ξ is also primitive. Since the construction of the cyclic table of $GF(p, P(x))$ depends only on $P(x)$ rather than on the specific ξ, it follows that either all the zeros of $P(x)$ are primitive or none of them is primitive. Thus the polynomial $x^3 + x^2 + 1$ is primitive over $\mathbb{Z}_2$, whereas $x^4 + x^3 + x^2 + x + 1$ is not. Finding a primitive polynomial is not an easy task. However, once such a polynomial has been found, it is easy to find all the other primitive polynomials of the same degree. Consider the polynomial $x^2 + x + 2$ which is known to be primitive over $\mathbb{Z}_3$. If σ is its associated Galois imaginary, then, by Proposition 7.4.3, the elements σ^3, σ^5, and σ^7 are all the other primitive elements of $GF(3, x^2 + x + 2)$. By Proposition 7.10, σ and σ^3 are the zeros of the same irreducible polynomial, as are σ^5 and $(\sigma^5)^3 = \sigma^{15} = \sigma^7$. Thus the monic primitive quadratic polynomials over $\mathbb{Z}_3$ are

$$(x - \sigma)(x - \sigma^3) = x^2 - (\sigma + \sigma^3)x + \sigma\sigma^3 = x^2 - (\sigma + 2\sigma + 2)x + \sigma^4$$
$$= x^2 + x + 2$$

and

$$(x - \sigma^5)(x - \sigma^7) = x^2 - (\sigma^5 + \sigma^7) + \sigma^5\sigma^7 = x^2 - (2\sigma + \sigma + 1)x + \sigma^{12}$$
$$= x^2 + 2x + 2$$

EXERCISES 7.4

1. Use the fact that $x^2 + 4x + 2$ is a primitive polynomial to determine all the monic primitive quadratic polynomials over $\mathbb{Z}_5$.
2. Verify that the polynomial $x^2 - x + 3$ is primitive over $\mathbb{Z}_7$. Write out its cyclic table, and use it to find all the other seven monic quadratic polynomials that are primitive over $\mathbb{Z}_7$.

3. Determine all the monic primitive cubic polynomials over $\mathbb{Z}_3$.
4. Prove that if $2^v - 1$ is a prime number, then every polynomial of degree v that is irreducible over $\mathbb{Z}_2$ is also primitive.
5. Prove that if p is a prime other than 2 and $P(x)$ is an irreducible polynomial over $\mathbb{Z}_p$ of degree greater than 1, then $\text{GF}(p, P(x))$ has some nonprimitive elements besides 0 and 1.
6. Suppose the polynomial $a_0 x^n + a_1 x^{n-1} + \cdots + a_{n-1} x + a_n$ $(a_0, a_n \neq 0)$ is primitive over a field F. Prove that the polynomial $a_n x^n + a_{n-1} x^{n-1} + \cdots + a_1 x + a_0$ is also primitive over F.
7. Prove that for every element α of $\text{GF}(p, P(x))$, there is a positive integer k such that $\alpha, \alpha^p, \alpha^{p^2}, \ldots, \alpha^{p^{k-1}}$ are all distinct, and $\alpha^{p^k} = \alpha$.
8. For any $\alpha \in \text{GF}(p, P(x))$, let $M_\alpha(x) = (x - \alpha)(x - \alpha^p)(x - \alpha^{p^2}) \cdots (x - \alpha^{p^{k-1}})$ where k is as defined in Exercise 7. Prove that $M_\alpha(x) \in \mathbb{Z}_p[x]$.
9. Prove Proposition 7.11.

By Proposition 7.11, any polynomial that has α as its zero is divisible by the polynomial $M_\alpha(x)$ of Exercise 8. The polynomial $M_\alpha(x)$ is therefore called the (monic) minimal polynomial of α. Find the monic minimal polynomials for all the elements of each of the fields in Exercises 10–14.

10. $\text{GF}(2, x^3 + x + 1)$ 11. $\text{GF}(2, x^4 + x + 1)$
12. $\text{GF}(3, x^2 + 2x + 2)$ 13. $\text{GF}(5, x^2 + 4x + 2)$
14. $\text{GF}(7, x^2 - x + 3)$
15. Show that the conclusion of Proposition 7.11 does not need to be valid if $P(x)$ is reducible over $\mathbb{Z}_p$.

CHAPTER SUMMARY

Working by analogy with the complex numbers, Galois created a host of new fields, now bearing his name. The nonzero elements of these Galois fields are also roots of unity and as such have orders whose properties are indistinguishable from the orders of the complex and modular roots of unity. We proved the Primitive Element Theorem for these (and the modular) roots of unity. It was also pointed out that these fields are subject to some subtle relationships, since seemingly different fields may turn out, upon careful examination, to be identical.

Chapter Review Exercises

Mark the following true or false.

1. Let $P(x)$ be an irreducible quadratic over $\mathbb{Z}_7$, and let α be the associated Galois imaginary. Then there exist $a, b \in \mathbb{Z}_7$ such that $(3 + 4\alpha)(a + b\alpha) = 1$.
2. Let $P(x)$ be an irreducible quadratic over $\mathbb{Z}_7$, and let α be the associated Galois imaginary. Then there exist $a, b \in \mathbb{Z}_7$ such that $(2 + 4\alpha)(a + b\alpha) = 1 + 5\alpha$.
3. Let $P(x)$ be an irreducible cubic over $\mathbb{Z}_3$. Then the number of elements of $\text{GF}(3, P(x))$ is 30.

4. Let $P(x)$ be an irreducible polynomial over $\mathbb{Z}_p$. Then the equation $P(x) = 0$ has a solution in $\mathrm{GF}(p, P(x))$.
5. Let $P(x)$ be an irreducible polynomial of degree 4 over $\mathbb{Z}_5$. Then every element of $\mathrm{GF}(4, P(x))$ is a zero of the polynomial $x^{600} + x^{25} + x^7 + x$.
6. The polynomial $x^2 + 1$ is primitive over $\mathbb{Z}_2$.
7. There is an element r in $\mathbb{Z}_{31}$ such that the sequence $1, r, r^2, \ldots$ (mod 31) contains all the nonzero elements of $\mathbb{Z}_{31}$.
8. Let $\alpha \in \mathrm{GF}(2, x^5 + x^2 + 1)$. If $P(x) \in \mathbb{Z}_2[x]$ and $P(\alpha) = 0$, then $P(\alpha^2) = 0$.

New Terms

Cyclic table	134
Galois field	133
Galois imaginary	130
Galois polynomial	140
Order of element	135
Primitive element	134
Primitive polynomial	149

Supplementary Exercises

1. Write a computer script that lists the primitive monic polynomials of degree d.
2. Prove that for every positive integer d and for every prime p, there is a primitive polynomial of degree d over $\mathbb{Z}_p$.
3. Find a formula for the number of monic irreducible polynomials of degree d over $\mathbb{Z}_p$.
4. Write a computer script that will solve any polynomial equation $Q(x) = 0$ over any Galois field $\mathrm{GF}(p, P(x))$.
5. If F is any field and

$$P(x) = a_0 x^n + a_1 x^{n-1} + \cdots + a_{n-1} x + a_n \in F[x]$$

then the *derivative* of $P(x)$ is defined as the polynomial

$$P'(x) = n a_0 x^{n-1} + (n-1) a_1 x^{n-2} + \cdots + a_{n-1} \in F[x]$$

Prove that this derivative has the following properties:
(a) $[P(x) + Q(x)]' = P'(x) + Q'(x)$ for any $P(x), Q(x) \in F[x]$.
(b) $[cP(x)]' = cP'(x)$ for any $c \in F$ and $P(x) \in F[x]$.
(c) $[P(x)Q(x)]' = P'(x)Q(x) + P(x)Q'(x)$ for any $P(x), Q(x) \in F[x]$.
(d) $P(x)$ has repeated zeros in F only if the greatest common divisor of $P(x)$ and $P'(x)$ has degree at least 1.

6. Let p be any prime and n any positive integer. Prove that the polynomial $x^n - 1$ has repeated zeros in $\mathbb{Z}_p$ if and only if p is factor of n.
7. For how many primes p is 10 a primitive root (mod p)?

CHAPTER 8

Permutations

Motivated by Lagrange's solution of the quartic equation, we consider the general question of what happens to a multivariable function when its variables are permuted. Some surprising results are derived, and they lead us to a deeper examination of the notion of a permutation.

8.1 PERMUTING THE VARIABLES OF A FUNCTION (I)

The key to Lagrange's solution of the general quartic equation (Section 6.5) was the fact that when the variables of the polynomial $x_1x_2 + x_3x_4$ are permuted in all the possible ways so as to produce the 24 polynomials

$$x_1x_2 + x_3x_4, \quad x_1x_2 + x_4x_3, \quad x_1x_3 + x_2x_4, \quad x_1x_3 + x_4x_2,$$
$$x_1x_4 + x_2x_3, \quad x_1x_4 + x_3x_2, \quad x_2x_1 + x_3x_4, \quad x_2x_1 + x_4x_3,$$
$$x_2x_3 + x_1x_4, \quad x_2x_3 + x_4x_1, \quad x_2x_4 + x_1x_3, \quad x_2x_4 + x_3x_1,$$
$$x_3x_1 + x_2x_4, \quad x_3x_1 + x_4x_2, \quad x_3x_2 + x_1x_4, \quad x_3x_2 + x_4x_1,$$
$$x_3x_4 + x_1x_2, \quad x_3x_4 + x_2x_1, \quad x_4x_1 + x_2x_3, \quad x_4x_1 + x_3x_2,$$
$$x_4x_2 + x_1x_3, \quad x_4x_2 + x_3x_1, \quad x_4x_3 + x_1x_2, \quad x_4x_3 + x_2x_1$$

it follows from the commutativity of addition and multiplication that in fact only *three* distinct polynomials emerge, namely

$$x_1x_2 + x_3x_4, \quad x_1x_3 + x_2x_4, \quad \text{and} \quad x_1x_4 + x_2x_3$$

In general, when two polynomials $P(x, y, z, \ldots)$ and $Q(x, y, z, \ldots)$ can be obtained from each other by permuting their variables, these polynomials are said to be *variants* of each other. Thus $x_1x_2 + x_3x_4$ has the 24 variants listed above, whereas $x_1x_2 + x_3$ has the 6 variants:

$$x_1x_2 + x_3, \quad x_2x_1 + x_3, \quad x_1x_3 + x_2, \quad x_3x_1 + x_2, \quad x_2x_3 + x_1, \quad x_3x_2 + x_1$$

Two variants are said to be *distinct* if they differ as functions over $\mathbb{C}$. Intuition is a fairly reliable guide in this context. When necessary, however, appropriate substitutions can be used to verify distinctness. Thus the substitution of $1, 1, 0, 0$ for x_1, x_2, x_3, x_4, respectively, proves that the variants $x_1x_2 + x_3x_4$ and $x_1x_3 + x_2x_4$ are distinct, since they assume the respective values 1 and 0. Lagrange's observation can now be phrased as *the polynomial $x_1x_2 + x_3x_4$ has three distinct variants*. Similarly the polynomial $x_1x_2 + x_3$ also has three distinct variants. A function that has no two distinct variants is said to be *invariant*. The polynomials x_1x_2 and $x_1 + x_2 + x_3$ as well as all the elementary symmetric polynomials of Section 6.4 are examples of invariant functions.

Since Lagrange's solution of the quartic equation hinges on the existence of a polynomial of four variables with three distinct variants, it is reasonable, when attempting the solution of the fifth-degree equation, to look for polynomials in five variables that have four (or perhaps three) distinct variants. Surprisingly such polynomials do not exist.

Let us digress here and change the point of view somewhat. The five-variable polynomial

$$x_1^2 + x_2^2 + x_3^2 + x_4^2 + x_5^2$$

is clearly *invariant*. On the other hand, the polynomial

$$x_1 + x_2 + x_3 + x_4 - x_5$$

has five distinct variants, namely itself and the four polynomials

$$x_1 + x_2 + x_3 + x_5 - x_4 \quad x_1 + x_2 + x_4 + x_5 - x_3$$
$$x_1 + x_3 + x_4 + x_5 - x_2 \quad x_2 + x_3 + x_4 + x_5 - x_1$$

Similarly the polynomial

$$x_1x_2 + x_3x_4x_5$$

has $\binom{5}{2} = 10$ distinct variants, namely itself and the nine polynomials

$$x_1x_3 + x_2x_4x_5 \quad x_1x_4 + x_2x_3x_5 \quad x_1x_5 + x_2x_3x_4$$
$$x_2x_3 + x_1x_4x_5 \quad x_2x_4 + x_1x_3x_5 \quad x_2x_5 + x_1x_3x_4$$
$$x_3x_4 + x_1x_2x_5 \quad x_3x_5 + x_1x_2x_4 \quad x_4x_5 + x_1x_2x_3$$

However, is there a function of these five variables that has four distinct variants? A formal proof of the nonexistence of such a function is offered below in Corollary 8.9. The need for such a proof is underscored by the fact that the beginner's search for functions of five variables that have *two* distinct variants is very likely also to meet with failure. Such functions, however, do exist, and a method for constructing them is suggested in Exercise 36. A complete proof of the existence of such two-variant functions is offered in Proposition 8.11.

That there are no functions of *five* variables that have *four* (or *three*) distinct variants was first recognized by Paolo Ruffini who incorporated this observation into his unsuccessful attempt to prove the unsolvability of the general quintic equation by radicals. Ruffini's theorem regarding the number of distinct variants that a function of five variables can have was generalized in 1815 by Cauchy to the statement that appears as Theorem 8.8 below. This proposition was incorporated by Abel into his groundbreaking proof of the unsolvability of the general quintic equation by radicals. In 1847 Cauchy returned to this topic and proved that *if a function on $n \geq 5$ variables has fewer than n distinct variants, then that function has only 1 or 2 distinct variants.* Since this stronger version turned out to play no special role in the evolution of the theory of algebraic resolvability of equations, it is mentioned without proof. We will, however, prove Cauchy's 1815 theorem after providing some basic theoretical information about permutations.

EXERCISES 8.1

Find the number of distinct variants that the functions in Exercises 1–35 have.

1. $x_1 + x_2$
2. $x_1 - x_2$
3. $(x_1 - x_2)^2$
4. $\dfrac{x_1}{x_2}$
5. $\dfrac{x_1}{x_2} + \dfrac{x_2}{x_1}$
6. $\sin(x - y)$
7. $\cos(x - y)$
8. $x_1 + 5x_2$
9. $x_1 x_2^2$
10. $x_1^3 x_2^3 x_3^3$
11. $x_1 x_2 x_3^2$
12. $(x_1 + x_2 - x_3)^2$
13. $(x_1 + x_2)(x_1 + x_3)(x_2 + x_3)$
14. $(x_1 - x_2)(x_1 - x_3)(x_2 - x_3)$
15. $x_1 x_2 + x_3$
16. $\dfrac{x_1}{x_2} + x_3$
17. $x_1 x_2^2 x_3^3$
18. $x_1 x_2 x_3 x_4$
19. $x_1 x_2 x_3 x_4^3$
20. $x_1 x_2 x_3^5 x_4^5$
21. $x_1 x_2 - x_3 x_4$
22. $(x_1 x_2 - x_3 x_4)^2$
23. $\dfrac{x_1 + x_2}{x_3 x_4}$
24. $(x_1 - x_2)^2 + (x_3 - x_4)^2$
25. $x_1 x_2^2 x_3^3 x_4^4$
26. $(x_1 - x_2)(x_1 - x_3)(x_2 - x_3) x_4$
27. $(x_1 + x_2)(x_1 + x_3)(x_1 + x_4)(x_2 + x_3)(x_2 + x_4)(x_3 + x_4)$
28. $(x_1 - x_2)(x_1 - x_3)(x_1 - x_4)(x_2 - x_3)(x_2 - x_4)(x_3 - x_4)$
29. $x_1 x_2 x_3 x_4 x_5$
30. $\dfrac{x_1 x_2 x_3 x_4}{x_5}$

31. $\dfrac{x_1 x_2 x_3 + x_4}{x_5}$
32. $\dfrac{x_1 x_2 x_3}{x_4 x_5}$
33. $(x_1 x_2 + x_3 x_4) x_5$
34. $(x_1 - x_2)(x_1 - x_3)(x_2 - x_3) x_4 x_5$
35. $x_1 x_2^2 x_3^3 x_4^4 x_5^5$
36. Use the answers of Exercises 14 and 28 to create a function of five variables that has two distinct variants.

Prove that for any positive integer n there exists a function of n variables that has the number of distinct variants that is specified in Exercises 37–43.

37. 1
38. n
39. $\binom{n}{2}$, $n \geq 2$
40. $3 \binom{n}{3}$, $n \geq 3$
41. $3 \binom{n}{4}$, $n \geq 4$
42. $\binom{n}{k}$, $n \geq k \geq 0$
43. $k \binom{n}{k}$, $n \geq k \geq 1$

44. Show that for any positive integer n there is a function of n variables such that every two of its variants are distinct.

8.2 PERMUTATIONS

In the previous section we had several occasions to shuffle some variables and to observe the effect that this transformation had on a function of these variables. We now focus on the shuffles themselves. The mathematical name for such a shuffle is a *permutation*. More formally, a permutation of a set S is a function σ that assigns to each element x of S an element $y = \sigma(x)$ of S so that

1. if x_1 and x_2 are distinct elements of S, then $\sigma(x_1) \neq \sigma(x_2)$;
2. if y is any element of S, then there is an element x in S such that $y = \sigma(x)$.

We note in passing that when the underlying set S is finite, these two conditions are equivalent (Exercise 35), and so only one needs to be verified. The *identity permutation* that transforms each element to itself is denoted by Id. In the earlier days of permutation theory, each permutation was written as an array of two rows. The first of these rows listed the elements of S in some natural order, and the second row listed the corresponding values of σ. Thus the array $\begin{pmatrix} x_1 & x_2 & x_3 & x_4 \\ x_2 & x_3 & x_4 & x_1 \end{pmatrix}$ was used to denote the permutation σ such that

$$\sigma(x_i) = x_{i+1} \quad \text{(addition modulo 4)}$$

and whose effect is to convert the polynomial $x_1 x_2 + x_3 x_4$ to the polynomial

$x_2x_3 + x_4x_1$. Similarly the permutation that interchanges x_1 with x_3 and also interchanges x_2 with x_4 was denoted by

$$\begin{pmatrix} x_1 & x_2 & x_3 & x_4 \\ x_3 & x_4 & x_1 & x_2 \end{pmatrix}$$

The letter x above serves merely as a place holder and it is more efficient to eliminate it. Thus we will generally restrict our attention to permutations on a set $S = \{1, 2, \ldots, n\}$, and the above two permutations can be denoted by $\begin{pmatrix} 1 & 2 & 3 & 4 \\ 2 & 3 & 4 & 1 \end{pmatrix}$ and $\begin{pmatrix} 1 & 2 & 3 & 4 \\ 3 & 4 & 1 & 2 \end{pmatrix}$, respectively. This notation will be further improved, but first we pause to count the permutations of a given set.

■ **PROPOSITION 8.1.** *For every positive integer n the number of permutations of the set $S = \{1, 2, 3, \ldots, n\}$ is $n! = 1 \cdot 2 \cdot 3 \cdot \cdots \cdot n$.*

Proof. Let $\begin{pmatrix} 1 & 2 & 3 & \cdots & k & \cdots & n \\ a & b & c & \cdots & h & \cdots & j \end{pmatrix}$ denote the arbitrary permutation of $S = \{1, 2, 3, \ldots, n\}$. Then the symbol a can be replaced by any of the n elements of S. Once a has been chosen, b can be replaced by any of the $n - 1$ elements of $S - \{a\}$. Once b has been chosen, c can be replaced by any of the $n - 2$ elements of $S - \{a, b\}$. Proceeding in the same manner, it is clear that the second row of this arbitrary permutation can be filled out in $n(n - 1)(n - 2) \cdots 1 = n!$ ways so as to define a bona fide permutation of S. ■

Like all functions, permutations can be composed, and this composition is associative (Exercise 29). If ρ and σ are two permutations of the set S, then their composition is denoted by either $\rho \circ \sigma$ or simply by their juxtaposition $\rho\sigma$ where $\rho\sigma(a) = \rho(\sigma(a))$. Thus, if

$$\rho = \begin{pmatrix} 1 & 2 & 3 & 4 \\ 3 & 1 & 4 & 2 \end{pmatrix} \quad \text{and} \quad \sigma = \begin{pmatrix} 1 & 2 & 3 & 4 \\ 2 & 3 & 4 & 1 \end{pmatrix}$$

then

$$\rho \circ \sigma(1) = \rho\sigma(1) = \rho(\sigma(1)) = \rho(2) = 1$$
$$\rho \circ \sigma(2) = \rho\sigma(2) = \rho(\sigma(2)) = \rho(3) = 4$$
$$\rho \circ \sigma(3) = \rho\sigma(3) = \rho(\sigma(3)) = \rho(4) = 2$$
$$\rho \circ \sigma(4) = \rho\sigma(4) = \rho(\sigma(4)) = \rho(1) = 3$$

In fact

$$\rho \circ \sigma = \rho\sigma = \begin{pmatrix} 1 & 2 & 3 & 4 \\ 1 & 4 & 2 & 3 \end{pmatrix}$$

This situation is quite typical and merits an explicit statement and perhaps also a formal proof.

■ **PROPOSITION 8.2.** *If ρ and σ are permutations of the set S, then so is their composition $\rho\sigma$ a permutation of the set S.*

Proof. Suppose that x_1 and x_2 are distinct elements of S. Then, since both ρ and σ are known to be permutations, it follows from property 1 of permutations first that

$$\sigma(x_1) \neq \sigma(x_2)$$

and next that

$$\rho(\sigma(x_1)) \neq \rho(\sigma(x_2))$$

or

$$\rho\sigma(x_1) \neq \rho\sigma(x_2)$$

Thus the composition $\rho\sigma$ also satisfies property 1. Similarly, if y is any element of S, then by property 2 of ρ, there exists an element z of S such that $\rho(z) = y$, and by property 2 of σ, there exists an element x of S such that $\sigma(x) = z$. Combining these two we see that

$$\rho\sigma(x) = \rho(\sigma(x)) = \rho(z) = y$$

so that the composition $\rho\sigma$ also has property 2. Thus $\rho\sigma$ is a permutation of S. ■

If σ is any permutation, then we define $\sigma^0 = \text{Id}$, $\sigma^1 = \sigma$, $\sigma^2 = \sigma\sigma$, and if k is any positive integer, σ^k is the composition of k σ's. Accordingly, if $\sigma = \begin{pmatrix} 1 & 2 & 3 & 4 & 5 \\ 2 & 3 & 4 & 5 & 1 \end{pmatrix}$, then $\sigma^2 = \begin{pmatrix} 1 & 2 & 3 & 4 & 5 \\ 3 & 4 & 5 & 1 & 2 \end{pmatrix}$ and $\sigma^3 = \begin{pmatrix} 1 & 2 & 3 & 4 & 5 \\ 4 & 5 & 1 & 2 & 3 \end{pmatrix}$. We now describe yet another, more efficient, way of writing down permutations of finite sets. We first define a *cycle*, or a *cyclic permutation* as a permutation of the form $\begin{pmatrix} a & b & c & \cdots & g & h \\ b & c & d & \cdots & h & a \end{pmatrix}$ and agree to write it in any of the forms

$$(a\ b\ c\ \cdots\ g\ h) = (b\ c\ \cdots\ g\ h\ a) = (c\ d\ \cdots\ h\ a\ b) = \cdots = (h\ a\ b\ \cdots\ g)$$

If k is any positive integer, then a *k-cycle* is a cycle that contains k elements. Thus (3 5 7) is a 3-cycle and (1 8 7 2 5) is a 5-cycle. Suppose next that σ is an arbitrary permutation and that a is an arbitrary element of the underlying set S. Consider the sequence

$$\sigma^0(a) = a, \quad \sigma^1(a) = \sigma(a), \quad \sigma^2(a), \quad \sigma^3(a), \ldots$$

Since S is finite, this infinite sequence must contain repetitions. Let k be the first exponent for which there exists another exponent $m > k$ such that

$$\sigma^k(a) = \sigma^m(a)$$

The exponent k must in fact be 0, since otherwise we would have

$$\sigma(\sigma^{k-1}(a)) = \sigma^k(a) = \sigma^m(a) = \sigma(\sigma^{m-1}(a))$$

and by property 1 of permutations we would have

$$\sigma^{k-1}(a) = \sigma^{m-1}(a)$$

contradicting the minimality of k. Hence the elements of

$$a, \sigma(a), \sigma^2(a), \sigma^3(a), \ldots, \sigma^{k-1}(a) \tag{1}$$

are all distinct and $\sigma(\sigma^{k-1}(a)) = a$. If we define σ_a to be the cycle

$$(a \ \sigma(a) \ \sigma^2(a) \ \sigma^3(a) \ \cdots \ \sigma^{k-1}(a))$$

then it is clear that σ and σ_a agree on all the elements of (1). If list (1) does not exhaust all the elements permuted by σ, let b be an element that does not appear in list (1), and let h be the least positive integer such that $\sigma^h(b) = b$. We then define

$$\sigma_b = (b \ \sigma(b) \ \sigma^2(b) \ \sigma^3(b) \ \cdots \ \sigma^{h-1}(b))$$

If this process is repeated until all the elements permuted by σ are exhausted, we have cyclic permutations $\sigma_a, \sigma_b, \ldots$ such that

$$\sigma = \sigma_a \sigma_b \cdots \tag{2}$$

Thus

$$\begin{pmatrix} 1 & 2 & 3 & 4 & 5 & 6 & 7 & 8 & 9 \\ 9 & 5 & 1 & 3 & 7 & 6 & 2 & 4 & 8 \end{pmatrix} = (1 \ 9 \ 8 \ 4 \ 3)(6)(2 \ 5 \ 7)$$

8.2 Permutations

We will refer to (2) as the *disjoint cycle decomposition* of σ. Note that the order in which the individual cycles in the disjoint cycle form of a permutation are written is arbitrary. Similarly it is only the *cyclic* order of the elements that appear in a cycle that is significant. Thus

$$(1\ 2\ 3\ 4)(5\ 6\ 7)(8\ 9) = (5\ 6\ 7)(1\ 2\ 3\ 4)(8\ 9) = (8\ 9)(5\ 6\ 7)(1\ 2\ 3\ 4)$$
$$= (3\ 4\ 1\ 2)(8\ 9)(6\ 7\ 5)$$

If this process gives rise to a cycle of length 1 that cycle is generally omitted. Accordingly

$$\begin{pmatrix} 1 & 2 & 3 & 4 & 5 & 6 \\ 3 & 2 & 4 & 6 & 5 & 1 \end{pmatrix} = (1\ 3\ 4\ 6)$$

It is clear that if

$$\rho = (a_1 a_2 a_3 \cdots a_{n-1} a_n) \quad \text{and} \quad \sigma = (a_n a_{n-1} \cdots a_3 a_2 a_1)$$

then

$$\rho\sigma = \sigma\rho = \text{Id}$$

We say that ρ and σ are *inverses* of each other and write

$$\sigma = \rho^{-1} \quad \text{and} \quad \rho = \sigma^{-1}$$

Even when ρ is not necessarily cyclic, it has an inverse, and this inverse is easily described. To see this, let

$$\sigma = \sigma_1 \sigma_2 \cdots \sigma_{n-1} \sigma_n$$

be the disjoint cycle decomposition of σ. If we now set

$$\rho = \sigma_n^{-1} \sigma_{n-1}^{-1} \cdots \sigma_2^{-1} \sigma_1^{-1}$$

then

$$\sigma\rho = \sigma_1 \sigma_2 \cdots \sigma_{n-1} \sigma_n \sigma_n^{-1} \sigma_{n-1}^{-1} \cdots \sigma_2^{-1} \sigma_1^{-1} = \sigma_1 \sigma_2 \cdots \sigma_{n-1} \sigma_{n-1}^{-1} \cdots \sigma_2^{-1} \sigma_1^{-1}$$
$$= \cdots = \text{Id}$$

Similarly

$$\rho\sigma = \text{Id}$$

Thus the inverse of the permutation

$$(1\ 9\ 8\ 4\ 3)(6)(2\ 5\ 7) \text{ is } (7\ 5\ 2)(6)(3\ 4\ 8\ 9\ 1)$$

It is clear from this description of inverses that every permutation has a unique inverse. If we set $\sigma^{-m} = (\sigma^{-1})^m$ for every nonnegative integer, then the powers of permutations obey the usual exponential rules (Exercise 28). The following lemma, however, is not quite so obvious:

■ **LEMMA 8.3.** *If ρ and σ are permutations of the same set, then*

$$(\rho\sigma)^{-1} = \sigma^{-1}\rho^{-1}$$

Proof. The proof is a straightforward application of the associativity of the composition of permutations. Note that

$$(\rho\sigma)(\sigma^{-1}\rho^{-1}) = \rho(\sigma\sigma^{-1})\rho^{-1} = \rho\text{Id}\rho^{-1} = \rho\rho^{-1} = \text{Id}$$

and

$$(\sigma^{-1}\rho^{-1})(\rho\sigma) = \sigma^{-1}(\rho^{-1}\rho)\sigma = \sigma^{-1}\text{Id}\sigma = \sigma^{-1}\sigma = \text{Id}$$

Thus $\sigma^{-1}\rho^{-1}$ fulfills the requisite conditions for being the inverse of $\rho\sigma$. ■

The analysis of the effects of permutations is often facilitated by factoring them into the composition of "smaller" permutations. A *transposition* is a permutation that interchanges only two elements, leaving all the others fixed. Thus every transposition has the form $(a\ b)$. It is clear that in an informal sense the transpositions are the smallest nontrivial permutations. The equations

$$(1\ 2\ 3) = (1\ 2)(2\ 3) \quad \text{and} \quad (1\ 2\ 3\ 4) = (1\ 2)(2\ 3)(3\ 4)$$

are instances of nontranspositions expressed as the composition of transpositions. This is always possible.

■ **PROPOSITION 8.4.** *Every permutation of a finite set is the composition of some transpositions.*

Proof. We already know that every permutation of a finite set is the composition of cyclic permutations, and hence it suffices to show that every cyclic permutation can be expressed as the composition of some transpositions. This, however, is easily accomplished as follows:

$$(a_1 a_2 a_3 \cdots a_n) = (a_1 a_2)(a_2 a_3) \cdots (a_{n-1} a_n) \qquad ■$$

Thus

$$(1\ 5\ 3)(2\ 4\ 8\ 9)(a\ b\ c\ d\ e) = (1\ 5)(5\ 3)(2\ 4)(4\ 8)(8\ 9)(a\ b)(b\ c)(c\ d)(d\ e)$$

It should be stressed that such expressions are not unique as illustrated by the equations

$$(1\ 2\ 3) = (1\ 2)(2\ 3) = (1\ 3)(1\ 2) = (2\ 3)(1\ 3) = (1\ 2)(3\ 4)(2\ 4)(3\ 4)$$
$$= (2\ 3)(1\ 3)(2\ 3)(3\ 4)(2\ 4)(3\ 4)$$

It is clear that if $\rho = (a_1 a_2 \cdots a_k)$ is any cyclic permutation then $\rho^k = \mathrm{Id}$. Consequently, if m is any common multiple of the lengths of $\sigma_1, \sigma_2, \ldots, \sigma_k$ in the disjoint cycle factorization of the arbitrary permutation $\sigma = \sigma_1 \sigma_2 \cdots \sigma_k$, then $\sigma^m = \mathrm{Id}$. Thus

$$((1\ 2\ 3)(4\ 5))^6 = \mathrm{Id}$$

The order $o(\sigma)$ of the permutation σ is the least positive integer m such that $\sigma^m = \mathrm{Id}$. The above considerations make it clear that every permutation of a finite set has a finite order. Exercise 31 asserts that the order of any permutation equals the least common multiple of the lengths of its disjoint cyclic factors.

EXERCISES 8.2

Express the permutations in Exercises 1–4 in the disjoint cycle form.

1. $\begin{pmatrix} 1 & 2 & 3 & 4 & 5 & 6 & 7 & 8 & 9 \\ 9 & 3 & 4 & 7 & 1 & 5 & 2 & 6 & 8 \end{pmatrix}$
2. $\begin{pmatrix} 1 & 2 & 3 & 4 & 5 & 6 & 7 & 8 & 9 \\ 5 & 3 & 4 & 7 & 1 & 6 & 2 & 9 & 8 \end{pmatrix}$
3. $\begin{pmatrix} 1 & 2 & 3 & 4 & 5 & 6 & 7 & 8 & 9 \\ 9 & 8 & 7 & 6 & 5 & 4 & 3 & 2 & 1 \end{pmatrix}$
4. $\begin{pmatrix} 1 & 2 & 3 & 4 & 5 & 6 & 7 & 8 & 9 \\ 9 & 7 & 5 & 3 & 1 & 8 & 6 & 4 & 2 \end{pmatrix}$

5. List the permutations of $\{1, 2\}$ in disjoint cycle form.
6. List the permutations of $\{1, 2, 3\}$ in disjoint cycle form.
7. List the permutations of $\{1, 2, 3, 4\}$ in disjoint cycle form.

Given the permutations $\rho = (1\ 2\ 3\ 4)(5\ 6\ 7)(8\ 9)$, $\sigma = (1\ 9\ 8\ 6\ 5)(2\ 3\ 4\ 7)$, express the permutations in Exercises 8–14 in disjoint cycle form, and compute their orders.

8. $\rho\sigma$
9. $\sigma\rho$
10. $\rho\sigma\rho$
11. $\rho\sigma\rho\sigma$
12. $\rho\sigma\rho^{-1}$
13. $\sigma\rho\sigma^{-1}$
14. $\rho^2\sigma^3$

Express the permutations in Exercises 15–18 as a composition of transpositions.

15. $(1\ 2\ 3)(4\ 5\ 6\ 7)(8\ 9)$
16. $(1\ 4\ 2\ 9)(5\ 6\ 3\ 7)$
17. $\begin{pmatrix} 1 & 2 & 3 & 4 & 5 & 6 & 7 & 8 & 9 \\ 9 & 3 & 4 & 7 & 1 & 5 & 2 & 6 & 8 \end{pmatrix}$
18. $\begin{pmatrix} 1 & 2 & 3 & 4 & 5 & 6 & 7 & 8 & 9 \\ 9 & 1 & 5 & 3 & 7 & 8 & 6 & 2 & 4 \end{pmatrix}$

19. Prove that if $n \geq 2$, then every permutation of $\{1, 2, \ldots, n\}$ can be expressed as the composition of transpositions of the form (1 a), $a = 2, 3, \ldots, n$.

20. Show that if $n \geq 4$, then every permutation of $\{1, 2, \ldots, n\}$ is expressible as the composition of 4-cycles.

21. Show that if $n \geq k \geq 2$ and k is even, then every permutation of $\{1, 2, \ldots, n\}$ is expressible as the composition of k-cycles.

22. Prove that if σ and ρ are permutations of $\{1, 2, \ldots, n\}$, then

$$o(\sigma) = o(\rho\sigma\rho^{-1})$$

23. (a) Prove that if any cycle of the permutation σ has the form

$$(a_1 a_2 a_3 \cdots a_k)$$

and if ρ is any other permutation, then $\rho\sigma\rho^{-1}$ has a corresponding cycle of the form

$$(\rho(a_1)\ \rho(a_2)\ \rho(a_3)\ \cdots\ \rho(a_k))$$

(b) Conclude that for each positive integer k, σ and $\rho\sigma\rho^{-1}$ have the same number of k-cycles.

(c) Show that if the two permutations σ and τ possess the same number of k-cycles for each positive integer k, then there is a permutation ρ such that $\tau = \rho\sigma\rho^{-1}$.

The number of cycles in the disjoint cycle form of the permutation σ on $\{1, 2, \ldots, n\}$ is denoted by $\|\sigma\|$. Thus $\|(1\ 2\ 3)(4\ 5)(6\ 7\ 8\ 9)\| = 3$ if $n = 9$, and $\|Id\| = n$ regardless of the value of n.

24. Let σ be any permutation, and let τ be any transposition on the same set. Prove that $\|\sigma\tau\| = \|\sigma\| \pm 1$.

25. Prove that if σ and ρ are any two permutations, then $\|\rho\sigma\rho^{-1}\| = \|\sigma\|$.

26. Prove that if σ and ρ are any two permutations, then $\|\sigma\rho\| = \|\rho\sigma\|$.

27. Prove that every permutation of $\{1, 2, \ldots, n\}$ is expressible as the composition of two permutations of $\{1, 2, \ldots, n\}$ that have order at most 2.

28. Prove that if σ is any permutation and m and n are any integers, then $\sigma^m \sigma^n = \sigma^{m+n}$ and $(\sigma^m)^n = \sigma^{mn}$.

29. Let S, T, U, V be sets, and let f be a function from S to T, g a function from T to U, and h a function from U to V. If the composition $g \circ f$ is defined via $g \circ f(x) = g(f(x))$, prove that

$$h \circ (g \circ f) = (h \circ g) \circ f$$

30. Find two permutations ρ and σ that have relatively prime orders and for which $o(\rho\sigma) \neq o(\rho)o(\sigma)$.

31. Suppose that $\sigma = \sigma_1 \sigma_2 \cdots \sigma_k$ is the disjoint cycle form of σ, and suppose that each factor σ_i contains m_i elements of S for $i = 1, 2, \ldots, k$. Prove that $o(\sigma)$ is the least common multiple of $m_1, m_2, \ldots, m_k$.

32. Prove that the order of every permutation on $\{1, 2, \ldots, n\}$ is a proper factor of $n!$ if $n > 2$.

33. For any integers $k \geq 0$ and $n > 0$, let $s(n, k)$ denote the number of permutations of $\{1, 2, \ldots, n\}$ whose disjoint cycle decomposition has exactly k cycles. Prove that for $k \geq 1$ and $n > 1$,

$$s(n, k) = s(n - 1, k - 1) + (n - 1)s(n - 1, k)$$

34. Show that the average number of cycles in the disjoint cycle decomposition of all the permutations on $\{1, 2, \ldots, n\}$ is

$$1 + \frac{1}{2} + \frac{1}{3} + \cdots + \frac{1}{n}$$

35. Prove that if the set S is finite and σ is a function of S into itself, then the following conditions are equivalent:
    (a) If x_1 and x_2 are distinct elements of S, then $\sigma(x_1) \neq \sigma(x_2)$.
    (b) If y is any element of S, then there is an element x in S such that $y = \sigma(x)$.

36. Show, by means of examples, that when S is an infinite set, neither of the conditions of Exercise 35 need entail the other.

8.3 PERMUTING THE VARIABLES OF A FUNCTION (II)

We now return to the issue of the number of distinct variants of a given function of several variables. We begin by formalizing the notion of interchanging the variables of a function. If $f = f(x_1, x_2, \ldots, x_n)$ is any function of n variables, and if σ is any permutation of the indices, then we define

$$\sigma f = f(x_{\sigma(1)}, x_{\sigma(2)}, \ldots, x_{\sigma(n)})$$

If $f = x_1 x_2 + x_3 x_4$ and σ is the cyclic permutation (1 2 3 4), then

$$\sigma f = x_2 x_3 + x_4 x_1$$

The following lemma is clear:

■ **LEMMA 8.5.** *If f and g are two functions of the variables $x_1, x_2, \ldots, x_n$ such that $f = g$, and if σ is any permutation of $\{1, 2, \ldots, n\}$, then $\sigma f = \sigma g$.*

The next two observations are easy, but fundamental. Their justification relies on the fact that in the composition $\rho\sigma$ it is σ that acts first on their permuted elements, and its action is followed by that of ρ.

■ **LEMMA 8.6.** *If f is any function of the variables $x_1, x_2, \ldots, x_n$, and if ρ and σ are any two permutations of $\{1, 2, \ldots, n\}$, then*

$$(\sigma\rho)f = \sigma(\rho f)$$

■ **COROLLARY 8.7.** *If $f = \sigma f$, then $f = \sigma^{-1} f$.*

Proof. If $f = \sigma f$, then, by the above two lemmas

$$\sigma^{-1} f = \sigma^{-1}(\sigma f) = (\sigma^{-1}\sigma)f = (\text{Id})f = f \qquad ■$$

If it so happens that $\sigma f = f$, we say that σ *leaves f unchanged.* Thus, $(1\ 3)(2\ 4)$ leaves $x_1 x_2 + x_3 x_4$ unchanged. We are now ready to state and prove this section's main theorem.

■ **THEOREM 8.8 (Cauchy).** *Let f be a function of n variables, and let p be any prime such that $p \leq n$. If f has $k < p$ distinct variants, then k is either 1 or 2.*

Proof. Let f and $k < p \leq n$ be as in the statement of the theorem. We first show that if σ is any permutation of order p of $\{1, 2, \ldots, n\}$, then $\sigma f = f$. Consider the p functions

$$f, \sigma f, \sigma^2 f, \ldots, \sigma^{p-1} f$$

Since f has only $k < p$ different variants, it follows that there exist two distinct integers $r, s, 0 \leq r < s < p$ such that

$$\sigma^s f = \sigma^r f \quad \text{or} \quad \sigma^{s-r} f = f$$

It is therefore clear that

$$\sigma^{a(s-r)} f = f \qquad \text{for every integer } a$$

Since $0 < s - r < p$, $s - r$ is relatively prime to p so that there exist integers A, B such that

$$A(s - r) + Bp = 1$$

But then, bearing in mind that $\sigma^p = \text{Id}$,

$$f = \sigma^{A(s-r)} f = \sigma^{-Bp+1} f = \sigma(\sigma^{-Bp} f) = \sigma(\text{Id}^{-B} f) = \sigma f$$

Thus we have proved that if σ is any permutation of order p, then $\sigma f = f$.

8.3 Permuting the Variables of a Function (II)

Next we show that the application of any two transpositions to the variables of f also leaves f unchanged. Let α and β be the two specific permutations

$$\alpha = (1\ 2\ 3\ 4\ \cdots\ p) \quad \text{and} \quad \beta = (p\ p-1\ \cdots\ 4\ 2\ 3\ 1)$$

Since they both have order p, both leave f unchanged. Consequently their composition

$$\beta\alpha = (1\ 3\ 2) = (3\ 2)(2\ 1)$$

also leaves f unchanged. By a similar argument, the permutation

$$(2\ 4\ 3) = (4\ 3)(3\ 2)$$

also leaves f unchanged, as must their composition

$$(4\ 3)(3\ 2)(3\ 2)(2\ 1) = (4\ 3)(2\ 1)$$

Since there was nothing special about the choice of the variables x_1, x_2, x_3, x_4, we now know that f is unchanged whenever its variables are permuted by two, or any even number of consecutive transpositions.

Finally, we demonstrate that if ρ and σ are any permutations expressible as the composition of an odd number of transpositions, then $\rho f = \sigma f$. Since it is already known that every permutation is expressible as the composition of some number of transpositions, this will conclude the proof of the theorem. Note that it suffices to show that for any such σ, $(1\ 2)f = \sigma f$. However, since σ is the product of an odd number of transpositions it follows that $(1\ 2)\sigma$ is the product of an even number of transpositions; therefore, by the previous argument

$$f = (1\ 2)\sigma f$$

and now

$$(1\ 2)f = (1\ 2)(1\ 2)\sigma f = \sigma f$$

Thus, in general, every variant σf of f equals either f or $(1\ 2)f$, depending on whether σ is expressible as the composition of an even or an odd number of transpositions. Of course it could happen that f and $(1\ 2)f$ might be equal, in which case f is invariant. In other words, k is either 2 or 1. ∎

While the above theorem was first stated and proved by Cauchy in 1815, the proof we gave is based on that which appears in Abel's 1826 memoir. A translation of Abel's proof is presented in Appendix C.

COROLLARY 8.9 (Ruffini). *There exists no function of 5 variables that has either 3 or 4 distinct variants.*

EXERCISES 8.3

For which of the values of n and k in Exercises 1–26 does there exist a function of n variables that has k distinct variants? Justify your answers.

1. $n = 1, k = 1$
2. $n = 2, k = 1$
3. $n = 2, k = 2$
4. $n = 3, k = 1$
5. $n = 3, k = 2$
6. $n = 3, k = 3$
7. $n = 4, k = 1$
8. $n = 4, k = 2$
9. $n = 4, k = 3$
10. $n = 4, k = 4$
11. $n = 5, k = 1$
12. $n = 5, k = 2$
13. $n = 5, k = 3$
14. $n = 5, k = 4$
15. $n = 5, k = 5$
16. $n = 5, k = 10$
17. $n = 6, k = 1$
18. $n = 6, k = 3$
19. $n = 6, k = 4$
20. $n = 6, k = 6$
21. $n = 7, k = 6$
22. n arbitrary, $k = n$
23. $n \geqslant 2, k = \binom{n}{2}$
24. $n \geqslant 3, k = 3\binom{n}{3}$
25. $n \geqslant 4, k = 3\binom{n}{4}$
26. $n \geqslant r \geqslant 0, k = \binom{n}{r}$

8.4 THE PARITY OF A PERMUTATION

It was seen above that every permutation can be expressed as the composition of transpositions. Moreover, in the last paragraphs of the proof of Cauchy's Theorem 8.8, the issue of the parity of the number of factors in this expression became significant. It was noted earlier that any permutation can be factored into transpositions in many ways, as illustrated by the equations

$$(1\ 2\ 3) = (1\ 3)(1\ 2) = (1\ 3)(1\ 4)(2\ 4)(1\ 4) = (1\ 3)(1\ 2)(2\ 4)(1\ 2)(2\ 4)(1\ 4)$$

However, as indicated by this example, the parity of the number of transpositions in any such factorization of a given permutation is fixed, a fact that we now set out to prove.

For every integer $n \geqslant 2$ we define the *discriminant* Δ_n as the polynomial

$$\Delta_n = (x_1 - x_2)(x_1 - x_3) \cdots (x_1 - x_n)(x_2 - x_3) \cdots (x_{n-1} - x_n)$$

Accordingly

$$\Delta_2 = x_1 - x_2, \quad \Delta_3 = (x_1 - x_2)(x_1 - x_3)(x_2 - x_3)$$
$$\Delta_4 = (x_1 - x_2)(x_1 - x_3)(x_1 - x_4)(x_2 - x_3)(x_2 - x_4)(x_3 - x_4)$$

8.4 The Parity of a Permutation

It will now be demonstrated, as promised in Section 8.1, that for each integer $n \geq 2$ the discriminant Δ_n has two distinct variants. This fact, in turn, will be used to draw some interesting conclusions regarding the parities of permutations. The brunt of the work is contained in the proof of the following observation:

■ **LEMMA 8.10.** *If τ is any transposition, then*

$$\tau \Delta_n = -\Delta_n$$

Proof. Suppose first that $\tau = (i \; i+1)$ with $1 \leq i < n$. The effect of τ on Δ_n is to replace the segment

$$(x_i - x_{i+1})(x_i - x_{i+2}) \cdots (x_i - x_n)(x_{i+1} - x_{i+2}) \cdots (x_{i+1} - x_n)$$

with

$$(x_{i+1} - x_i)(x_{i+1} - x_{i+2}) \cdots (x_{i+1} - x_n)(x_i - x_{i+2}) \cdots (x_i - x_n)$$

Since $(x_{i+1} - x_i) = -(x_i - x_{i+1})$, it follows that for $\tau = (i \; i+1)$ we do indeed have

$$\tau \Delta_n = -\Delta_n$$

Next let τ be an arbitrary transposition $(a \; b)$ with $a < b$. Since

$$(a \; b) = (a \; a+1 \; a+2 \cdots b-1 \; b)(b-1 \; b-2 \cdots a+1 \; a)$$
$$= (a \; a+1)(a+1 \; a+2) \cdots$$
$$(b-2 \; b-1)(b-1 \; b)(b-1 \; b-2)(b-2 \; b-3) \cdots$$
$$(a+2 \; a+1)(a+1 \; a) \qquad (3)$$

and since the right-hand side of (3) consists of an odd number (specifically, $2(b-a) - 1$) transpositions of the form $(i \; i+1)$, it follows from the first part of the proof that in this case too

$$\tau \Delta_n = -\Delta_n \qquad ■$$

A permutation is said to be *even* (or *odd*) according as it is expressible as the composition of an even (or odd) number of transpositions. Thus every transposition is necessarily odd, whereas every 3-cycle of the form $(a \; b \; c)$ is even since $(a \; b \; c) = (a \; b)(b \; c)$. Similarly

$$(1 \; 2 \; 3)(4 \; 5 \; 6 \; 7) = (1 \; 2)(2 \; 3)(4 \; 5)(5 \; 6)(6 \; 7)$$

is an odd permutation, whereas

$$(1 \; 2 \; 3 \; 4 \; 5)(6 \; 7)(8 \; 9 \; a \; b) = (1 \; 2)(2 \; 3)(3 \; 4)(4 \; 5)(6 \; 7)(8 \; 9)(9 \; a)(a \; b)$$

is an even permutation. It follows from Proposition 8.4 that every permutation is necessarily either even or odd, or both. However, it follows from Lemma 8.10 that

$$\sigma \Delta_n = \begin{cases} \Delta_n & \text{if } \sigma \text{ is even} \\ -\Delta_n & \text{if } \sigma \text{ is odd} \end{cases}$$

Since $\Delta_n \neq -\Delta_n$ (Exercise 27), we conclude that no permutation σ can be both odd and even. This is an important fact that deserves being stated as a proposition.

■ **PROPOSITION 8.11.** *A permutation σ of $\{1, 2, \ldots, n\}$, $n \geq 2$, is either even or odd according as $\sigma \Delta_n = \Delta_n$ or $-\Delta_n$. Consequently no permutation can be both even and odd, and Δ_n has only two distinct variants.*

An alternate proof of this proposition is indicated in Exercise 17. The *parity* of a permutation is its evenness or oddness.

Since it was already seen in the proof of Proposition 8.4 that every k-cycle can be expressed as the composition of $k - 1$ transpositions, the parity of a permutation can be easily computed from its disjoint cycle decomposition. Specifically, if $\sigma_1 \sigma_2 \cdots \sigma_m$ is the disjoint cycle decomposition of σ, where each σ_i is a k_i-cycle, then σ can be expressed as the composition of $\Sigma_{i=1}^{m}(k_i - 1)$ transpositions. Thus the parity of σ is identical with the parity of the integer $\Sigma_{i=1}^{m}(k_i - 1)$. In particular, the parity of (1 2 3 4)(5 6 7) is the same as the parity of $(4 - 1) + (3 - 1) = 5$, which is odd.

This notion of parity can be used to give another convincing example of the utility of permutations in proving negative results. The well known 15-puzzle consists of 15 square pieces, numbered 1 through 15, that are placed inside a larger square frame, as indicated in Figure 8.1. A *legitimate move* consists in the sliding of a neighboring piece into the empty space. Figure 8.1 describes the effect of several successive legitimate moves. One now faces the challenge of rearranging the pieces into any prescribed configuration by means of legitimate moves alone. It turns out that some configurations, such as the configuration R (for reverse) of Figure 8.2, are in fact unattainable by means of legitimate moves. Parities of permutations can be used to give a rigorous proof of this fact, and we will do so here.

Let I be the initial configuration of the 15-puzzle as described in Figure 8.1. To every prescribed configuration X of the 15-puzzle we assign a permutation P_X of the set $\{1, 2, 3, \ldots, 15, b\}$ by

1. thinking of the empty space as just another piece labeled b (for blank),
2. setting $P_X(i)$ to be the label of the piece that occupies in X the same location that i has in I.

8.4 The Parity of a Permutation

1	2	3	4
5	6	7	8
9	10	11	12
13	14	15	

I

1	2	3	4
5	6	7	8
9	10	11	12
13	14		15

A

1	2	3	4
5	6	7	8
9	10		12
13	14	11	15

B

1	2	3	4
5	6		8
9	10	7	12
13	14	11	15

C

1	2	3	4
5	6	8	
9	10	7	12
13	14	11	15

D

1	2	3	4
5	6	8	12
9	10	7	
13	14	11	15

E

1	2	3	4
5	6	8	12
9	10		7
13	14	11	15

F

1	2	3	4
5	6	8	12
9		10	7
13	14	11	15

G

Figure 8.1. Legitimate moves for the 15-puzzle.

For the configurations of Figure 8.1,

$$P_I = \text{Id}$$
$$P_A = (15\ b) \qquad\qquad P_B = (15\ 11\ b)$$
$$P_C = (7\ b\ 15\ 11) \qquad\qquad P_D = (7\ 8\ b\ 15\ 11)$$
$$P_E = (7\ 8\ 12\ b\ 15\ 11) \qquad\qquad P_F = (7\ 8\ 12)(b\ 15\ 11)$$
$$P_G = (7\ 8\ 12)(10\ b\ 15\ 11)$$

Note that the sliding of some square labeled x into the empty space is tantamount to the transposition $(x\ b)$, and hence we have the following lemma:

■ **LEMMA 8.12.** *If the configuration Y is obtained from the configuration X by sliding the piece x into the empty space, then*

$$P_Y = (x\ b)P_X$$

In Figure 8.1, F is obtained from E by sliding 7 into the empty space, and indeed

$$P_F = (7\ 8\ 12)(b\ 15\ 11) = (7\ b)(7\ 8\ 12\ b\ 15\ 11) = (7\ b)P_E$$

It follows from Lemma 8.12 that if a configuration X is obtained from the initial configuration I by a sequence of m moves, then P_X can be expressed as the composition of m transpositions.

We will now show that the configuration R of Figure 8.2 is unattainable by legitimate moves. Observe that

$$P_R = (1\ 15)(2\ 14)(3\ 13)(4\ 12)(5\ 11)(6\ 10)(7\ 9)$$

15	14	13	12
11	10	9	8
7	6	5	4
3	2	1	

R

Figure 8.2. An unattainable configuration.

8.4 The Parity of a Permutation 171

Figure 8.3. A coloring of the 15-puzzle.

is an *odd* permutation, since it is expressed here as the composition of 7 transpositions. On the other hand, since the empty space occupies the same positions in the initial configuration I and in R, it follows that a sequence of legitimate moves leading from I to R must consist of an *even* number of moves. One way to justify this assertion is to note that since such a sequence of moves terminates with b in its original location, the number of vertical moves must have been even, as must have been the number of horizontal moves. An alternate justification is obtained by coloring the underlying framework in the checkerboard pattern of Figure 8.3. Each legitimate move then changes the color showing in the empty space. Since the empty space returned to its original position, this sequence must consist of an even number of moves. Either way, Proposition 8.11 guarantees that the odd permutation P_R is not expressible by means of an even number of transpositions and hence *the configuration R is not attainable by legitimate moves*.

A configuration X is said to be *standard* if the empty space occupies the same position in X as it does in I, that is, if $P_X(b) = b$. It is clear that the above argument proves the unattainability of any standard configuration X for which P_X is an odd permutation.

■ **PROPOSITION 8.13.** *A standard configuration X is attainable from the initial configuration I if and only if P_X is an even permutation.*

Sketch of Proof. If X is a standard configuration with P_X odd, then the argument that was applied to prove the unattainability of R above accomplishes the same goal for X.

The proof of the converse is based on the observation (Exercise 29) that regardless of the values of the a_i's, one of the two configurations of Figure 8.4 is necessarily attainable by legitimate moves. Moreover the permutations P_X and P_Y associated with these configurations are related by

$$P_Y = (a_{14} \ a_{15}) P_X$$

172 Permutations

```
┌────┬────┬────┬────┐        ┌────┬────┬────┬────┐
│ a₁ │ a₂ │ a₃ │ a₄ │        │ a₁ │ a₂ │ a₃ │ a₄ │
├────┼────┼────┼────┤        ├────┼────┼────┼────┤
│ a₅ │ a₆ │ a₇ │ a₈ │        │ a₅ │ a₆ │ a₇ │ a₈ │
├────┼────┼────┼────┤        ├────┼────┼────┼────┤
│ a₉ │a₁₀ │a₁₁ │a₁₂ │        │ a₉ │a₁₀ │a₁₁ │a₁₂ │
├────┼────┼────┼────┤        ├────┼────┼────┼────┤
│a₁₃ │a₁₄ │a₁₅ │    │        │a₁₃ │a₁₅ │a₁₄ │    │
└────┴────┴────┴────┘        └────┴────┴────┴────┘
          X                            Y
```

Figure 8.4. Either X or Y is attainable.

Consequently, if P_X is even then P_Y is odd, and hence Y is unattainable. Since we have argued that either X or Y must be attainable, it follows that X must be attainable. ■

EXERCISES 8.4

Determine the parities of the permutations in Exercises 1–4.

TABLE 8.1. Parities of permutations

∘	Even	Odd
Even	Even	Odd
Odd	Odd	Even

1. $\begin{pmatrix} 1 & 2 & 3 & 4 & 5 & 6 & 7 & 8 & 9 \\ 9 & 3 & 4 & 7 & 1 & 5 & 2 & 6 & 8 \end{pmatrix}$
2. $\begin{pmatrix} 1 & 2 & 3 & 4 & 5 & 6 & 7 & 8 & 9 \\ 5 & 3 & 4 & 7 & 1 & 6 & 2 & 9 & 8 \end{pmatrix}$
3. $\begin{pmatrix} 1 & 2 & 3 & 4 & 5 & 6 & 7 & 8 & 9 \\ 9 & 8 & 7 & 6 & 5 & 4 & 3 & 2 & 1 \end{pmatrix}$
4. $\begin{pmatrix} 1 & 2 & 3 & 4 & 5 & 6 & 7 & 8 & 9 \\ 9 & 7 & 5 & 3 & 1 & 8 & 6 & 4 & 2 \end{pmatrix}$

5. List all the even permutations of $\{1, 2\}$ in disjoint cycle form.
6. List all the even permutations of $\{1, 2, 3\}$ in disjoint cycle form.
7. List all the even permutations of $\{1, 2, 3, 4\}$ in disjoint cycle form.
8. List all the odd permutations of $\{1, 2\}$ in disjoint cycle form.
9. List all the odd permutations of $\{1, 2, 3\}$ in disjoint cycle form.
10. List all the odd permutations of $\{1, 2, 3, 4\}$ in disjoint cycle form.
11. Prove that the effect of composition on the parities of permutations is described by Table 8.1 above.

12. Prove that if ρ and σ are any two permutations then
    (a) $\sigma\rho$ and $\rho\sigma$ have the same parities,
    (b) σ and $\rho\sigma\rho^{-1}$ have the same parities.
13. Prove that every even permutation is expressible as the composition of 3-cycles and that this is not true for odd permutations.
14. Let $k > 1$ be an odd positive integer. Prove that every even permutation of $\{1, 2, \ldots, n\}$, $n \geq k$, is expressible as the composition of k-cycles and that this is not true for odd permutations.
15. Prove that for $n \geq 3$ every even permutation of $\{1, 2, \ldots, n\}$ is expressible as a composition of 3-cycles of the form
    (a) $(1\ a\ b)$ (b) $(1\ 2\ a)$
16. Determine for which n there exist cyclic permutations σ and τ of $\{1, 2, \ldots, n\}$ such that $\sigma\tau = (1\ 2\ 3\ \cdots\ n)$. Prove your answer.
17. Use Exercise 8.2.24 to prove Proposition 8.11.
18. Which of the configurations of the 15-puzzle in Figure 8.5 are attainable by legitimate moves?

6	11	4	15
3	14	7	10
8	5	12	1
13	2	9	

a

9	11	4	8
3	14	6	10
15	5	12	1
13	2	7	

b

1	15	5	12
8	11	2	14
3	4	9	6
10	7	13	

c

1	8	9	12
15	11	2	10
13	5	4	6
14	7	3	

d

Figure 8.5. Some configurations of the 15-puzzle.

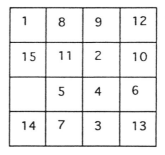

Figure 8.6. Some configurations of the 15-puzzle.

19. Formulate and prove a necessary and sufficient condition for an arbitrary (not necessarily standard) configuration to be attainable by legitimate moves.
20. Which of the configurations of Figure 8.6 is attainable?
21. How many of the standard configurations of the 15-puzzle are attainable and how many are not? Prove your answer.
22. Prove that for $n > 1$, exactly half of the permutations of $\{1, 2, \ldots, n\}$ are even.

For the values of n and k specified in Exercises 23–26, does there exist a function of n variables that has k variants? Justify your answers.

23. $n \geqslant 3, k = 2n$
24. $n \geqslant 2, k = 2\binom{n}{2}$
25. $n \geqslant 3, k = 2\binom{n}{3}$
26. $n > r + 1 \geqslant 2, k = 2\binom{n}{r}$

27. Let σ be a permutation of $\{1, 2, \ldots, n\}$. Prove that the parity of the permutation σ is the same as the parity of the integer $n - \|\sigma\|$.
28. Prove that for $n \geqslant 2$, $\Delta_n \neq -\Delta_n$.
29. Complete the proof of Proposition 8.12 by showing that regardless of the values of the a_i's, one of the two configurations of Figure 8.4 is necessarily attainable by legitimate moves.

CHAPTER SUMMARY

Motivated by Lagrange's solution of the general quartic equation, we studied the number of distinct variants that a multivariable function can have. It was shown in Theorem 8.8 that there are some strong and surprising limitations on the number of such variants. The proof called for a deep analysis of the structure of permutations. The same proof also led to the formulation of the notion of the parity of a permutation. Finally, these new tools, developed with functions in mind, were applied toward the resolution of the popular 15-puzzle.

Chapter Review Exercises

Mark the following true or false.

1. The number of distinct variants of the function $x_1(x_2 - x_3)^2$ is 6.
2. The number of permutations of $\{a, b, c, d, e\}$ is 120.
3. (1 2 3)(4 7 6 5)(1 7 6 2 3)(4 5) = (1 6 3 2)(5 7).
4. The permutation (7 1 4 3 6 2)(5 9)(8) is expressible as the composition of transpositions.
5. If $f = x_1 + 2x_2 + 3x_3 + 4x_4 + 5x_5$, then $(1\ 2\ 3\ 4\ 5)f = (3\ 4\ 5\ 1\ 2)f$.
6. $(1\ 2\ 3)(2\ 3\ 4)f = (1\ 2)(3\ 4)f$.
7. There is a function of 8 variables that has 4 variants.
8. There is a function of 8 variables that has 2 variants.
9. The 15-puzzle configuration of Figure 8.7 is attainable by legitimate moves.

New Terms

Cycle	157
Discriminant	166
Disjoint cycle decomposition	159
Even permutation	167
15-puzzle	168
k-cycle	158
Legitimate move in 15-puzzle	168
Odd permutation	167
Parity of permutation	168
Permutation	155
Transposition	160
Variant	153

2	1	3	4
5	6	7	8
9	10	11	12
13	14	15	

Figure 8.7. A configuration of the 15-puzzle.

176 Permutations

Supplementary Exercises

If k and n are any two positive integers, the (k, n)-puzzle is similar to the 15-puzzle, except that it consists of a rectangular array of squares containing k rows and n columns. Accordingly the (4, 4)-puzzle is identical with the 15-puzzle analyzed in this chapter.

1. Given any initial configuration of the (4, 5)-puzzle, describe all the configurations attainable from it.
2. Given any initial configuration of the (5, 5)-puzzle, describe all the configurations attainable from it.
3. For any positive integers k and n, given any initial configuration of the (k, n)-puzzle, describe all the configurations attainable from it.
4. The *cylindrical (k, n)-puzzle* is obtained from the traditional version by adding some moves that allow for the pieces to cross the vertical boundaries of the board. Figure 8.8 illustrates some legitimate moves for this cylindrical version of the puzzle. The reason for the *cylindrical* appellation is that it is possible to obtain a physical interpretation of the legitimacy of the new moves by bending the puzzle into a cylinder and gluing its vertical edges so that in this new form the pieces always do slide into adjacent spaces. Given any initial configuration of the cylindrical (k, n)-puzzle, decide which configurations are attainable from it.
5. The *toroidal (k, n) puzzle* is obtained from the cylindrical puzzle by allowing the pieces to slide into the empty space across the horizontal boundaries as well (Figure 8.9). Given any initial configuration of the toroidal (k, n)-puzzle, decide which

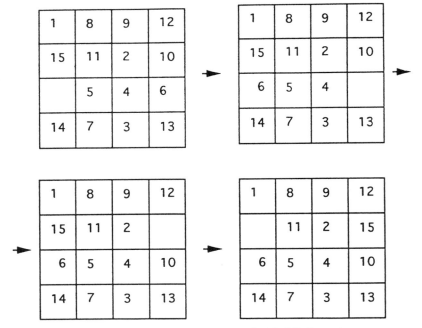

Figure 8.8. Legitimate moves for the cylindrical (2, 2)-puzzle.

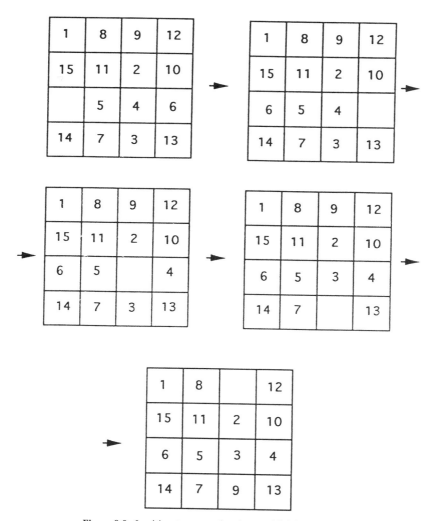

Figure 8.9. Legitimate moves for the toroidal (2, 2)-puzzle.

configurations are attainable from it. This game can be visualized as taking place on a torus (Figure 8.10).

6. The *Moebius* (k, n)-*puzzle* is obtained from the traditional one by allowing for the filling of the blanks across the vertical boundaries, with the additional twist that the blank can be filled only by boundary pieces that are diametrically opposite to it (Figure 8.11). As its name implies, this version can be visualized on the Moebius strip of Figure 8.12. For any initial configuration of the Moebius (k, n)-puzzle, determine the configurations that are attainable from it.

178 Permutations

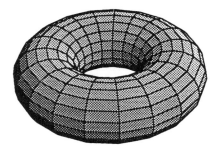

Figure 8.10. The torus.

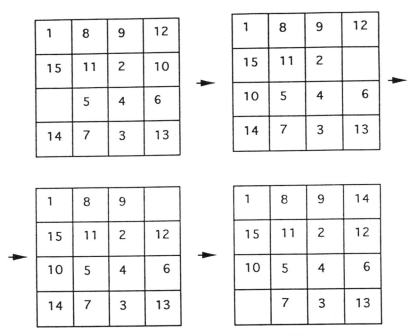

Figure 8.11. Legitimate moves on the Moebius (2, 2)-puzzle.

7. The *Klein-bottle* (k, n)-*puzzle* allows for cylindrical moves across its horizontal boundaries and Moebius-type moves across its vertical boundaries (Figure 8.13). For any initial configuration of the Klein-bottle (k, n)-puzzle, determine the configurations that are attainable from it. This game can be visualized (with some difficulty) on the Klein-bottle of Figure 8.14.

8. Generalize the (k, n)-puzzle to other planar versions.

Chapter Summary 179

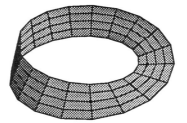

Figure 8.12. The Moebius strip.

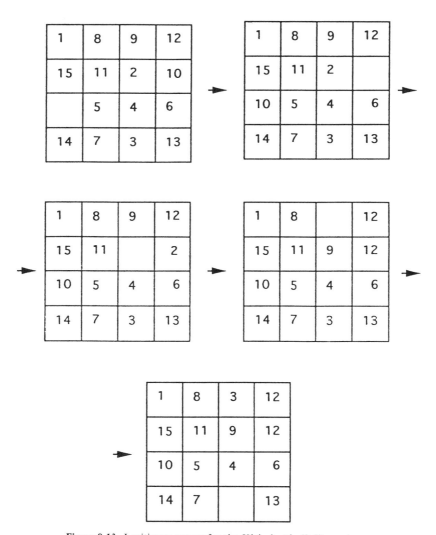

Figure 8.13. Legitimate moves for the Klein-bottle (2, 2)-puzzle.

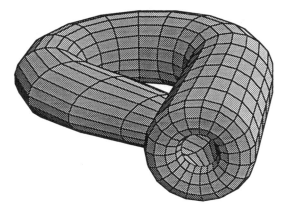

Figure 8.14. The Klein-bottle.

9. Formulate a three-dimensional version of the 15-puzzle and solve it.
10. Formulate a three-dimensional version of the toroidal (k, n)-puzzle and solve it.
11. Formulate a three-dimensional version of the Klein-bottle (k, n)-puzzle and solve it.
12. Formulate and solve d-dimensional analogs of all the versions of the (k, n)-puzzle for every positive integer d.
13. Prove Cauchy's Theorem that a function of $n \geqslant 5$ variables that has fewer than n distinct variants must have either 1 or 2 distinct variants.
14. Let n, k be two positive integers. Which permutations of $\{1, 2, \ldots, n\}$ are expressible as the composition of k cyclic permutations of the same set?
15. Is it true that for any three integers $k, l, m > 1$, there exist permutations ρ, σ such that ρ, σ, and $\rho\sigma$ have orders k, l, m, respectively?

CHAPTER 9

Groups

We now extract the notion of a group from the variety of algebraic structures that have been discussed in the previous chapters. This leads to a natural classification problem and information is provided toward the classification of some elementary groups.

9.1 PERMUTATION GROUPS

A few years after Abel proved that the general quintic equation was not solvable by radicals the young Galois discovered a general criterion for determining whether any given equation was solvable by radicals. The only two specific examples that Galois gave were Abel's aforementioned theorem and Gauss's work on the cyclotomic equation, which was discussed in some detail in Chapters 2 and 3. Galois observed that permutations (he sometimes called them substitutions) played a crucial role in both proofs. As was seen in Chapter 8, these permutations are quite explicit in Abel's work. They are less evident in Gauss's proof, and the best that can be said within the confines of this book is that in the special case discussed in Section 2.4 the crucial permutation is

$$\sigma = (\zeta \ \zeta^3 \ \zeta^9 \ \zeta^{10} \ \zeta^{13} \ \zeta^5 \ \zeta^{15} \ \zeta^{11} \ \zeta^{16} \ \zeta^{14} \ \zeta^8 \ \zeta^7 \ \zeta^4 \ \zeta^{12} \ \zeta^2 \ \zeta^6)$$

where ζ is any primitive 17th root of 1. Of course σ can also be thought of as the algebraically defined function

$$\sigma(\alpha) = \alpha^3$$

defined on the imaginary 17th roots of 1, but we will regard it as a purely formal permutation. This permutation bears an obvious relation to the sum

$$\zeta + \zeta^3 + \zeta^9 + \zeta^{10} + \zeta^{13} + \zeta^5 + \zeta^{15} + \zeta^{11} + \zeta^{16} + \zeta^{14}$$
$$+ \zeta^8 + \zeta^7 + \zeta^4 + \zeta^{12} + \zeta^2 + \zeta^6$$

that was used to prove the ruler and compass constructibility of the regular 17-sided polygon.

The work of Lagrange, Gauss, and Abel eventually led Galois to formulate the notion of a *group of permutations*, or *permutation group*. Given a set S, a group of permutations of S is a set G of permutations of S such that

1. G contains the identity permutation;
2. if σ is in G, so is σ^{-1};
3. if ρ and σ are in G, so is their composition $\rho\sigma$.

We have already encountered several groups of permutations. For each positive integer n, the group of all the permutations of the set $\{1, 2, \ldots, n\}$ is called the *symmetric group* and is denoted by S_n. That S_n satisfies requirements 1–3 follows from the fact that it consists of all of the permutations of $\{1, 2, \ldots, n\}$.

The list of permutations

$$\text{Id}, \quad (1\ 2), \quad (3\ 4), \quad (1\ 2)(3\ 4)$$

which consists of all the permutations that leave the polynomial

$$x_1 + x_2 - x_3 - x_4$$

unchanged, is also a group of permutations. To see this, it need only be observed that each of these permutations is its own inverse and that the composition of any two distinct nonidentity elements is equal to the third nonidentity element. This is no coincidence.

■ **PROPOSITION 9.1.** *If f is any function of $x_1, x_2, \ldots, x_n$, then the set $S_{n,f}$ of all the permutations of these variables that leave f unchanged is a group of permutations.*

Proof. This follows from Lemma 8.6 and Corollary 8.7. ■

This proposition of course provides us with a host of groups of permutations. Thus, if $f = x_1 + x_2 + x_3$, then $S_{3,f} = S_3$. If $f = x_1 x_2^2 x_3^3$, then $S_{3,f} = \{\text{Id}\}$. For $f = x_1 x_2 + x_3$, $S_{3,f} = \{\text{Id}, (1\ 2)\}$, and finally, for

$$f = x_1 x_2^2 x_3^3 + x_2 x_3^2 x_1^3 + x_3 x_1^2 x_2^3$$

we have

$$S_{3,f} = \{\text{Id}, (1\ 2\ 3), (1\ 3\ 2)\}$$

The converse of Proposition 9.1 also holds.

9.1 Permutation Groups

■ **PROPOSITION 9.2.** *Let G be any group of permutations of* $\{1, 2, \ldots, n\}$. *Then there is a polynomial function f such that* $G = S_{n,f}$.

Proof. It is clear that the polynomial $g = x_1 x_2^2 x_3^3 \cdots x_n^n$ is such that $S_{n,g} = \{\text{Id}\}$. Suppose now that $G = \{\sigma_1, \sigma_2, \ldots, \sigma_k\}$ is a listing of the elements of G. Set

$$f = \sigma_1 g + \sigma_2 g + \cdots + \sigma_k g$$

Since G is a group, it follows that for every $\sigma, \sigma_i \in G$, $\sigma \sigma_i$ also belongs to G. Moreover $\sigma \sigma_i = \sigma \sigma_j$ if and only if $\sigma_i = \sigma_j$, if and only if $i = j$, so

$$\sigma \sigma_1, \sigma \sigma_2, \ldots, \sigma \sigma_k$$

also constitute a listing of the elements of G. Hence, for any $\sigma \in G$,

$$\sigma f = \sigma \sigma_1 g + \sigma \sigma_2 g + \cdots + \sigma \sigma_k g = \sigma_1 g + \sigma_2 g + \cdots + \sigma_k g = f$$

so $\sigma \in S_{n,f}$. Thus we have shown that $S_{n,f} \supset G$.

Conversely, suppose that $\sigma \in S_{n,f}$. Then σ must transform each summand of f into another summand of f. In particular, for some $j = 1, 2, \ldots, k$,

$$\sigma \sigma_1 g = \sigma(\sigma_1 g) = \sigma_j g \quad \text{or} \quad \sigma_j^{-1} \sigma \sigma_1 g = g$$

Since $S_{n,g} = \{\text{Id}\}$, it follows that $\sigma_j^{-1} \sigma \sigma_1 = \text{Id}$, and so

$$\sigma = \sigma_j \text{Id} \sigma_1^{-1} = \sigma_j \sigma_1^{-1} \in G$$

Thus $S_{n,f} = G$. ■

If $G = \{\text{Id}, (1\ 2\ 3), (1\ 3\ 2)\}$, then $g = x_1 x_2^2 x_3^3$ and

$$f = x_1 x_2^2 x_3^3 + x_2 x_3^2 x_1^3 + x_3 x_1^2 x_2^3$$

The group of permutations G that Galois associated with the cyclotomic equation

$$x^{17} - 1 = 0$$

consists of all the powers of the permutation

$$\sigma = (\zeta\ \zeta^3\ \zeta^9\ \zeta^{10}\ \zeta^{13}\ \zeta^5\ \zeta^{15}\ \zeta^{11}\ \zeta^{16}\ \zeta^{14}\ \zeta^8\ \zeta^7\ \zeta^4\ \zeta^{12}\ \zeta^2\ \zeta^6)$$

Put differently

$$G = \{\text{Id}, \sigma, \sigma^2, \sigma^3, \ldots, \sigma^{15}\}$$

where

$$\sigma^2 = (\zeta \;\; \zeta^9 \;\; \zeta^{13} \;\; \zeta^{15} \;\; \zeta^{16} \;\; \zeta^8 \;\; \zeta^4 \;\; \zeta^2)(\zeta^3 \;\; \zeta^{10} \;\; \zeta^5 \;\; \zeta^{11} \;\; \zeta^{14} \;\; \zeta^7 \;\; \zeta^{12} \;\; \zeta^6)$$
$$\vdots$$
$$\sigma^{16} = \text{Id}$$

This generalizes as follows:

■ **PROPOSITION 9.3.** *Let σ be any permutation of $\{1, 2, \ldots, n\}$, and let d be the order of σ. Then the set*

$$\langle \sigma \rangle = \{\text{Id}, \sigma, \sigma^2, \ldots, \sigma^{d-1}\}$$

is a group of permutations.

Proof. The identity permutation Id belongs to $\langle \sigma \rangle$ by definition. Since $\sigma^d = \text{Id}$, it follows that

$$(\sigma^k)^{-1} = \sigma^{-k} = \sigma^{d-k} \qquad \text{for } k = 0, 1, 2, \ldots, d-1$$

and so every element of $\langle \sigma \rangle$ has its inverse in $\langle \sigma \rangle$. Finally, if σ^k and σ^m are two arbitrary elements of $\langle \sigma \rangle$, then so is

$$\sigma^k \sigma^m = \sigma^{k+m}$$

an element of $\langle \sigma \rangle$. Thus $\langle \sigma \rangle$ is indeed a group of permutations. ■

Accordingly,

$$\langle (1\ 2\ 3)(4\ 5) \rangle = \{\text{Id}, (1\ 2\ 3)(4\ 5), (1\ 3\ 2), (4\ 5), (1\ 2\ 3), (1\ 3\ 2)(4\ 5)\}$$

and

$$\langle (1\ 2\ 3\ 4\ 5\ 6) \rangle = \{\text{Id}, (1\ 2\ 3\ 4\ 5\ 6), (1\ 3\ 5)(2\ 4\ 6), (1\ 4)(2\ 5)(3\ 6),$$
$$(1\ 5\ 3)(2\ 6\ 4), (1\ 6\ 5\ 4\ 3\ 2)\}$$

are groups of permutations. The *alternating group* A_n consists of the set of all the even permutations of n symbols. Thus

$A_1 = \{\text{Id}\}, \quad A_2 = \{\text{Id}\},$
$A_3 = \{\text{Id}, (1\ 2\ 3), (1\ 3\ 2)\}$
$A_4 = \{\text{Id}, (1\ 2)(3\ 4), (1\ 3)(2\ 4), (1\ 4)(2\ 3), (1\ 2\ 3), (1\ 3\ 2),$
$\qquad (1\ 2\ 4), (1\ 4\ 2), (1\ 3\ 4), (1\ 4\ 3), (2\ 3\ 4), (2\ 4\ 3)\}$

9.1 Permutation Groups

The formal proof of the fact that A_n is indeed a group of permutations is relegated to Exercise 17.

There is a host of groups of permutations that are defined geometrically in terms of symmetries of configurations rather than algebraic symbols. Consider the rectangle of Figure 9.1 whose vertices are labeled 1, 2, 3, 4. It has two obvious symmetries, with respect to the x- and y-axes. The first of these interchanges the vertices 1 and 4 and also 2 and 3. Thus it can be denoted by the permutation (1 4)(2 3). Similarly the symmetry with respect to the y-axis induces the permutation (1 2)(3 4) on the vertices. The central symmetry that the rectangle has with respect to the origin induces the permutation (1 3)(2 4) on the vertices. Each of these three permutations is its own inverse, and the composition of any two distinct ones equals the third. It therefore follows that if we add Id as a trivial symmetry, then the set

$$K = \{\text{Id}, (1\ 2)(3\ 4), (1\ 3)(2\ 4), (1\ 4)(2\ 3)\}$$

is a group of permutations. This group is known as the *Klein 4-group*, after the mathematician Felix Klein whose *Erlanger Programm* of 1872 set the tone for the investigation of the relationship between geometry and group theory for generations to come.

Before we go on to describe some more geometrical groups of permutations, it is necessary to firm up the notion of a symmetry. For the purposes of this discussion, the geometrical configurations in question are all assumed to be centered at the origin of a three-dimensional coordinate system, and a symmetry of the configuration is a rotation of the ambient space about an axis through the origin that leaves the position of the configuration unchanged. Thus, from our point of view, a symmetry with respect to the x-axis results from a 180° rotation of space about the x-axis. This rotation clearly transforms the rectangle of Figure 9.1 right back onto itself. The central symmetry of the rectangle with respect to the origin comes from a 180° rotation of space about the z-axis, which is not drawn in the figure. It is clear that by this definition

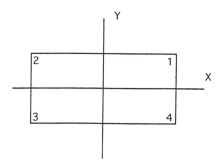

Figure 9.1. The symmetries of the rectangle.

every configuration possesses at least one symmetry, namely the trivial one that is defined by the identity transformation of space. Since each symmetry transforms the configuration onto itself, each such symmetry must necessarily permute the vertices of the configuration, as was the case in the above rectangle. We refer to these permutations as *vertex symmetries*.

■ **PROPOSITION 9.4.** *The vertex symmetries of any geometrical configuration form a group of permutations.*

Proof. As was noted above, the identity permutation is a vertex symmetry of any configuration. Since the inverse of any rotation is also a rotation, it follows that the inverse of any vertex symmetry is also a vertex symmetry. Finally, since the composition of any two rotations whose axes intersect at the origin is known to be another such rotation, it follows that the composition of any two vertex symmetries is also a vertex symmetry. ■

We emphasize here that if σ is the vertex permutation that describes the rotation R, then the rotation R replaces the vertex v with the vertex $\sigma(v)$. The square of Figure 9.2 has eight symmetries: the four that it has as a rectangle, two more that result from clockwise and counterclockwise 90° rotations about the z-axis, and two more that result from 180° rotations about the diagonals 13 and 24, respectively. These last four induce the following respective vertex permutations:

$$(1\ 2\ 3\ 4), (1\ 4\ 3\ 2), (2\ 4), (1\ 3)$$

Thus the vertex symmetries of the square constitute the permutation group

$$D_4 = \{\text{Id}, (1\ 2)(3\ 4), (1\ 3)(2\ 4), (1\ 4)(2\ 3), (1\ 2\ 3\ 4), (1\ 4\ 3\ 2), (1\ 3), (2\ 4)\}$$

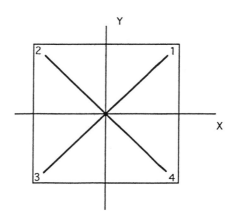

Figure 9.2. The symmetries of the square.

In general, the group of vertex symmetries of the regular n-gon is called the *dihedral group* D_n.

We turn next to some interesting vertex symmety groups defined by solid configurations. The regular tetrahedron of Figure 9.3, with vertices 1, 2, 3, 4 has four faces each of which is an equilateral triangle. Each altitude (the line joining a vertex to the center of the opposite face) serves as the axis of two nontrivial $\pm 120°$ rotations, thus contributing a total of eight vertex symmetries:

(1 2 3), (1 3 2), (1 2 4), (1 4 2), (1 3 4), (1 4 3), (2 3 4), (2 4 3)

In addition the line joining the midpoints of the two edges 23 and 14 serves as the axis of a 180° rotation that defines the vertex symmetry

(2 3)(1 4)

with the analogous lines defining the additional vertex symmetries

(1 3)(2 4)   and   (1 2)(3 4)

It is now clear that the vertex symmetries of the tetrahedron constitute a by now familiar group, namely the alternating group A_4 that consists of all the even permutations of $\{1, 2, 3, 4\}$.

This is the time to note that we have excluded some symmetries that others might, and sometimes do, include. Specifically, it could be argued that the tetrahedron of Figure 9.3 possesses a symmetry with respect to the plane that contains the edge 14 and bisects the edge 23. The decision to restrict our attention only to symmetries that are realizable as rotations in three-dimensional space was arbitrary and based on pedagogical grounds.

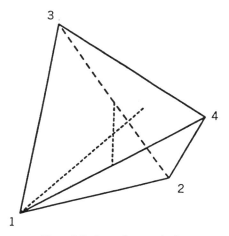

Figure 9.3. A regular tetrahedron.

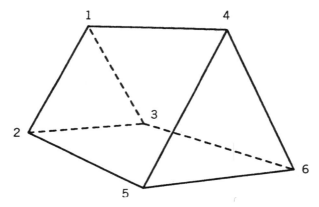

Figure 9.4. A triangular prism.

Consider the triangular prism of Figure 9.4, whose lateral sides are equilateral triangles. Its group of vertex symmetries is

{Id, (1 2 3)(4 5 6), (1 3 2)(4 6 5), (1 5)(2 4)(3 6), (1 4)(2 6)(3 5),

(1 6)(2 5)(3 4)}

EXERCISES 9.1

Which of the sets of permutations of $\{1, 2, 3, 4, 5\}$ in Exercises 1–6 form a group? Justify your answer.

1. All the even permutations.
2. All the odd permutations.
3. All the transpositions.
4. All the permutations that leave 3 fixed.
5. All the permutations that interchange 2 and 4.
6. All the permutations that map 1 to 5.

List the elements of the group $S_{n,f}$ for the function f in each of the Exercises 7–16.

7. $x_1 + x_2 + 2x_3$
8. $x_1 x_2 + x_2 x_3 + x_3 x_1$
9. $x_1 + x_2 + x_3 - x_4$
10. $(x_1 + x_2)(x_3 + x_4)$
11. $x_1(x_2 + x_3 - x_4)$
12. $(x_1 + x_2)x_3 x_4 x_5$
13. $(x_1 - x_2)(x_3 - x_4)$
14. $(x_1 + x_2 - x_3 - x_4)x_5$
15. $(x_1 + x_2 - x_3 - x_4)^2$
16. $\dfrac{x_1}{x_2} + \dfrac{x_3}{x_4}$

17. Prove that if $f = \Delta_n$, then $S_{n,f} = A_n$.
18. Describe the vertex symmetries of the cube of Figure 9.5.

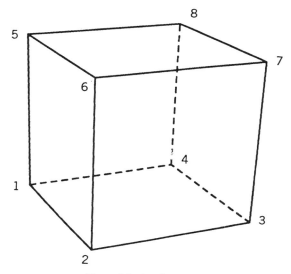

Figure 9.5. A cube.

19. Describe the vertex symmetries of the octahedron of Figure 9.6.
20. Describe, without listing them, all the vertex symmetries of the dodecahedron of Figure 9.7.
21. Describe, without listing them, all the vertex symmetries of the icosahedron of Figure 9.8.

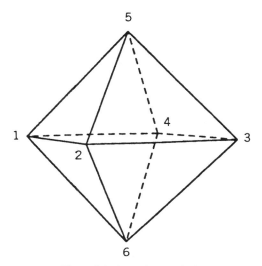

Figure 9.6. A regular octahedron.

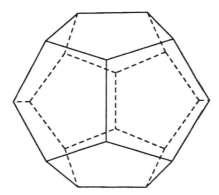

Figure 9.7. A regular dodecahedron.

Let $A=(1\ 2\ 3)$, $B=(2\ 4\ 3)$, $C=(1\ 2)(3\ 4)$, $D=(1\ 3)(2\ 4)$ *be four rotations of the tetrahedron of Figure 9.3. Find the axes and the angles of the rotations in Exercises 22–27.*

22. $A \circ B$ 23. $A \circ C$ 24. $B \circ A$
25. $C \circ B$ 26. $C \circ A$ 27. $D \circ C$

Let $A = (1\ 2\ 3\ 4)(5\ 6\ 7\ 8)$, $B = (1\ 8\ 6)(2\ 4\ 7)$, *and* $C = (1\ 2)(7\ 8)(4\ 6)(3\ 5)$ *be three rotations of the cube of Figure 9.5. Find the axes and the angles of the rotations in Exercises 28–33.*

28. $A \circ B$ 29. $A \circ C$ 30. $B \circ A$
31. $C \circ B$ 32. $C \circ A$ 33. $B \circ C$

34. List the elements of the dihedral group D_5 as permutations.
35. How many elements does the dihedral group D_n have?
36. How many of the elements of the dihedral group D_n have order 2?
37. Prove that for every positive integer k there is a polygon whose group of vertex symmetries contains k elements.

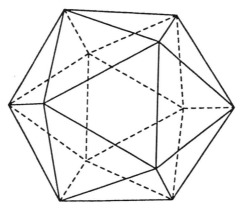

Figure 9.8. A regular icosahedron.

9.2 ABSTRACT GROUPS

Every group of permutation resembles the earlier algebraic structures of this text, such as the real numbers, the complex numbers, $\mathbb{Z}_p$, and the Galois fields, in that it involves a binary operation on a set of objects, this operation being of course the composition of permutations. Thus it is possible to associate with every group of permutations a multiplication table that greatly resembles the tables that were associated with $\mathbb{Z}_n$ for each positive integer n. Table 9.1 describes the compositions of the elements of the Klein 4-group K. Here as elsewhere, the product ab is to be found at the intersection of row a and column b.

Similarly the group of vertex symmetries of the square has Table 9.2 associated with it. It was the British mathematician A. Cayley who eventually extracted the notion of an abstract group from these tables (see Appendix E). Accordingly an *abstract group* consists of a set G and a binary operation · on its elements such that the following four properties are satisfied:

1. For any two elements a and b of G $a \cdot b$ is also in G.
2. There is an element 1_G of G such that $a \cdot 1_G = 1_G \cdot a = a$ for every element a of G.
3. For any elements a, b, c of G, $(a \cdot b) \cdot c = a \cdot (b \cdot c)$.
4. For every element a of G there is an element $a^\#$ such that $a \cdot a^\# = a^\# \cdot a = 1_G$.

Such an abstract group is denoted by $(G, \cdot)$, or sometimes by G alone, if the binary operation is understood. The element 1_G is called the *identity* of G, and the element $a^\#$ is called the *inverse* of a.

It is clear that every group of permutations is an abstract group, with composition as the binary operation in question. The identity permutation Id functions as the identity 1_G and σ^{-1} functions as $\sigma^\#$ for every permutation σ.

TABLE 9.1. The multiplication table of the Klein 4-group K.

	Id	(1 2)(3 4)	(1 3)(2 4)	(1 4)(2 3)
Id	Id	(1 2)(3 4)	(1 3)(2 4)	(1 4)(2 3)
(1 2)(3 4)	(1 2)(3 4)	Id	(1 4)(2 3)	(1 3)(2 4)
(1 3)(2 4)	(1 3)(2 4)	(1 4)(2 3)	Id	(1 2)(3 4)
(1 4)(2 3)	(1 4)(2 3)	(1 3)(2 4)	(1 2)(3 4)	Id

TABLE 9.2. The multiplication table of D_4.

	Id	(1 2 3 4)	(1 3)(2 4)	(1 4 3 2)	(1 3)	(2 4)	(1 2)(3 4)	(1 4)(2 3)
Id	Id	(1 2 3 4)	(1 3)(2 4)	(1 4 3 2)	(1 3)	(2 4)	(1 2)(3 4)	(1 4)(2 3)
(1 2 3 4)	(1 2 3 4)	(1 3)(2 4)	(1 4 3 2)	Id	(1 4)(2 3)	(1 2)(3 4)	(1 3)	(2 4)
(1 3)(2 4)	(1 3)(2 4)	(1 4 3 2)	Id	(1 2 3 4)	(2 4)	(1 3)	(1 4)(2 3)	(1 2)(3 4)
(1 4 3 2)	(1 4 3 2)	Id	(1 2 3 4)	(1 3)(2 4)	(1 2)(3 4)	(1 4)(2 3)	(2 4)	(1 3)
(1 3)	(1 3)	(1 2)(3 4)	(2 4)	(1 4)(2 3)	Id	(1 3)(2 4)	(1 2 3 4)	(1 4 3 2)
(2 4)	(2 4)	(1 4)(2 3)	(1 3)	(1 2)(3 4)	(1 3)(2 4)	Id	(1 4 3 2)	(1 2 3 4)
(1 2)(3 4)	(1 2)(3 4)	(2 4)	(1 4)(2 3)	(1 3)	(1 2 3 4)	(1 4 3 2)	Id	(1 3)(2 4)
(1 4)(2 3)	(1 4)(2 3)	(1 3)	(1 2)(3 4)	(2 4)	(1 4 3 2)	(1 2 3 4)	(1 3)(2 4)	Id

It is also clear that the set of the real numbers, together with the operation of addition, constitutes a group. Here 0 functions as the group identity and $a^\# = -a$ for all real a. This group will be denoted by $(\mathbb{R}, +)$. Similarly $(\mathbb{Z}, +)$, $(\mathbb{Q}, +)$, and $(\mathbb{C}, +)$ denote groups whose underlying sets are the integers, the rationals, and the complex numbers, respectively. On the other hand, the integers under multiplication do not constitute a group, since very few integers have multiplicative inverses. Nor do the rational, the real, or the complex numbers constitute a group under multiplication, since in each case 0 fails to have a multiplicative inverse.

We have already encountered many other groups in this book. Recalling that $\sqrt[n]{1}$ denotes the set of all the n complex roots of 1, we note that $(\sqrt[n]{1}, \cdot)$ is a group wherein $\cdot$ denotes the ordinary multiplication of complex numbers. The identity element of this group is 1. The inverse $\zeta^\#$ of the root ζ is simply ζ^{-1}, also an element of $\sqrt[n]{1}$, and it is clear that if ζ and η are any two elements of $\sqrt[n]{1}$, then so is their product $\zeta\eta$, since

$$(\zeta\eta)^n = \zeta^n \eta^n = 1$$

For any integer n, addition modulo n defines a group $(\mathbb{Z}_n, +)$, with 0 acting as the group identity and $a^\# = -a$. However, when p is a prime integer, modular arithmetic can be used to define another, less expected, collection of groups. If p is a prime number, set

$$\mathbb{Z}_p^* = \{1, 2, 3, \ldots, p-1\}$$

Then we know that each element of $\mathbb{Z}_p^*$ has a multiplicative inverse in $\mathbb{Z}_p$. Consequently $(\mathbb{Z}_p^*, \cdot)$ is a group with identity 1, where $\cdot$ denotes multiplication modulo p. This collection of groups can be considerably enlarged as follows: For each positive integer n, let $\mathbb{Z}_n^*$ denote the set of positive integers that are both smaller than n and relatively prime to it. For example,

$$\mathbb{Z}_6^* = \{1, 5\}, \quad \mathbb{Z}_8^* = \{1, 3, 5, 7\}, \quad \mathbb{Z}_{10}^* = \{1, 3, 7, 9\}$$

Corollary 4.2 and Lemma 4.4 guarantee that $(\mathbb{Z}_n^*, \cdot)$ is indeed a group.

If F is any field, then $(F, +)$ also forms a group. Note that if $F =$ GF(2, $P(x)$), then $a^\# = a$ for each a in F. Again, if F^* denotes all the nonzero elements of F, then $(F^*, \cdot)$ is also a group, where $\cdot$ denotes the multiplication operation in the field.

If F is any field, then $F[x]$, the set of polynomials with coefficients in F is a group with respect to addition. Similarly, if n is any positive integer and $F[x, \leq n]$ denotes all the polynomials in $F[x]$ that have degree at most n together with the zero polynomial, then $F[x, \leq n]$ is also a group with respect to addition. Specifically,

$$\mathbb{Z}_2[x, \leq 1] = \{0, 1, x, 1 + x\}$$

Some groups can be defined directly by means of a multiplication table. The group whose multiplication table appears below is called the *Quaternion* group. It is clear that in this multiplication table, 1 functions as an identity and $a^\# = e$, $b^\# = f$, and so on. The direct verification of the associativity of this multiplication table calls for several hundreds of computations. More efficient techniques are available, but they fall outside the confines of this text.

A group is said to be *commutative* (or *abelian*) if for any two of its elements a and b we have

$$ab = ba$$

Thus the groups $(\mathbb{Z}_n, +)$, $GF(p, P(x))$, $\mathbb{Z}_n^*$, $\mathbb{Z}$, $\mathbb{C}$, are all commutative. On the other hand, if $n \geqslant 3$, then the group S_n is not commutative, since in each such group

$$(1\ 2)(2\ 3) = (1\ 2\ 3) \neq (3\ 2\ 1) = (2\ 3)(1\ 2)$$

A digression on the nature of multiplication tables might be in order here. It is clear that (the interior of) the multiplication table of a group with n elements consists of an n-by-n array. It is customary to list the rows and the columns of the array in the same order, with the row and column that correspond to the identity element appearing first. Each row and each column of the multiplication table constitutes a permutation of the elements of the group. A table that possesses all these properties is called a *Latin square*. Thus the multiplication table of every group is a Latin square. The converse is not

TABLE 9.3. The multiplication table of the Quaternion group.

	1	a	b	c	d	e	f	g
1	1	a	b	c	d	e	f	g
a	a	d	c	f	e	1	g	b
b	b	g	d	a	f	c	1	e
c	c	b	e	d	g	f	a	1
d	d	e	f	g	1	a	b	c
e	e	1	g	b	a	d	c	f
f	f	c	1	e	b	g	d	a
g	g	f	a	1	c	b	e	d

true. Most Latin squares do not come from groups, and Exercises 31–35 contain additional information on this subject.

The definition of an abstract group stipulates the existence of inverses but says nothing about the possible existence of multiple inverses. The next proposition shuts the door on this possibility.

■ **PROPOSITION 9.5.** *If G is a group and $a \in G$, then a has exactly one inverse in G.*

Proof. Suppose that both b and c are inverses of a in G, that is,

$$ba = ab = 1_G = ca = ac$$

Then

$$b = b1_G = b(ac) = (ba)c = 1_G c = c \qquad ■$$

The next proposition about inverses will prove useful later.

■ **PROPOSITION 9.6.** *If a and b are elements of the group G, then*

$$(ab)^{\#} = b^{\#} a^{\#}$$

Proof. Several applications of the Associative Law yield

$$(b^{\#} a^{\#})(ab) = b^{\#}(a^{\#} a)b = b^{\#} 1_G b = b^{\#} b = 1_G$$

and

$$(ab)(b^{\#} a^{\#}) = a(bb^{\#})a^{\#} = a1_G a^{\#} = aa^{\#} = 1_G$$

Thus $b^{\#} a^{\#}$ acts like an inverse of ab, and hence, by the previous proposition, it must be the inverse of ab, that is,

$$b^{\#} a^{\#} = (ab)^{\#} \qquad ■$$

If a is any element of a group G, and n is any positive integer, then we define a^n as the product of n a's. Thus

$$a^1 = a, \quad a^2 = aa, \quad a^3 = aaa, \ldots$$

If we also define

$$a^0 = 1_G \quad \text{and} \quad a^{-n} = (a^{\#})^n$$

then it easily verified (Exercise 36) that, just like the powers of real numbers, the powers of abstract group elements satisfy the conditions

$$a^m a^n = a^{m+n} \quad \text{and} \quad (a^m)^n = a^{mn}$$

for any two integers m and n.

EXERCISES 9.2

Each of Exercises 1–14 specifies a set and a binary operation. In which cases do these form a group? If not, explain why not.

1. All the even elements of $\mathbb{Z}_{1000}$ under addition.
2. All the odd elements of $\mathbb{Z}_{1000}$ under addition.
3. All the even elements of $\mathbb{Z}_{1000}$ under multiplication.
4. All the odd elements of $\mathbb{Z}_{1000}$ under multiplication.
5. All the even elements of $\mathbb{Z}_{64}$ under multiplication.
6. All the odd elements of $\mathbb{Z}_{64}$ under multiplication.
7. All the integers under subtraction.
8. All the integers under addition.
9. All the integers under multiplication.
10. All the positive real numbers under addition.
11. All the positive real numbers under multiplication.
12. All the positive real numbers under division.
13. All the polynomials over $\mathbb{Z}_2$ under addition.
14. All the polynomials over $\mathbb{Z}_2$ under multiplication.

In each of the groups in Exercises 15–27 pair each element with its inverse.

15. $(\mathbb{Z}_4, +)$
16. $(\mathbb{Z}_5^*, \cdot)$
17. $\sqrt[4]{1}$
18. K
19. $\mathbb{Z}_2[x, \leq 1]$
20. S_3
21. $(\mathbb{Z}_5, +)$
22. $\mathbb{Z}_2[x, \leq 2]$
23. $(\mathbb{Z}_6^*, \cdot)$
24. $\sqrt[6]{1}$
25. $(\mathbb{Z}_6, +)$
26. $\mathbb{Z}_3[x, \leq 1]$
27. The Quaternion group.
28. Prove that if G is a group for which $(ab)^\# = a^\# b^\#$ for every pair of elements a, b, then G is a commutative group.
29. Prove that if G is a group in which $a^2 = \text{Id}$ for each a in G, then G is commutative.
30. What geometrical feature characterizes the multiplication table of commutative groups?
31. Explain why the Latin square below is not the multiplication table of a group.

$$\begin{array}{ccccc} 1 & 2 & 3 & 4 & 5 \\ 2 & 1 & 5 & 3 & 4 \\ 3 & 4 & 1 & 5 & 2 \\ 4 & 5 & 2 & 1 & 3 \\ 5 & 3 & 4 & 2 & 1 \end{array}$$

32. Explain why the Latin square below is not the multiplication table of a group.

$$\begin{array}{cccccc} a & b & c & d & e & f \\ b & c & a & e & f & d \\ c & a & b & f & d & e \\ d & e & f & a & b & c \\ e & f & d & c & a & b \\ f & d & e & b & c & a \end{array}$$

33. Prove that every 1-by-1 and every 2-by-2 Latin square is the multiplication table of a group.
34. Prove that every 3-by-3 Latin square is the multiplication table of a group.
35. Prove that every 4-by-4 Latin square is the multiplication table of a group.
36. Prove that if a is an element of the abstract group G, and m and n are arbitrary integers, then $a^m a^n = a^{m+n}$ and $(a^m)^n = a^{mn}$.
37. Prove that every group has a unique identity element.

9.3 ISOMORPHISMS OF GROUPS AND ORDERS OF ELEMENTS

Since abstract groups are defined in terms of their multiplication tables, it makes sense to identify abstract groups that have identical tables. With this in mind, let us examine the tables below. Table 9.4 is an abbreviation of the table associated with the Klein 4-group. Table 9.5 represents $(\mathbb{Z}_4, +)$. Table 9.6 represents a mystery group, as yet unidentified. Table 9.7 represents the group $(F, +)$ where F is the Galois field $GF(2, x^2 + x + 1)$.

Tables 9.4 and 9.7 are readily recognized as being essentially one and the same. In each table the diagonal entry is the group identity, and in each table the group multiplication of any two distinct nonidentity elements equals the third nonidentity element. Tables 9.5 and 9.6 are also essentially the same. To see this it is merely necessary to switch the columns and rows of Table 9.6 that correspond to the elements A and B so that this table takes the form displayed in Table 9.8.

When the symbols e, B, A, C of Table 9.8 are replaced by 0, 1, 2, 3, respectively, Table 9.5 is obtained, thus showing that Tables 9.5 and 9.6 are only superficially different.

Is it possible that some such switching of rows and columns and rewriting of symbols could transform Table 9.4 to Table 9.5? The answer is no, and a reason for this can be found in the diagonals of these tables. Notice that the diagonal of Table 9.4 contains only the group identity, whereas the diagonal of Table 9.5 contains another element besides the identity. Now it is clear that no matter how the symbols of Table 9.4 are relabeled, the diagonal will always contain only that symbol that stands for the group identity. Moreover, even

TABLE 9.4. The Klein 4-group.

	Id	a	b	c
Id	Id	a	b	c
a	a	Id	c	b
b	b	c	Id	a
c	c	b	a	Id

TABLE 9.5. Addition modulo 4.

	0	1	2	3
0	0	1	2	3
1	1	2	3	0
2	2	3	0	1
3	3	0	1	2

TABLE 9.6. A mystery group.

	e	A	B	C
e	e	A	B	C
A	A	e	C	B
B	B	C	A	e
C	C	B	e	A

TABLE 9.7. Addition in $GF(2, x^2+x+1)$.

	0	1	α	$1+\alpha$
0	0	1	α	$1+\alpha$
1	1	0	$1+\alpha$	α
α	α	$1+\alpha$	0	1
$1+\alpha$	$1+\alpha$	α	1	0

when the row and column of any element are extracted and moved (in a consistent manner) to a new location, the diagonal still only contains the group identity. Hence Tables 9.4 and 9.5 are different in an essential way.

We formalize this notion of sameness with the term of *isomorphism*. Two groups $(G, \cdot)$ and $(H, \oplus)$ are said to be *isomorphic* provided that their elements

TABLE 9.8. A rewriting of the mystery group.

	e	B	A	C
e	e	B	A	C
B	B	A	C	e
A	A	C	e	B
C	C	e	B	A

can be matched up so that when the elements of G in a table of $(G, \cdot)$ are replaced with the corresponding elements of H, then the table of $(G, \cdot)$ is transformed into a multiplication table for $(H, \oplus)$. In other words, the two groups $(G, \cdot)$ and $(H, \oplus)$ are isomorphic if there exists a function

$$f: G \to H$$

such that

1. $f(a) = f(b)$ if and only if $a = b$,
2. for every $h \in H$ there is a $g \in G$ such that $f(g) = h$,
3. $f(a \cdot b) = f(a) \oplus f(b)$.

Requirement 1 says that f assigns distinct elements of H to distinct elements of G. Requirement 2 says that every element of H is assigned to some element of G. When dealing with finite groups, which are our main concern here, these two conditions are redundant in the sense that each implies the other. For infinite groups this need not be the case (see Exercises 20, 21).

To understand the last requirement 3, note that if in a multiplication table for $(G, \cdot)$ each element a is replaced by $f(a)$, and if the result is a multiplication table for $(H, \oplus)$, then the entry $a \cdot b$ in the a row and b column will be replaced by $f(a \cdot b)$. However, because this is now the entry in the $f(a)$ row and $f(b)$ column of the multiplication table of $(H, \oplus)$, it must also equal $f(a) \oplus f(b)$ (see Figure 9.9). Hence

$$f(a \cdot b) = f(a) \oplus f(b)$$

The function f itself is called an *isomorphism*. Contrary to its usage in

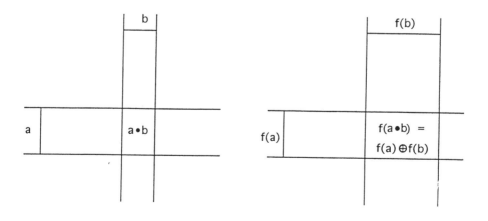

Figure 9.9. Requirement 3 for isomorphisms.

elementary calculus, the word function is used here in its most abstract sense, that of a mere association. Thus the function

$$f(\text{Id}) = 0, \quad f(a) = 1, \quad f(b) = \alpha, \quad f(c) = 1 + \alpha$$

is an isomorphism of the groups whose tables appear in Tables 9.4 and 9.7. The function

$$f'(\text{Id}) = 0, \quad f'(a) = \alpha, \quad f'(b) = 1 + \alpha, \quad f'(c) = 1$$

is another isomorphism of the groups whose tables appear in Tables 9.4 and 9.7. Similarly

$$g(e) = 0, \quad g(A) = 2, \quad g(B) = 1, \quad g(C) = 3$$

is an isomorphism of the groups whose tables appear in Tables 9.6 and 9.5. However, the function

$$g'(e) = 0, \quad g'(A) = 1, \quad g'(B) = 2, \quad g'(C) = 3$$

is not an isomorphism. It violates requirement 3 of isomorphisms, since

$$g'(A \cdot A) = g'(e) = 0$$

and

$$g'(A) + g'(A) = 1 + 1 \neq 0$$

If the groups G and H are isomorphic to each other, then this fact is denoted by writing

$$G \cong H$$

It is clear that if two finite groups are isomorphic, then they must have the same number of elements. This leads us to define the *order* of a finite group as the number of its elements. So $(\mathbb{Z}_n, +)$ has order n, and S_n has order $n!$. In his note of 1878, reproduced in Appendix E, Cayley already mentions the problem of classifying all the isomorphism types of groups. The difficulty of this task is underscored by the fact that Cayley asserts erroneously that, up to isomorphism, there are 3 groups of order 6. It will be proved in Chapter 11 that every group of order 6 is isomorphic either to $(\mathbb{Z}_6, +)$ or to S_3.

One useful tool for distinguishing between nonisomorphic groups is the property of commutativity. It is easily verified (Exercise 19) that for $n \geqslant 3$ the dihedral group D_n is not commutative, as is the case for the Quaternion group. Thus the commutative group $(\mathbb{Z}_8, +)$ is not isomorphic to either D_4 or to the Quaternion group.

9.3 Isomorphisms of Groups and Orders of Elements

Another tool for distinguishing between nonisomorphic groups is provided by the notion of the order of an element of a group. In analogy with the notion of the order of a permutation, we define the *order* $o(a)$ of an element a of an abstract group G as the least positive integer n such that

$$a^n = 1_G$$

If no such integer n exists, then we say that the element a has *infinite order*. The element d of the Quaternion group has order 2, whereas each of the elements a, b, c, e, f, g has order 4. The identity element always has order 1, and it is clearly the only group element that can have order 1. On the other hand, the element 2 of $\mathbb{Z}$ has infinite order, since $2 + 2 + 2 + \cdots + 2$ is never $0 = 1_{\mathbb{Z}}$.

Suppose now that the groups $(G, \cdot)$ and $(H, \oplus)$ are isomorphic. Since this is tantamount to saying that they have the same multiplication tables, it follows that corresponding elements of G and H must have the same orders. Returning to the Quaternion group and the dihedral group D_4, the first has exactly one element of order 2, whereas the latter has five such elements. Consequently these two groups are not isomorphic.

Group theoretic order enjoys the same properties as do the orders of roots of unity and permutations. Proposition 9.7 below is practically identical with Proposition 7.4 above and for that reason no proof is offered.

■ **PROPOSITION 9.7.** *Let g and h be any elements of a finite group G. Then*

1. $g^n = 1_G$ *if and only if n is a multiple of $o(g)$,*
2. $g^a = g^b$ *if and only if $a - b$ is a multiple of $o(g)$,*
3. $o(g^k) = \dfrac{n}{(k, n)}$ *if $o(g) = n$,*
4. $o(gh) = o(g)o(h)$ *if $o(g)$ and $o(h)$ are relatively prime and $gh = hg$.*

That the assumption $gh = hg$ is necessary in part 4 of the above proposition can be seen by choosing $g = (1\ 2)(3\ 4)$, $h = (1\ 4\ 5)$. Here

$$o(gh) = o((1\ 2)(3\ 4)(1\ 4\ 5)) = o((1\ 3\ 4\ 5\ 2)) = 5 \neq 2 \cdot 3 = o(g)o(h)$$

This assumption is of course automatically satisfied in the context of fields wherein Proposition 7.4 was stated.

EXERCISES 9.3

1. Prove that every two groups of order 2 are isomorphic to each other.
2. Prove that every two groups of order 3 are isomorphic to each other.
3. Are every two groups of order 4 isomorphic to each other?

TABLE 9.9. A group table.

	a	b	c	d	e
a	a	b	c	d	e
b	b	c	d	e	a
c	c	d	e	a	b
d	d	e	a	b	c
e	e	a	b	c	d

TABLE 9.10. A group table.

	A	B	C	D	E
A	A	B	C	D	E
B	B	D	A	E	C
C	C	A	E	B	D
D	D	E	B	C	A
E	E	C	D	A	B

4. Prove that the groups whose multiplication tables are in Tables 9.9, 9.10 are isomorphic to each other.
5. Prove that the groups whose multiplication tables are in Tables 9.10, 9.11 are isomorphic to each other.
6. Prove that the groups whose multiplication tables are in Tables 9.9, 9.11 are isomorphic to each other.
7. Explain why the cube and the octahedron of Figure 9.5 have isomorphic vertex groups.
8. A double pyramid is formed by joining the vertices of a regular pentagon to two points, one directly above and one directly below the pentagon's geometrical center. Explain why the group of vertex symmetries of this solid is isomorphic to the dihedral group D_5.

An isomorphism of a group with itself is called an automorphism.

9. Prove that the function $f(x) = 3x$ is an automorphism of $(\mathbb{Z}_{100}, +)$.
10. Prove that if k and n are relatively prime positive integers, then the function $f(x) = kx$ is an automorphism of $(\mathbb{Z}_n, +)$.

TABLE 9.11. A group table.

	α	β	γ	δ	ε
α	α	β	γ	δ	ε
β	β	ε	α	γ	δ
γ	γ	α	δ	ε	β
δ	δ	γ	ε	β	α
ε	ε	δ	β	α	γ

11. Is the function $f(x) = x^3$ an automorphism of $(\mathbb{Z}_{100}, +)$?
12. Is the function $f(x) = x^{-1}$ an automorphism of $(\mathbb{Z}_{17}^*, \cdot)$?
13. Prove that the function $f(x) = x^p$ is an automorphism of $(GF(p, P(x)), +)$.
14. Prove that the function $f(x) = x^p$ is an automorphism of $(GF(p, P(x))^*, \cdot)$.
15. For any element x of the group G, let f_x be the function from G into itself defined by

$$f_x(a) = xax^{-1} \qquad \text{for all } a \in G$$

Prove that f_x is an automorphism of G.

16. For any element x of the group G, let h_x be the function from G into itself defined by

$$h_x(a) = xax \qquad \text{for all } a \in G$$

Prove that h_x is an automorphism of G if and only if $x = x^{-1}$.

17. Prove that every group of even order contains an element of order 2.
18. Let G be a finite commutative group. The *exponent* of G is the least common multiple of all orders of all the elements of G. Prove that G has an element whose order equals the exponent of G.
19. Prove that for $n \geqslant 3$ the dihedral group D_n is not commutative.
20. For the group $G = (\mathbb{Z}, +)$, find a function f from G to G that satisfies conditions 1 and 3 but not condition 2.
21. Find a function f of the positive integers into themselves that satisfies condition 2 for isomorphisms but does not satisfy condition 1.
22. Prove that if a and b are elements of a group G, then $o(a) = o(bab^{-1})$.

9.4 SUBGROUPS AND THEIR ORDERS

A copy of the group

$$K = \{\text{Id}, (1\ 2)(3\ 4), (1\ 3)(2\ 4), (1\ 4)(2\ 3)\}$$

consisting of the vertex symmetries of the rectangle, is clearly contained in the group

$$D_4 = \{\text{Id}, (1\ 2)(3\ 4), (1\ 3)(2\ 4), (1\ 4)(2\ 3), (1\ 2\ 3\ 4), (1\ 4\ 3\ 2), (1\ 3), (2\ 4)\}$$

which consists of the vertex symmetries of the square. This relationship is formalized by the notion of a *subgroup*. If $(G, \cdot)$ is an abstract group, and if H is a subset of G such that $(H, \cdot)$ is a group in its own right, then $(H, \cdot)$ is said to be a subgroup of $(G, \cdot)$. Thus $(K, \circ)$ is a subgroup of $(D_4, \circ)$. This is generally abbreviated to say that K is a subgroup of D_4. Similarly each of the groups $(\mathbb{Z}, +), (\mathbb{Q}, +), (\mathbb{R}, +), (\mathbb{C}, +)$ is a subgroup of the next. It is clear that if G is a group and H is a subset of G, then H is a subgroup of G if and only if the following conditions hold:

1. 1_G is in H.
2. If a and b are in H, so is ab.
3. If a is in H, then so is its inverse a^{-1}.

Thus $\{0, 2, 4\}$ is a subgroup of $\mathbb{Z}_6$. If β is the Galois imaginary associated with the irreducible polynomial $x^3 + x^2 + 1$ over $\mathbb{Z}_2$ and $F = \mathrm{GF}(2, x^3 + x^2 + 1)$, then each of the following sets defines a subgroup of $(F, +)$:

$$\{0, 1, \beta, 1 + \beta\}$$
$$\{0, 1, \beta^2, 1 + \beta^2\}$$
$$\{0, \beta, \beta^2, \beta + \beta^2\}$$

The alternating group A_n which consists of all the even permutations of n symbols is a subgroup of the symmetric group S_n which consists of all the permutations on n symbols. The group of symmetries of the cube is a subgroup of the group S_8 of all the permutations of the eight vertices of the cube. The group $\sqrt[6]{1} = \{1, -\omega^2, \omega, -1, \omega^2, -\omega\}$ contains both $\sqrt{1} = \{1, -1\}$ and $\sqrt[3]{1} = \{1, \omega, \omega^2\}$ as subgroups. If f is any function of n variables, then the group $S_{n,f}$ that consists of all the permutations that leave f unchanged is a subgroup of the symmetric group S_n.

Every group contains two obvious subgroups — itself and the *trivial subgroup* $\{1_G\}$ that consists of G's identity element alone. Any other subgroup of G is said to be *proper*.

The following theorem is quite possibly the most important theorem of group theory. In this form it was first stated and proved by Camille Jordan in his book *Traité des Substitutions*. Because he modestly attributed it to Lagrange who in fact had only proved the limited version of Corollary 9.13, this theorem nowadays bears the latter's name.

■ **THEOREM 9.8 (Lagrange).** *If G is a finite group, then the order of any subgroup of G is a divisor of the order of G.*

Since the proof of this theorem relies on the notion of a coset, a concept that is all but explicit in Lagrange's original proof, the proof is preceded by a discussion of this concept. If

$$H = \{h_1, h_2, h_3, \ldots\}$$

is any subgroup of the group $(G, \cdot)$ and if a is any element of the original group G, we define

9.4 Subgroups and Their Orders

$$a \cdot H = \{a \cdot h_1, a \cdot h_2, a \cdot h_3, \ldots\}$$

and call $a \cdot H$ a *coset* of H. If $G = (\mathbb{Z}_{12}, +)$ and $H = \{0, 4, 8\}$, then

$$0 + H = \{0, 4, 8\} = H = 4 + H = 8 + H$$
$$1 + H = \{1, 5, 9\} = 5 + H = 9 + H$$
$$2 + H = \{2, 6, 10\} = 6 + H = 10 + H$$
$$3 + H = \{3, 7, 11\} = 7 + H = 11 + H$$

These cosets are pictured in Figure 9.10. Another example begins with the polynomial $f = x_1 + x_2 - x_3 - x_4$ and its group

$$H = S_{4,f} = \{\text{Id}, (1\ 2), (3\ 4), (1\ 2)(3\ 4)\}$$

which is by definition a subgroup of S_4. Here

$$(1\ 2\ 3\ 4)H = \{(1\ 2\ 3\ 4), (1\ 3\ 4), (1\ 2\ 3), (1\ 3)\}$$
$$= (1\ 3\ 4)H = (1\ 2\ 3)H = (1\ 3)H$$
$$(1\ 2\ 4\ 3)H = \{(1\ 2\ 4\ 3), (1\ 4\ 3), (1\ 2\ 4), (1\ 4)\}$$
$$= (1\ 4\ 3)H = (1\ 2\ 4)H = (1\ 4)H$$
$$(1\ 3\ 2\ 4)H = \{(1\ 3\ 2\ 4), (1\ 4)(2\ 3), (1\ 3)(2\ 4), (1\ 4\ 2\ 3)\}$$
$$= (1\ 4)(2\ 3)H = (1\ 3)(2\ 4)H = (1\ 4\ 2\ 3)H$$
$$(1\ 3\ 4\ 2)H = \{(1\ 3\ 4\ 2), (2\ 3\ 4), (1\ 3\ 2), (2\ 3)\}$$
$$= (2\ 3\ 4)H = (1\ 3\ 2)H = (2\ 3)H$$
$$(1\ 4\ 3\ 2)H = \{(1\ 4\ 3\ 2), (2\ 4\ 3), (1\ 4\ 2), (2\ 4)\}$$
$$= (2\ 4\ 3)H = (1\ 4\ 2)H = (2\ 4)H$$

These sets are pictured in Figure 9.11.

0	4	8
1	5	9
2	6	10
3	7	11

Figure 9.10. The cosets of $\{0, 4, 8\}$ in $\mathbb{Z}_{12}$.

Id	(1 2)	(3 4)	(1 2)(3 4)
(1 2 3 4)	(1 3 4)	(1 2 3)	(1 3)
(1 2 4 3)	(1 4 3)	(1 2 4)	(1 4)
(1 3 2 4)	(1 4)(2 3)	(1 3)(2 4)	(1 4 2 3)
(1 3 4 2)	(2 3 4)	(1 3 2)	(2 3)
(1 4 3 2)	(2 4 3)	(1 4 2)	(2 4)

Figure 9.11. The cosets of $\{\text{Id}, (1\ 2), (3\ 4), (1\ 2)(3\ 4)\}$ in S_4.

The patterns that are indicated in the above examples hold in general.

■ **PROPOSITION 9.9.** *Let H be a subgroup of the group G. Then*

  1. *if H is finite, every two cosets of H have the same number of elements;*
  2. *every two distinct cosets of H are in fact disjoint.*

Proof. For part 1, it suffices to show that if H has m elements, then every coset of H also has m elements. Suppose that

$$H = \{h_1, h_2, h_3, \ldots, h_m\}$$

then

$$aH = \{ah_1, ah_2, ah_3, \ldots, ah_m\}$$

Moreover, if $ah_i = ah_j$, then

$$h_i = a^{-1}ah_i = a^{-1}ah_j = h_j$$

and so distinct elements of H give rise to distinct elements of aH. Thus H and aH contain the same number of elements.

For part 2, suppose that the two cosets aH and bH share some element. In other words, suppose that there exist $h, k \in H$ such that

$$ah = bk$$

or

$$a = bkh^{-1}$$

Since H is a group in its own right, all the elements of the product $kh^{-1}H$ are back in H. Hence

$$kh^{-1}H \subset H$$

and we can conclude that

$$aH = b(kh^{-1})H \subset bH$$

A symmetrical argument leads to the inclusion $bH \subset aH$, and hence we have

$$aH = bH$$

Thus, if any two cosets of H share an element, they must in fact be equal. In other words, distinct cosets must be disjoint. ∎

Proof of Theorem 9.8. If H is subgroup of the finite group G then, by Proposition 9.9, the cosets of H in G constitute a partition of the elements of G into sets all of which have the same cardinality as H. Consequently the order of H must divide the order of G. ∎

Section 7.2 already contains such a computation of cosets. Specifically, it was demonstrated there that if γ is the Galois imaginary associated with $\text{GF}(2, x^4 + x^3 + x^2 + x + 1)$, then the cosets of the subgroup

$$\langle \gamma \rangle = \{1, \gamma, \gamma^2, \gamma^3, 1 + \gamma + \gamma^2 + \gamma^3\}$$

of the multiplicative group $\text{GF}^*(2, x^4 + x^3 + x^2 + x + 1)$ consist of $\langle \gamma \rangle$ itself as well as the two sets

$$\{1 + \gamma, \quad \gamma + \gamma^2, \quad \gamma^2 + \gamma^3, \quad 1 + \gamma + \gamma^2, \quad \gamma + \gamma^2 + \gamma^3\}$$
$$\{1 + \gamma^2, \quad \gamma + \gamma^3, \quad 1 + \gamma + \gamma^3, \quad 1 + \gamma^3, \quad 1 + \gamma^2 + \gamma^3\}$$

Cosets also played crucial, though implicit, roles elsewhere in this book. They appear in the proof of Galois's Theorem 7.5. The sums used by Gauss in the proof of Theorem 2.13 are also cosets in disguise. Thus the exponents of summands of A in this proof constitute the order 8 subgroup $\{1, 9, 13, 15, 16, 8, 4, 2\}$ of $\mathbb{Z}_{17}^*$ and those of B are its only other coset. Again, the exponents of the summands of C constitute the subgroup $\{1, 13, 16, 4\}$ of $\mathbb{Z}_{17}^*$ and those of D, E, and F are its cosets. Similarly, if z is any complex number, then the argument of z is in fact a coset of the subgroup

$$\langle 360° \rangle = \{\ldots, -720°, -360°, 0°, 360°, 720°, \ldots\}$$

of $\mathbb{R}$.

The number of cosets that the subgroup H has in G is called the *index* of H in G and is denoted by $[G:H]$. The corollary below follows directly from Proposition 9.9, seeing as $H = 1_G H$ is a coset.

■ **COROLLARY 9.10.** *If H is a subgroup of the finite group G, then*

$$[G:H] = \frac{\text{order of } G}{\text{order of } H}$$

The next corollary greatly facilitates the task of deciding when two elements belong to the same coset. It says that a and b determine the same coset of H if and only if $a^{-1}b \in H$.

■ **COROLLARY 9.11.** *Let H be a subgroup of G and let a and b be two elements of G. Then the following are equivalent:*

1. $aH = bH$.
2. *There exists an element c of G such that $a, b \in cH$.*
3. $a^{-1}b \in H$.

Proof. $1 \Rightarrow 2$. Suppose that $aH = bH$. Since

$$a = a1_G \in aH \quad \text{and} \quad b = b1_G \in bH = aH$$

it follows that both a and b belong to aH.

$2 \Rightarrow 3$. Suppose that there is an element c of G such that $a, b \in cH$. In other words, suppose that there exist $h, k \in H$ such that $a = ch$ and $b = ck$. Then

$$a^{-1}b = (ch)^{-1}(ck) = h^{-1}c^{-1}ck = h^{-1}k \in H$$

$3 \Rightarrow 1$. Suppose that $a^{-1}b \in H$, or, in other words,

$$a^{-1}b = h \quad \text{for some } h \in H$$

Then $b = ah$, and so the cosets aH and bH both contain the element $ah = b1_G$. By Proposition 9.9.2, $aH = bH$. ■

Not surprisingly, the significance of a coset depends on the meaning of both the ambient group and the defining subgroup. In the case of the subgroup

$$H = S_{4, x_1 + x_2 - x_3 - x_4} = \{\text{Id}, (1\ 2), (3\ 4), (1\ 2)(3\ 4)\}$$

of the symmetric group $G = S_4$, H consists of all the permutations of $1, 2, 3, 4$ that leave the polynomial $f = x_1 + x_2 - x_3 - x_4$ unchanged. Here the cosets of H turn out to be in a one-to-one correspondence with the distinct variants of f. Thus the elements of the coset

$$(1\ 2\ 3\ 4)H = \{(1\ 2\ 3\ 4), (1\ 3\ 4), (1\ 2\ 3), (1\ 3)\}$$

all change f to the polynomial $x_2 + x_3 - x_4 - x_1$, the elements of the coset

$$(1\ 2\ 4\ 3)H = \{(1\ 2\ 4\ 3), (1\ 4\ 3), (1\ 2\ 4), (1\ 4)\}$$

all change f to the polynomial $x_2 + x_4 - x_1 - x_3$, and so on. In general, we have the following proposition whose verification is relegated to Exercise 58:

■ **PROPOSITION 9.12.** *Let f be any function of the variables $x_1, x_2, \ldots, x_n$. Then the two elements ρ and σ of S_n are in the same coset of $S_{n,f}$ if and only if*

$$\rho f = \sigma f$$

We next point out what Lagrange actually proved.

■ **COROLLARY 9.13.** *If f is a function of n variables and m is the number of distinct variants of f, then m is a divisor of $n!$.*

Proof. According to Proposition 9.12, $m = [S_n : S_{n,f}]$, so by Corollary 9.10, m is a divisor of the order of S_n which is $n!$. ■

Another setting wherein the cosets of a group have an interesting interpretation is that of the vertex symmetries of the tetrahedron. Let G be this group of vertex symmetries, and let H be the subgroup that consists of all the vertex symmetries that leave the vertex 4 unchanged:

$$G = \{\text{Id}, (1\ 2)(3\ 4), (1\ 3)(2\ 4), (1\ 4)(2\ 3), (1\ 2\ 3), (1\ 3\ 2), (1\ 2\ 4),$$
$$(1\ 4\ 2), (1\ 3\ 4), (1\ 4\ 3), (2\ 3\ 4), (2\ 4\ 3)\}$$

and

$$H = \{\text{Id}, (1\ 2\ 3), (1\ 3\ 2)\}$$

Then the cosets of H are

$$\{\text{Id}, (1\ 2\ 3), (1\ 3\ 2)\}, \quad \{(1\ 2)(3\ 4), (2\ 4\ 3), (1\ 4\ 3)\}$$
$$\{(1\ 3)(2\ 4), (1\ 4\ 2), (2\ 3\ 4)\}, \quad \{(1\ 4)(2\ 3), (1\ 3\ 4), (1\ 2\ 4)\}$$

Note that the permutations of the first coset all fix 4, the permutations of the second coset all transform 4 to 3, those of the third coset all transform 4 to 2, and the permutations of the last coset all transform 4 to 1. There is a general principle in operation here, and it can be found in Exercise 40.

EXERCISES 9.4

1. Find all the subgroups of $(\mathbb{Z}_m, +)$ for $m = 1, 2, \ldots, 10$.
2. Find all the subgroups of S_n for $n = 1, 2, 3$.
3. Show that S_4 contains subgroups of orders 1, 2, 3, 4, 6, 8, 12, and 24.
4. Show that S_4 contains two nonisomorphic subgroups of order 4.
5. Show that S_5 contains subgroups of orders 6, 8, 10, 12.
6. Find all the subgroups of $(GF(2, x^2 + x + 1), +)$.
7. Find all the subgroups of $(GF(2, x^3 + x^2 + 1), +)$.
8. Find all the subgroups of D_n for $n = 3, 4, 5$.
9. Find all the subgroups of the Quaternion group.
10. Prove that A_4 does not have a subgroup of order 6.

Compute the cosets of the subgroup H of G for the groups specified in Exercises 11–22.

11. $G = (\mathbb{Z}_{15}, +)$, $H = \{0, 3, 6, 9, 12\}$
12. $G = (\mathbb{Z}_{15}, +)$, $H = \{0, 5, 10\}$
13. $G = (\mathbb{Z}_{18}, +)$, $H = \{0, 9\}$
14. $G = (\mathbb{Z}_{24}, +)$, $H = \{0, 6, 12, 18\}$
15. $G = S_3$, $H = \{\text{Id}, (1\ 2)\}$
16. $G = S_3$, $H = \{\text{Id}, (1\ 2\ 3), (1\ 3\ 2)\}$
17. $G = A_4$, $H = K$
18. $G = A_4$, $H = \{\text{Id}, (1\ 2\ 3), (1\ 3\ 2)\}$
19. $G = $ Quaternion group, $H = \{1, d\}$
20. $G = $ Quaternion group, $H = \{1, a, d, e\}$
21. $G = D_n$, $H = \{1, \rho, \rho^2, \ldots, \rho^{n-1}\}$, where ρ is the counterclockwise rotation by the angle $\dfrac{2\pi}{n}$
22. $G = D_{2n}$, $H = \{1, \alpha\}$, where α is the 180° rotation about any diagonal of the underlying polygon
23. For which values of k can a function of 3 variables have k distinct variants? Justify your answer.
24. For which values of k can a function of 4 variables have k distinct variants? Justify your answer.
25. For which of the following values of k can a function of 5 variables have k distinct variants? Justify your answer.
    (a) $k = 1, 2, 3, 4, 5$
    (b) $k = 10, 11, 12, 13, 14$
26. For which values of $k = 1, 2, 3, 4, 5, 6, 7$ can a function of 7 variables have k distinct variants? Justify your answer.
27. Prove that for each positive integer n there exists a function of n variables that has $(n-1)!$ variants.

For which of the values of m in Exercises 28–37 does S_5 contain a subgroup of order m? Justify your answers.

28. 50
29. 40
30. 30
31. 24

32. 18 **33.** 15 **34.** 10 **35.** 8
36. 6 **37.** 5

38. Find a divisor d of $6! = 720$ such that S_6 does not have a subgroup of order d.

39. Prove that for every positive integer $n > 1$, the set A_n of even permutations of $\{1, 2, \ldots, n\}$ is a subgroup of S_n. Show that A_n contains exactly half the elements of S_n.

40. Let G be a group of permutations of the set $\{1, 2, \ldots, n\}$, and let G_1 be the set of all those permutations σ of G such that $\sigma(1) = 1$.
   (a) Prove that G_1 is a subgroup of G.
   (b) Show that two elements ρ and σ of G belong to the same coset of G_1 in G if and only if $\rho(1) = \sigma(1)$.

41. Suppose that A and B are subgroups of G. Prove that $A \cap B$ is also a subgroup of G. Is the same true for $A \cup B$?

42. Suppose that H is a subgroup of G and g is some element of G. Prove that the set

$$gHg^{-1} = \{ghg^{-1} \mid h \in H\}$$

is also a subgroup of G and is isomorphic to H.

The centralizer Z_a of the element a of the group G consists of the set of all the elements of G that commute with a:

$$Z_a = \{x \in G \mid xa = ax\}$$

43. Prove that for any a in G, Z_a is a subgroup of G.

44. Compute the centralizer of each element of the following groups:
   (a) $(\mathbb{Z}_n, +)$ (b) S_3
   (c) A_4 (d) Quaternion group

The center $Z(G)$ of the group G consists of the set of all the elements of G that commute with all the elements of G:

$$Z(G) = \{x \in G \mid xa = ax \text{ for all } a \in G\}$$

45. Prove that the center of the group G is a subgroup of G.

Find the centers of the groups in Exercises 46–51.

46. $(\mathbb{Z}_n, +)$ **47.** Quaternion group
48. S_3 **49.** A_4
50. D_4 **51.** S_n

52. Let G be a permutation group on $\{1, 2, \ldots, n\}$, and let H consist of all the even permutations of G. Prove that if H does not equal G, then its order equals half the order of G.

53. Prove that S_4 does not contain a subgroup isomorphic to the Quaternion group.

54. Prove that if $n > 4$ and p is a prime such that $p \leq n$, then S_n contains no subgroup of index k for any k such that $2 < k < p$.

55. Let $P(x)$ be a polynomial of degree v that is irreducible over $\mathbb{Z}_p$, where p is prime, and let $M(x)$ be the minimal polynomial of some $\alpha \in \mathrm{GF}(p, P(x))$. Prove that the degree of $M(x)$ is a divisor of the degree of $P(x)$.
56. Let p be a prime number and n a positive integer. Prove that every group of order p^n has an element of order p.
57. Let f be a function of 4 variables that has 3 distinct variants. Prove that a certain pair of these variables can be interchanged without changing the value of the function. For example, if $f = xy + zw$, then $\{z, w\}$ constitutes such a pair.
58. Prove Proposition 9.12.
59. Prove that A_5 does not contain a subgroup that is isomorphic to S_4.
60. Prove that A_6 does not contain a subgroup that is isomorphic to S_5.
61. Prove that condition 3 for subgroups is redundant when G is finite.
62. Prove that if the union of two subgroups of G is a group, then one of those subgroups contains the other.

9.5 CYCLIC GROUPS AND SUBGROUPS

If a is any element of the group G, we let $\langle a \rangle$ denote the set of all the integer powers of a. Since

$$a^0 = 1_G$$
$$a^m a^n = a^{m+n} \quad \text{for all integers } m, n$$
$$(a^m)^\# = a^{-m}$$

it follows that $\langle a \rangle$ is a subgroup of G. The subgroup $\langle a \rangle$ is said to be *generated* by a. If σ is a permutation, then there is nothing new about this notation; it was already used in the same sense in that more restricted context. If $G = \mathbb{Z}$, then $\langle 2 \rangle$ consists of all the even integers. If $G = (\mathbb{Z}_{2n}, +)$ for some positive integer n, then

$$\langle 2 \rangle = \{0, 2, 4, \ldots, 2n - 2\} \quad \text{and} \quad \langle n \rangle = \{0, n\}$$

On the other hand, in $\mathbb{Z}_5$,

$$\langle 2 \rangle = \{0, 2, 4, 1, 3\} = \mathbb{Z}_5$$

This of course generalizes to the fact that $\langle 2 \rangle = \mathbb{Z}_m$ whenever m is an odd integer. If ζ is the Galois imaginary associated with the irreducible polynomial $x^4 + x^3 + 1$ over $\mathbb{Z}_2$, then ζ is primitive so that it has order 15 in F^*, where $F = \mathrm{GF}(2, x^4 + x^3 + 1)$. Consequently

$$\langle \zeta^3 \rangle = \langle 1, \zeta^3, \zeta^6, \zeta^9, \zeta^{12} \rangle$$

and

$$\langle \zeta^5 \rangle = \langle 1, \zeta^5, \zeta^{10} \rangle$$

If it so happens that a is an element of the group G such that $\langle a \rangle = G$, then we say that a is a *generator* of G. Thus 1 is always a generator of $(\mathbb{Z}_n, +)$, every odd element of $\mathbb{Z}_8$ is a generator of $\mathbb{Z}_8$, and each of the elements $\zeta, \zeta^2, \zeta^4, \zeta^7, \zeta^8, \zeta^{11}, \zeta^{13}, \zeta^{14}$ is a generator of the multiplicative group F^* of the Galois field above. A group that is generated by one of its elements is said to be *cyclic*. Thus $(\mathbb{Z}_n, +)$ is cyclic for each n, since, as noted above, it has 1 as a generator. The Primitive Element Theorem 7.8 asserts that the multiplicative group of the nonzero elements of every Galois field is cyclic. On the other hand, the additive group of every Galois field $GF(p, P(x))$ is not cyclic for $v \geq 2$, since it has order p^v and every nonzero element has order p. Similarly the group S_n is not cyclic for $n > 2$, since every cyclic group is necessarily commutative.

The cyclic groups are considered to be the simplest of all groups, and they can be classified in a very simple manner.

■ **THEOREM 9.14.** *Every two cyclic groups of the same order are isomorphic.*

Proof. Let $(G, \cdot)$ and $(H, \oplus)$ be two cyclic groups of order n. Suppose that they are generated by the elements a and b, respectively:

$$G = \{1_G, a, a^2, a^3, \ldots, a^{n-1}\} \quad \text{and} \quad H = \{1_H, b, b^2, b^3, \ldots, b^{n-1}\}$$

Then the function $f(a^k) = b^k$, $k = 0, 1, 2, \ldots, n-1$ is an isomorphism of G and H because

$$f(a^k \cdot a^m) = f(a^{k+m}) = b^{k+m} = b^k \oplus b^m = f(a^k) \oplus f(a^m)$$

where the exponents are added modulo n.

The proof of the theorem for infinite cyclic groups is relegated to Exercise 27. ■

It follows from this theorem that for a fixed positive integer n, the groups $(\sqrt[n]{1}, \cdot)$, $(\mathbb{Z}_n, +)$, $\langle (1 \; 2 \cdots n) \rangle$ are all isomorphic to one another. Similarly, if $P(x)$ is irreducible of degree v over $\mathbb{Z}_p$, then the group $(GF^*(p, P(x)), \cdot)$ is isomorphic to $(\mathbb{Z}_{p^v-1}, +)$. In particular, $(\mathbb{Z}_p^*, \cdot)$ is isomorphic to $(\mathbb{Z}_{p-1}, +)$.

As was mentioned above, one of the main tasks of abstract group theory is the classification of all groups up to isomorphism. Since every two isomorphic groups necessarily have the same order, this can be rephrased as looking for the classification of all groups of a fixed order n. The next proposition resolves this classification problem when n is a prime number.

■ **PROPOSITION 9.15.** *Every group of a prime order p is cyclic and is therefore isomorphic to* $(\mathbb{Z}_p, +)$.

Proof. Let G be a group of order p where p is a prime integer. Let a be any nonidentity element of G. If k is the order of a, then the subgroup $\langle a \rangle$ also has order k. However, by Theorem 9.8, k must divide p, which is prime. Since $k > 1$, it follows that $k = p$, so $\langle a \rangle = G$. ■

Clearly every group of order 5 is necessarily isomorphic to $(\mathbb{Z}_5, +)$. Curiously this fact seems to have eluded Cayley in his 1878 paper. Another surprising consequence of this proposition is that every group of prime order is necessarily commutative, since it is isomorphic to a commutative group. The following consequence is also very useful:

■ **PROPOSITION 9.16.** *If G is a group of finite order n, and if a is any element of G, then* $a^n = 1_G$. *Consequently the order of a is a divisor of the order of G.*

Proof. Let G be a group of order n, and let a be an element of G. The subgroup $\langle a \rangle$ has order $o(a)$. By Theorem 9.8, $o(a)$ is therefore a divisor of n. It now follows from Proposition 9.7.1 that $a^n = 1_G$. ■

This proposition yields new proofs of Fermat's Theorem 5.4 and Galois's Theorem 7.5 (Exercise 12).

We conclude this section by pointing out that the information we have obtained so far also allows us to classify all the groups of order 4 up to isomorphism.

■ **PROPOSITION 9.17.** *If G is a group of order 4, then it is isomorphic to either* $(\mathbb{Z}_4, +)$ *or K.*

Sketch of Proof. If G has an element of order 4, then it is cyclic and hence it is isomorphic to $(\mathbb{Z}_4, +)$. Otherwise, every nonidentity element of G has order 2, meaning that the diagonal entries of the multiplication table of G are all 1_G. It is now easily verified that the multiplication table of G must be identical with that of K. The details are relegated to Exercise 1. ■

With Proposition 9.17 available, we have classified all the groups of order at most 5 up to isomorphism. Those of orders 1, 2, 3, 5 are isomorphic to $(\mathbb{Z}_1, +)$, $(\mathbb{Z}_2, +)$, $(\mathbb{Z}_3, +)$, $(\mathbb{Z}_5, +)$, respectively, whereas every group of order 4 is isomorphic to either $(\mathbb{Z}_4, +)$ or K. A sense of the enormity of the task of classifying all finite groups can be obtained from the amount of theory that was required by the classification of these small groups alone.

EXERCISES 9.5

1. Complete the proof of Proposition 9.17.

For each of the groups in Exercises 2–11, decide whether it is isomorphic to $(\mathbb{Z}_4, +)$ *or to K.*

2. {Id, (1 2), (3 4), (1 2)(3 4)}
3. $(\mathbb{Z}_5^*, \cdot)$
4. $\langle(1\ 2\ 3\ 4)\rangle$
5. $\langle(1\ 2\ 3\ 4)(5\ 6)\rangle$
6. $S_{4,f}$ where $f = x_1 + 2x_2 + 2x_3 + x_4$
7. $(\mathbb{Z}_8, \cdot)$
8. $(\mathrm{GF}(2, x^2 + x + 1), +)$
9. $(\mathbb{Z}_{10}^*, \cdot)$
10. $\sqrt[4]{1}$
11. $\mathbb{Z}_2[x, \leq 1]$
12. Use Proposition 9.16 to give a new proof of Fermat's Theorem 5.4 and Galois's Theorem 7.5.
13. Prove that every subgroup of every cyclic group is also cyclic.
14. Suppose that G is a group of order 187. Prove that if two subgroups of G have the same order, then they are isomorphic.

Find the largest value of k for which the groups in Exercises 15–23 contain a cyclic subgroup of order k.

15. $\sqrt[n]{1}$
16. $(\mathrm{GF}(p, P(x)), +)$
17. D_{10}
18. S_5
19. A_5
20. S_{10}
21. A_{10}
22. $(\mathbb{Z}_{37}^*, \cdot)$
23. $(\mathbb{Z}_{16}^*, \cdot)$
24. Prove that a group has exactly 1 subgroup if and only if it is isomorphic to $(\mathbb{Z}_1, +)$.
25. Prove that a group has exactly 2 subgroups if and only if it is isomorphic to $(\mathbb{Z}_p, +)$ for some prime p.
26. Prove that a group has exactly 3 subgroups if and only if it is isomorphic to $(\mathbb{Z}_{p^2}, +)$ for some prime p.
27. Complete the proof of Theorem 9.14 (the infinite case).
28. Prove that if m is relatively prime to n, then $m^{\phi(n)} \equiv 1 \pmod{n}$, where $\phi(n)$ is the Euler ϕ function.

9.6 CAYLEY'S THEOREM

The first groups to be examined by mathematicians were groups of permutations. It was not until a century had past that Cayley pointed out that every group is determined up to isomorphism by its multiplication table, and that therefore this table could be used to define the notion of an abstract group. At the same time Cayley noted that this innovation did not introduce any genuinely new structures into the study of groups, for, he said, every abstract group can be shown to be isomorphic to a group of permutations. Cayley did not formally prove this assertion; he contented himself with an example. His short note on the subject is included as Appendix E. Cayley's assertion will be formally stated and proved below as Theorem 9.18, but we first paraphrase Cayley's ideas in more modern terminology. Table 9.12 contains the multiplication table of the symmetric group S_3 with $a = (1\ 2)$, $b = (3\ 2\ 1)$, $c = (1\ 3)$, $d = (1\ 2\ 3)$, and $e = (2\ 3)$.

Reverting to our original notation for permutations, we associate with each element x a two-rowed array P_x, whose first row is Id $a\ b\ c\ d\ e$ and whose second row is that row of Table 9.12 that corresponds to the element x. Thus

$$P_{\text{Id}} = \begin{pmatrix} \text{Id} & a & b & c & d & e \\ \text{Id} & a & b & c & d & e \end{pmatrix} \qquad P_a = \begin{pmatrix} \text{Id} & a & b & c & d & e \\ a & \text{Id} & c & b & e & d \end{pmatrix}$$

$$P_b = \begin{pmatrix} \text{Id} & a & b & c & d & e \\ b & e & d & a & \text{Id} & c \end{pmatrix} \qquad P_c = \begin{pmatrix} \text{Id} & a & b & c & d & e \\ c & d & e & \text{Id} & a & b \end{pmatrix}$$

$$P_d = \begin{pmatrix} \text{Id} & a & b & c & d & e \\ d & c & \text{Id} & e & b & a \end{pmatrix} \qquad P_e = \begin{pmatrix} \text{Id} & a & b & c & d & e \\ e & b & a & d & c & \text{Id} \end{pmatrix}$$

Note that

$$P_a P_d = \begin{pmatrix} \text{Id} & a & b & c & d & e \\ a & \text{Id} & c & b & e & d \end{pmatrix} \begin{pmatrix} \text{Id} & a & b & c & d & e \\ d & c & \text{Id} & e & b & a \end{pmatrix}$$

$$= \begin{pmatrix} \text{Id} & a & b & c & d & e \\ e & b & a & d & c & \text{Id} \end{pmatrix} = P_e = P_{ad}$$

and

$$P_d P_e = \begin{pmatrix} \text{Id} & a & b & c & d & e \\ d & c & \text{Id} & e & b & a \end{pmatrix} \begin{pmatrix} \text{Id} & a & b & c & d & e \\ e & b & a & d & c & \text{Id} \end{pmatrix}$$

$$= \begin{pmatrix} \text{Id} & a & b & c & d & e \\ a & \text{Id} & c & b & e & d \end{pmatrix} = P_a = P_{de}$$

TABLE 9.12. The multiplication table of S_3.

	Id	a	b	c	d	e
Id	Id	a	b	c	d	e
a	a	Id	c	b	e	d
b	b	e	d	a	Id	c
c	c	d	e	Id	a	b
d	d	c	Id	e	b	a
e	e	b	a	d	c	Id

In other words, the function that assigns to each element x the corresponding permutation P_x behaves just like an isomorphism. It is in fact always an isomorphism, and that is the gist of Cayley's assertion.

■ **THEOREM 9.18 (Cayley).** *Every group is isomorphic to a group of permutations.*

Proof. Let G be any group. To each element x of G, we assign a permutation P_x of the elements of G that transforms each element a to xa:

$$P_x(a) = xa \qquad \text{for all } x, a \in G$$

The function P_x is a permutation because $xa = xb$ if and only if $a = b$ and because $P_x(x^{-1}y) = y$. Let H be the set of permutations thus obtained:

$$H = \{P_x \mid x \in G\}$$

To see that H constitutes a group of permutations, we observe first that for any two elements P_x and P_y of H, we have

$$(P_x \circ P_y)(a) = P_x(P_y(a)) = P_x(ya) = xya = P_{xy}(a)$$

so

$$P_x \circ P_y = P_{xy} \qquad (1)$$

Hence the composition of any two elements of H is in H, and the inverse of P_x is $P_{x^{-1}}$ which is also in H. Thus H is a group of permutations.

Suppose that x and y are distinct elements of G. Then

$$P_x(\text{Id}) = x \,\text{Id} = x \neq y = y \,\text{Id} = P_y(\text{Id})$$

and so P_x and P_y are distinct elements of H. Thus the function $f(x) = P_x$ matches up all the elements of G and H. That f is in fact an isomorphism follows from (1) since

$$f(xy) = P_{xy} = P_x \circ P_y = f(x) \circ f(y) \qquad ■$$

EXERCISES 9.6

1. Write out the Cayley representation P_x for every element x of $(\mathbb{Z}_2, +)$.
2. Write out the Cayley representation P_x for every element x of $(\mathbb{Z}_3, +)$.
3. Write out the Cayley representation P_x for every element x of $(\mathbb{Z}_4, +)$.
4. Write out the Cayley representation P_x for every element x of K.
5. Write out the Cayley representation P_x for every element x of $(\mathbb{Z}_5, +)$.

6. Write out the Cayley representation P_x for every element x of the Quaternion group.
7. Write out the Cayley representation P_x for every element x of $(\mathbb{Z}_6, +)$.
8. Prove that for any finite group G and for any element x of G, all the cycles in the disjoint cycle representation of P_x have the same length. Explain why this common cycle length is necessarily a divisor of the order of G.
9. A set with a binary operation whose multiplication table is a Latin square is called a *loop*. The definition of the permutation P_x applies to loops as well as to groups. Prove that a loop X is a group if the set of permutations $P_x, x \in X$, is a group (under composition).

CHAPTER SUMMARY

We have defined groups, both concretely, as permutations, and abstractly, in terms of axioms. Many of the algebraic systems examined in the earlier chapters are examples of such groups, and this leads to a natural classification problem. The notion of isomorphism was defined to formalize the notion of "sameness" of groups. Some inroads were made on this difficult, if not impossible, classification problem by recognizing cyclic groups and using them to show that any two groups of the same prime order are necessarily isomorphic. Finally, we presented Cayley's Theorem which asserts that every abstract group is in fact isomorphic to some group of permutations.

Chapter Review Exercises

Mark the following true or false.

1. The set of permutations that leave the function $(x_1 + x_2)(x_3 + x_4)$ unchanged is a group.
2. Every permutation belongs to some group.
3. There is no function f such that $S_{n,f} = D_n$.
4. No group of permutations is an abstract group.
5. $\sqrt[17]{1}$ is an abstract group.
6. $\sqrt[17]{1}$ is a permutation group.
7. The inverse of (1 2 3 4 5)(6 7 8 9) is (5 4 3 2 1)(9 8 7 6).
8. Every two groups of order 4 are isomorphic.
9. If the element a of the group G has order 24, then $o(a^{20}) = 6$.
10. S_6 has a subgroup of order 6.
11. S_6 has a subgroup of order 7.
12. If H is a subgroup of G, then every coset of H in G is also a subgroup of G.
13. If a subgroup of S_n has order n, then its index is $(n-1)!$.
14. The permutations (1 2) and (1 3) belong to the same coset of A_5 in S_5.
15. Let $f = x_1 x_2 + x_3 + x_4 + x_5$. Then the permutations (1 3)(2 4 5) and (1 4 5 2 3) belong to the same coset of $S_{5,f}$.

16. There is a function of 10 variables that has 11 distinct variants.
17. Every two groups of order 17 are isomorphic.
18. S_{10} has no element of order 11.
19. Given any three groups of order 4, some two of them are isomorphic to each other.
20. The group $(\mathbb{Z}_{25}, +)$ is isomorphic to some group of permutations.

New Terms

Abstract group	191
Alternating group	184
Commutative group	194
Coset	205
Cyclic group	213
Dihedral group	187
Generator	213
Index of subgroup	208
Inverse	191
Isomorphism of groups	198
Latin square	194
Loop	218
Multiplication table	191
Order of element	201
Order of group	200
Permutation group	182
Quaternion group	194
Subgroup	203
Symmetric group	182
Symmetry	185
Vertex symmetry group	186

Supplementary Exercises

1. Write a computer script that will decide whether or not a given multiplication table is a Latin square.
2. Write a computer script that will decide whether or not a given multiplication table describes a group.
3. Write a computer script that will list all the subgroups of a group that is given by its multiplication table.
4. Write a computer script that will list all the subgroups of S_n for as many values of n as possible.

5. Classify the subgroups of S_n by isomorphism type for as many values of n as possible.
6. Decide whether the groups whose multiplication tables appear below are isomorphic.

$$\begin{array}{cccccc} A & B & C & D & E & F \\ B & C & D & E & F & A \\ C & D & E & F & A & B \\ D & E & F & A & B & C \\ E & F & A & B & C & D \\ F & A & B & C & D & E \end{array} \qquad \begin{array}{cccccc} a & b & c & d & e & f \\ b & c & a & e & f & d \\ c & a & b & f & d & e \\ d & e & f & a & b & c \\ e & f & d & b & c & a \\ f & d & e & c & a & b \end{array}$$

7. Characterize all the groups that have exactly 4 subgroups.
8. Characterize all the groups that have exactly 5 subgroups.
9. Write a computer program that will list all the cosets of a subgroup of a group.
10. For each positive integer n, what are the n-dimensional analogs of the 5 three-dimensional regular solids? Describe their vertex symmetry groups.

CHAPTER 10

Quotient Groups and Their Uses

We will define an operation on groups that is in some ways analogous to division of integers. This operation will provide us with a rigorous method for constructing the Galois fields of Chapter 7 as well as a host of new fields. Finally, Galois's Theorem on the resolvability of algebraic equations will be described.

10.1 QUOTIENT GROUPS

The notion of a set as a point is one of the recurrent themes of modern mathematics. Loosely speaking, the idea is to create new structures from old ones by considering sets of elements of the old structure as elements of the new structure. In the context of group theory, this process takes the following form. Let $(G, \cdot)$ be a group, and let S and T be any two subsets of G. We define

$$S \cdot T = \{a \cdot b \mid a \in S \text{ and } b \in T\}$$

and

$$S^{-1} = \{a^{-1} \mid a \in S\}$$

If $G = (\mathbb{Z}_9, +)$, $S = \{1, 3, 5\}$, and $T = (3, 7)$, then

$$S + T = \{1 + 3, 1 + 7, 3 + 3, 3 + 7, 5 + 3, 5 + 7\} = \{4, 8, 6, 1, 8, 3\}$$
$$= \{1, 3, 4, 6, 8\}$$

and
$$S^{-1} = \{-1, -3, -5\} = \{4, 6, 8\}$$

Given a group $(G, \cdot)$, this process defines a binary operation on all the subsets of G. The associativity of the operation $\cdot$ on the elements of G entails its associativity as an operation on sets as well (Exercise 35). Let us examine this operation on some collections of cosets. Suppose that $(G, \cdot) = (\mathbb{Z}_{12}, +)$, H is the subgroup $\{0, 3, 6, 9\}$, and let the cosets of H be labeled

$$H_0 = H = \{0, 3, 6, 9\}$$
$$H_1 = 1 + H = \{1, 4, 7, 10\}$$
$$H_2 = 2 + H = \{2, 5, 8, 11\}$$

Then, for example,

$$H_1 + H_2 = \{1+2, 1+5, 1+8, 1+11, 4+2, 4+5, 4+8, 4+11,$$
$$7+2, 7+5, 7+8, 7+11, 10+2, 10+5, 10+8, 10+11\}$$
$$= \{3, 6, 9, 0, 6, 9, 0, 3, 9, 0, 3, 6, 0, 3, 6, 9\}$$
$$= \{0, 3, 6, 9\} = H_0$$

In fact the sum of any two of the cosets of this subgroup H is again a coset of H. Table 10.1 summarizes the results of the operation of addition on the cosets of H and makes it clear that the cosets of H form a new group that is isomorphic to $\mathbb{Z}_3$.

Let us examine the Quaternion group G of Table 9.3. Since $d^2 = 1$, it follows that $H = \{1, d\}$ is a subgroup of G. Its cosets are

$$1H = dH = \{1, d\}, \quad aH = eH = \{a, e\}$$
$$bH = fH = \{b, f\}, \quad cH = gH = \{c, g\}$$

Note that

$$(aH)(bH) = \{ab, af, eb, ef\} = \{c, g, g, c\} = \{c, g\} = cH$$

TABLE 10.1. A quotient group of $(\mathbb{Z}_{12}, +)$.

	H_0	H_1	H_2
H_0	H_0	H_1	H_2
H_1	H_1	H_2	H_0
H_2	H_2	H_0	H_1

TABLE 10.2. A quotient group of the Quaternion group.

	H	aH	bH	cH
H	H	aH	bH	cH
aH	aH	H	cH	bH
bH	bH	cH	H	aH
cH	cH	bH	aH	H

Table 10.2 displays the result of multiplying any two of the cosets of the subgroup $H = \{1, d\}$ of the Quaternion group. It is clear that the multiplication Table 10.2 is isomorphic with that of the Klein 4-group of Table 9.4. We hasten to point out that the cosets of a subgroup do not always form a group. If $G = S_3$ and $H = \{\text{Id}, (1\ 2)\}$, then

$$(1\ 3)H = \{(1\ 3), (1\ 2\ 3)\}$$

but

$$[(1\ 3)H][(1\ 3)H] = \{(1\ 3)(1\ 3), (1\ 3)(1\ 2\ 3), (1\ 2\ 3)(1\ 3), (1\ 2\ 3)(1\ 2\ 3)\}$$
$$= \{\text{Id}, (1\ 2), (2\ 3), (1\ 3\ 2)\}$$

which is clearly not a coset of H. This counterexample notwithstanding, the previous two examples indicate that the cosets of a subgroup form a group of their own often enough for this phenomenon to merit attention. The following lemma acknowledges the fact that in one very special case the product of two cosets is necessarily a coset:

■ **LEMMA 10.1.** *If H is any subgroup of G, then $HH = H$.*

Proof. Since H is a subgroup, the product of any two of its elements is in H, and hence

$$HH \subset H$$

On the other hand, since $1_G \in H$, it also follows that

$$HH \supset H1_G = H$$

Hence $HH = H$. ■

Let us examine the question of just when the product of two cosets is necessarily a coset. For the two cosets aH and bH of the subgroup H to have another coset, say, cH, as their product, this coset cH would have to contain the element ab, since a is in aH and b is in bH. Hence, since the coset that contains ab is abH, we would have

$$(aH)(bH) = cH = abH$$

or, upon multiplying both sides by $b^{-1}a^{-1}$,

$$b^{-1}HbH = H$$

Hence, for any $h \in H$,

$$b^{-1}hb = b^{-1}hb1_G \in b^{-1}HbH = H$$

Consequently, if the product of every pair of cosets is again a coset, then

$$b^{-1}hb \in H \qquad \text{for all } b \in G \text{ and } h \in H$$

This motivates the following definition. A subgroup H of G is said to be *normal* if for every $b \in G$ and every $h \in H$, $b^{-1}hb \in H$. Thus every subgroup of a commutative group G is normal, since in such groups

$$b^{-1}hb = hb^{-1}b = h \in H$$

If G is the Quaternion group, and if $H = \{1_G, d\}$ then clearly

$$x^{-1}1_G x = x^{-1}x = 1_G$$

for every element x of G. Moreover, since the column and the row of d in Table 9.3 are identical, it follows that for each element x of G,

$$dx = xd$$

and hence

$$x^{-1}dx = d$$

for all such x's. Thus $\{1, d\}$ is a normal subgroup of the Quaternion group. It is clear that a subgroup H of G is normal if and only if

$$b^{-1}Hb \subset H$$

for every element b of G.

■ **THEOREM 10.2.** *Let H be a subgroup of G. Then the following are equivalent:*

1. *H is a normal subgroup of G.*
2. $x^{-1}Hx = H$ *for every element x of G.*
3. $xHx^{-1} = H$ *for every element x of G.*
4. $Hx = xH$ *for every element x of G.*
5. $(xH)(yH) = xyH$ *for all* $x, y \in G$.
6. *The multiplication of the cosets of H forms a group.*

Proof. $1 \Rightarrow 2$. If H is a normal subgroup of G, then for every element x of G,

$$x^{-1}Hx \subset H$$

Consequently, if x is any element of G, two applications of the definition of normality yield

$$H \supset x^{-1}Hx \supset x^{-1}[(x^{-1})^{-1}Hx^{-1}]x = 1_G H 1_G = H$$

Consequently

$$x^{-1}Hx = H$$

$2 \Rightarrow 3$. Replacing x by x^{-1} clearly transforms the expression $x^{-1}Hx$ to xHx^{-1}.

$3 \Rightarrow 4$. If $xHx^{-1} = H$, then $xH = xH(x^{-1}x) = (xHx^{-1})x = Hx$.

$4 \Rightarrow 5$. Suppose that $x, y \in G$. Since $Hy = yH$, we get

$$(xH)(yH) = x(Hy)H = x(yH)H = xyHH = xyH$$

$5 \Rightarrow 6$. Let H be a subgroup of G such that $(xH)(yH) = xyH$ holds for every two of its cosets. Then

$$(xH)(1_G H) = x 1_G H = xH = 1_G xH = (1_G H)(xH)$$

so that the coset $H = 1_G H$ acts as the identity for this multiplication of cosets. In addition, for every coset xH, we have

$$(xH)(x^{-1}H) = xx^{-1}H = 1_G H = H = x^{-1}xH = (x^{-1}H)(xH)$$

In other words, the coset $x^{-1}H$ is the inverse of the coset xH. As was noted above, the multiplication of subsets of a group is always associative. It follows that the multiplication of the cosets of H is indeed a group.

$6 \Rightarrow 1$. This was already proved as part of the argument that motivated the definition of normal subgroups. ∎

It would be useful to have a quick method for deciding whether a given subgroup is in fact normal. Unfortunately, no such method is known. There is, however, a variety of helpful ad hoc techniques. It is clear from the definition of normality that both the whole group G and the trivial subgroup $\{1_G\}$ are normal subgroups of G. More interestingly, we have the following observation:

■ **PROPOSITION 10.3.** *If H is a subgroup of index 2 in the group G, then H is a normal subgroup of G.*

Proof. Assume for a contradiction that $xhx^{-1} \notin H$ for some $x \in G$, $h \in H$. Then $x \notin H$, so that $xhx^{-1} \in xH$, and there exists an element k in H such that

$$xhx^{-1} = xk$$

But then

$$x^{-1} = (xh)^{-1}(xk) = h^{-1}x^{-1}xk = h^{-1}k \in H$$

contradicting the fact that $x \notin H$. ∎

The alternating group A_4, which consists of all the even permutations in S_4, has twelve elements, and it therefore contains exactly half the elements of the symmetric group S_4. It follows that A_4 is a normal subgroup of S_4 which has only two cosets, itself and its complement in S_4.

If H is a subgroup of G such that no other subgroup of G has the same order as H, then H is necessarily a normal subgroup of G (Exercise 21). Thus, since d is the only element of the Quaternion group that has order 2, it follows that $\{1, d\}$ is the only subgroup of the Quaternion group that has order 2. It therefore is a normal subgroup of G.

If G is a commutative group, then for any subgroup H and for any element a of G, $a^{-1}Ha = Ha^{-1}a = H$, and so H is necessarily normal. This situation arises often enough to merit highlighting.

■ **PROPOSITION 10.4.** *Every subgroup of a commutative group is normal.*

The converse to this proposition is false. The Quaternion group is an example of a noncommutative group all of whose subgroups are normal (Exercise 12).

In the special case where G is a permutation group, Exercise 8.2.23 provides an occasionally useful criterion for recognizing normal subgroups. According to this exercise, *if ρ and σ are any two permutations in S_n, then $\rho\sigma\rho^{-1}$ and σ have the same number of k-cycles for each positive integer k.* Consequently the Klein 4-group

$$K = \{\text{Id}, (1\ 2)(3\ 4), (1\ 3)(2\ 4), (1\ 4)(2\ 3)\}$$

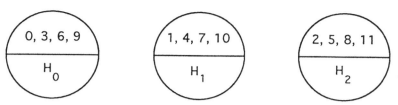

Figure 10.1. The elements of $(\mathbb{Z}_{12}, +)/\{0, 3, 9\}$.

is a normal subgroup of A_4, since K consists of the identity and all three of the elements of A_4 that consist of exactly two 2-cycles and nothing else.

The subgroup $H = \{\text{Id}, (1\ 2)\}$ of the symmetric group S_3 is not normal, since for $x = (2\ 3)$ we have

$$xHx^{-1} = (2\ 3)\{\text{Id}, (1\ 2)\}(2\ 3) = \{\text{Id}, (1\ 3)\} \neq H$$

If H is a normal subgroup of G, then the group formed by the cosets of H in G is called the *quotient* of G by H and is denoted by G/H. Each element of G/H is a coset of H in G, and so it is a subset of G. Figures 10.1 and 10.2 illustrate this relationship for the examples displayed in Tables 10.1 and 10.2.

The nature of the quotient group G/H is also not easily determined, and again only ad hoc arguments are available. In some cases, however, the quotient group G/H is easily identified. If $H = G$, then H has only one coset in G, namely itself, and so G/H is necessarily the trivial group with one element only. At the other end of the spectrum we have the possibility that $H = \{1_G\}$. In this case each coset of H consists of exactly one element of G, and in view of Theorem 10.2, G/H is isomorphic to G. A somewhat less trivial, though still surprisingly useful, observation deals with subgroups of index 2, namely subgroups H that have exactly two cosets. As pointed out in Proposition 10.3, such subgroups are necessarily normal. Moreover, as they determine only two cosets, their quotients contain only two elements, and so they are isomorphic to $(\mathbb{Z}_2, +)$. Accordingly S_4/A_4 is isomorphic to $(\mathbb{Z}_2, +)$.

Similarly, if H is a normal subgroup of index 3 in G, then G/H contains three elements, and so it must be isomorphic to $(\mathbb{Z}_3, +)$. Since it was pointed

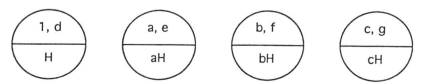

Figure 10.2. The elements of Quaternion group/$\{1, d\}$.

out above that K is a normal subgroup of A_4 and since K has index 3 in A_4, it follows that A_4/K isomorphic to $(\mathbb{Z}_3, +)$.

We offer two observations that can be helpful in identifying quotient groups:

■ **PROPOSITION 10.5.** *Let H be a normal subgroup of the group G. Then the order of aH in G/H is a divisor of the order of a in G.*

Proof. Suppose that a has order m in G. Then

$$(aH)^m = (aH)(aH) \cdots (aH) = a^m H = 1_G H = H$$

Since H is the identity element of the quotient group G/H, it follows from Proposition 9.7.1 that the order of aH is a divisor of m. ■

Consider the group $(F, +)$ where $F = \operatorname{GF}(2, x^4 + x + 1)$. This group has order 16 and is commutative. If δ is the Galois imaginary that is associated with this field, then $H = \{0, 1, \delta, 1 + \delta\}$ is a subgroup of order 4. Since $(F, +)$ is commutative, this subgroup is normal. Thus F/H is group of order 4. Since every nonidentity element of F has order 2, it follows from the above proposition that every nonidentity element of F/H also has order 2. Hence F/H is isomorphic to the Klein 4-group.

■ **PROPOSITION 10.6.** *If G is a cyclic group and H is a subgroup of G, then G/H is also cyclic.*

Proof. Let g be a generator of G. Since the typical element of G has the form g^k, it follows that the typical element of G/H has the form

$$g^k H = (gH)^k$$

Hence G/H is generated by the coset gH, and so G/H is necessarily cyclic. ■

Consider the group $(F^*, \cdot)$ where $F = \operatorname{GF}(3, x^2 + x + 2)$ which was discussed in detail in Section 7.1. This is a commutative group of order 8. If δ is the Galois imaginary associated with this field, then we already know that δ is primitive, so that δ is a generator of $(F^*, \cdot)$. Thus $(F^*, \cdot)$ is a cyclic group. This group has $H = \{1, \delta^4\}$ as a subgroup, which is necessarily normal. Since F^* was cyclic, so is F^*/H. Since the quotient F^*/H has order 4, it follows that F^*/H is isomorphic to $(\mathbb{Z}_4, +)$.

The quotient method for constructing groups provides us with a new perspective on modulo n arithmetic.

■ **COROLLARY 10.7.** *Let $n\mathbb{Z}$ denote the subgroup of the group of integers $\mathbb{Z}$ that is generated by the integer n. Then the quotient group $\mathbb{Z}/n\mathbb{Z}$ is isomorphic to $(\mathbb{Z}_n, +)$.*

Proof. By definition,

$$n\mathbb{Z} = \{0, \pm n, \pm 2n, \pm 3n, \ldots\}$$

and hence by Proposition 9.11 the cosets of H are $H, 1 + H, 2 + H, \ldots, n - 1 + H$. Since, by Proposition 10.6, $\mathbb{Z}/n\mathbb{Z}$ is cyclic, it follows from Theorem 9.15 that $\mathbb{Z}/n\mathbb{Z}$ is isomorphic to $(\mathbb{Z}_n, +)$. ∎

This corollary tells us that another way to define $\mathbb{Z}_n$ is to view each of its elements as a coset of $n\mathbb{Z}$ in $\mathbb{Z}$. Thus the element 2 in $\mathbb{Z}_5$ is to be regarded as a shorthand notation for the coset

$$2 + 5\mathbb{Z} = \{\ldots, -8, -3, 2, 7, 12, \ldots\}$$

of $5\mathbb{Z}$ in $\mathbb{Z}$.

However, the elements of $\mathbb{Z}_n$ are also subject to multiplication, and it behooves us to verify that this multiplication is consistent with the coset point of view. To do this, we *define* the product of the two cosets $a + n\mathbb{Z}$ and $b + n\mathbb{Z}$ in $\mathbb{Z}_n$ as the coset $ab + n\mathbb{Z}$. There is a potential problem with this glib definition. Note that

$$2 + 5\mathbb{Z} = 7 + 5\mathbb{Z} \quad \text{and} \quad 4 + 5\mathbb{Z} = 9 + 5\mathbb{Z}$$

Is the product of these two cosets to be $2 \cdot 4 + 5\mathbb{Z}$, $2 \cdot 9 + 5\mathbb{Z}$, $7 \cdot 4 + 5\mathbb{Z}$, or $7 \cdot 9 + 5\mathbb{Z}$? Fortunately, this question is moot, since $2 \cdot 4 \equiv 2 \cdot 9 \equiv 7 \cdot 4 \equiv 7 \cdot 9 \pmod{5}$, and so

$$2 \cdot 4 + 5\mathbb{Z} = 2 \cdot 9 + 5\mathbb{Z} = 7 \cdot 4 + 5\mathbb{Z} = 7 \cdot 9 + 5\mathbb{Z}$$

This is the case in general as well. If both $a - a'$ and $b - b'$ are divisible by n, then

$$ab - a'b' = b(a - a') + a'(b - b')$$

is also divisible by n, and hence $ab + n\mathbb{Z} = a'b' + n\mathbb{Z}$.

EXERCISES 10.1

For each of the pairs $\{G, H\}$ in Exercises 1–11, compute the cosets of H in G. Decide whether H is a normal subgroup of G. If it is normal, identify the isomorphism type of the quotient group G/H. If H is not a normal subgroup of G explain why not.

1. $G = (\mathbb{Z}_{16}, +)$, $H = \langle 4 \rangle$
2. $G = (\mathbb{Z}_9, +)$, $H = \langle 3 \rangle$
3. $G = (\mathbb{Z}_{15}, +)$, $H = \langle 5 \rangle$
4. $G = (\mathbb{Z}_{15}, +)$, $H = \langle 3 \rangle$

5. $G = S_3$, $H = \langle (1\ 2) \rangle$
6. $G = S_3$, $H = \langle (1\ 2\ 3) \rangle$
7. $G = \sqrt[6]{1}$, $H = \{1, -1\}$
8. $G = \sqrt[6]{1}$, $H = \{1, \omega, \omega^2\}$
9. $G = A_4$, $H = \langle (1\ 2\ 3) \rangle$
10. $G = (\mathbb{Z}_5^*, \cdot)$, $H = \langle 2 \rangle$
11. $G = (\mathbb{Z}_{16}^*, \cdot)$, $H = \langle 7 \rangle$

For each of the groups G in Exercises 12–20 determine all the nontrivial normal subgroups H of G, and identify G/H for such subgroups H.

12. $G = $ Quaternion group
13. $G = D_4$
14. $G = (\text{GF}(2, x^2 + x + 1), +)$
15. $G = (\text{GF}(2, x^3 + x^2 + 1), +)$
16. $G = (\text{GF}(3, x^2 + x + 2), +)$
17. $G = (\text{GF}^*(3, x^2 + x + 2), \cdot)$
18. $G = (\text{GF}(5, x^2 + 4x + 2), +)$
19. $G = A_4$
20. $G = D_5$
21. Suppose that H is a subgroup of G such that no other subgroup of G is isomorphic to H. Prove that H is a normal subgroup of G.
22. Suppose that a is an element of the group G such that no other element has the same order as a. Prove that a has order 1 or 2 and that $\langle a \rangle$ is a normal subgroup of G.
23. Prove that the center $Z(G)$ of the group G is a normal subgroup of G.

The element a of the group G is said to be conjugate to the element b if there exists an element x of G such that $xax^{-1} = b$. The set $C(a)$ consists of the set of all the elements of G that are conjugate to a and is called the conjugacy class of a.

24. Prove that a is conjugate to b if and only if b is conjugate to a.
25. Prove that if a is conjugate to b and b is conjugate to c, then a is conjugate to c.
26. Prove that if a and b are conjugate, then $C(a) = C(b)$.
27. Prove that if a and b are any two elements of a group G, then ab and ba are conjugate.
28. Describe the conjugacy class of each element of S_4.
29. Describe the conjugacy class of each element of S_n.
30. Prove that the number of elements in $C(a)$ equals $[G:Z_a]$ and is therefore a divisor of the order of G whenever G is finite.
31. Prove that if p is a prime number, then every group of order p^n, $n > 0$, has a nontrivial center.
32. Prove that if H is a normal subgroup of G, then $(xH)^{-1} = x^{-1}H$ for all x in G.
33. Suppose that H and K are normal subgroups of G such that $H \cap K = \{1_G\}$. Prove that $ab = ba$ whenever $a \in H$ and $b \in K$.

34. Prove that A_n is a normal subgroup of S_n and that $S_n/A_n \equiv (\mathbb{Z}_2, +)$.

35. Let $(G, \cdot)$ be any group. Prove that if A, B, C are any subsets of G, then $A \cdot (B \cdot C) = (A \cdot B) \cdot C$.

36. Prove that the intersection of two normal subgroups of G is a normal subgroup of G.

10.2 THE RIGOROUS CONSTRUCTION OF FIELDS

If H is a normal subgroup of G, then, loosely speaking, we say that G/H inherits its binary operation from that of G. Some well-known groups, however, are subject to additional operations, as is the case, for instance, for the groups $(\mathbb{Z}, +)$ and $(F[x], +)$ for any field F. We saw above that the multiplication of integers could be transferred to $(\mathbb{Z}_n, +)$ so as to convert it to arithmetic modulo n. It will now be demonstrated that a similar procedure can be employed to yield rigorous constructions of both the complex numbers of Chapter 2 and the Galois fields of Chapter 7, as well as a host of new fields.

Let F be an arbitrary field, and let $P(x)$ be any polynomial in $F[x]$. If we set $G = (F[x], +)$ and let H be the set of all the polynomials in $F[x]$ that are divisible by $P(x)$, then H is a subgroup of G. The reason for this is that $0 = 0P(x)$ is clearly a multiple of $P(x)$, and if $A(x)P(x)$ and $B(x)P(x)$ are any two multiples of $P(x)$, then so are

$$A(x)P(x) + B(x)P(x) = (A(x) + B(x))P(x)$$

and

$$-A(x)P(x) = (-A(x))P(x)$$

multiples of $P(x)$.

Since G is a commutative group it follows that H is necessarily a normal subgroup of G, and so there exists a quotient group G/H. We will show that when the polynomial $P(x)$ is irreducible the quotient group G/H can be converted to a field. The fields so obtained include the Galois fields as well as a variety of new fields. Consider the case where $F = \mathbb{Z}_2$ and $P(x) = x^2 + x + 1$ which is known to be irreducible over F. By Corollary 9.11.3, the polynomials $M(x)$ and $N(x)$ in $F[x]$ belong to the same coset of H if and only if

$$-N(x) + M(x) = M(x) - N(x) \in H$$

which is true if and only if

$$P(x) \quad \text{divides} \quad M(x) - N(x)$$

Since the difference of any two of the polynomials $0, 1, x, 1 + x$ has degree at most 1, it follows that the four cosets

$$0 + H, \quad 1 + H, \quad x + H, \quad 1 + x + H \tag{2}$$

are all distinct. Moreover, if $M(x)$ is any polynomial of $F[x]$, then, by Proposition 6.4, there exist polynomials $Q(x)$ and $R(x)$ such that

$$M(x) = Q(x)(x^2 + x + 1) + R(x) \tag{3}$$

and $R(x)$ is either 0 or else has degree at most 1. In other words, $R(x)$ is one of the polynomials

$$0, \quad 1, \quad x, \quad 1 + x$$

However, it is clear from (3) above that $M(x) - R(x)$ is divisible by $x^2 + x + 1$, and hence $M(x)$ and $R(x)$ belong to the same coset of H. Consequently every polynomial of $F[x]$ belongs to one of the cosets listed in (2). Thus G/H has four elements only. (It happens to be isomorphic to the Klein 4-group, but that is of no interest to us now.)

We will next convert G/H, which already has the additive operation,

$$(a + H) + (b + H) = (a + b) + H$$

to a field by defining a multiplication of its elements. Specifically, we define

$$(a + H) \cdot (b + H) = ab + H$$

These two operations on G/H are summarized in Tables 10.3 and 10.4. In creating these tables we wrote the entry for $(x + H) \cdot (x + H) = x^2 + H$ as $1 + x + H$, since these two cosets are identical. These tables are in fact identical with those of GF(2, $x^2 + x + 1$) (Tables 10.5 and 10.6) if the H is either suppressed or replaced by zero and the x is replaced by α.

TABLE 10.3. Addition in $\mathbb{Z}_2[x]/(x^2 + x + 1)$.

+	H	$1 + H$	$x + H$	$1 + x + H$
H	H	$1 + H$	$x + H$	$1 + x + H$
$1 + H$	$1 + H$	H	$1 + x + H$	$x + H$
$x + H$	$x + H$	$1 + x + H$	H	$1 + H$
$1 + x + H$	$1 + x + H$	$x + H$	$1 + H$	H

TABLE 10.4. Multiplication in $\mathbb{Z}_2[x]/(x^2+x+1)$.

·	H	$1+H$	$x+H$	$1+x+H$
H	H	H	H	H
$1+H$	H	$1+H$	$x+H$	$1+x+H$
$x+H$	H	$x+H$	$1+x+H$	$1+H$
$1+x+H$	H	$1+x+H$	$1+H$	$x+H$

TABLE 10.5. Addition in $GF(2, x^2+x+1)$.

+	0	1	α	$1+\alpha$
0	0	1	α	$1+\alpha$
1	1	0	$1+\alpha$	α
α	α	$1+\alpha$	0	1
$1+\alpha$	$1+\alpha$	α	1	0

TABLE 10.6. Multiplication in $GF(2, x^2+x+1)$.

·	0	1	α	$1+\alpha$
0	0	0	0	0
1	0	1	α	$1+\alpha$
α	0	α	$1+\alpha$	1
$1+\alpha$	0	$1+\alpha$	1	α

Thus our quotient G/H is none other than the Galois field $GF(2, x^2+x+1)$ in disguise. This laborious reconstruction of this Galois field has the advantage of mathematical rigor. Galois's construction of his fields presumed the existence of at least one zero for every irreducible polynomial. The only justification given by Galois for this assumption was the analogy with the complex numbers. The construction given here, on the other hand, makes no such assumptions.

The same method that was used above to construct $GF(2, x^2 + x + 1)$ can be applied in other situations to construct a wide variety of both old and new fields. Before describing this construction in its full generality, it is necessary to formalize some concepts. Two fields F and F' are said to be *isomorphic* if there is a function $f: F \to F'$ such that

1. $f(a_1) \neq f(a_2)$ for any distinct $a_1, a_2 \in F_0$
2. For each $b \in F'$ there is an $a \in F$ such that $f(a) = b$.
3. $f(a_1 + a_2) = f(a_1) + f(a_2)$ for any $a_1, a_2 \in F$.
4. $f(a_1 a_2) = f(a_1) f(a_2)$ for any $a_1, a_2 \in F$.

The function f is said to be an *isomorphism*. A comparison of this notion with that of the group isomorphism defined in Section 9.3 leads to the conclusion that every field isomorphism of F and F' constitutes both a group isomorphism of $(F, +)$ and $(F', +)$, on the one hand, and a group isomorphism of $(F^*, \cdot)$ and $(F'^*, \cdot)$, on the other. The function

$$f(r + s\alpha) = (r + sx) + H$$

constitutes an isomorphism of $GF(2, x^2 + x + 1)$ and the above-constructed quotient G/H. Properties 1 and 2 are clearly satisfied, whereas properties 3 and 4 follow from an examination of the above tables and the arguments:

$$\begin{aligned} f(r_1 + s_1\alpha + r_2 + s_2\alpha) &= f[(r_1 + r_2) + (s_1 + s_2)\alpha] \\ &= (r_1 + r_2) + (s_1 + s_2)x + H \\ &= (r_1 + s_1 x + H) + (r_2 + s_2 x + H) \\ &= f(r_1 + s_1\alpha) + f(r_2 + s_2\alpha) \end{aligned}$$

and, since $\alpha^2 = \alpha + 1$ and $x^2 + H = x + 1 + H$,

$$\begin{aligned} f((r_1 + s_1\alpha)(r_2 + s_2\alpha)) &= f[r_1 r_2 + (r_1 s_2 + r_2 s_1)\alpha + s_1 s_2 \alpha^2] \\ &= f[(r_1 r_2 + s_1 s_2) + (r_1 s_2 + r_2 s_1 + s_1 s_2)\alpha] \\ &= (r_1 r_2 + s_1 s_2) + (r_1 s_2 + r_2 s_1 + s_1 s_2)x + H \\ &= r_1 r_2 + (r_1 s_2 + r_2 s_1)x + s_1 s_2 x^2 + H \\ &= (r_1 + s_1 x)(r_2 + s_2 x) + H \\ &= (r_1 + s_1 x + H) \cdot (r_2 + s_2 x + H) \\ &= f(r_1 + s_1\alpha) \cdot f(r_2 + s_2\alpha) \end{aligned}$$

We now generalize this procedure to arbitrary fields and irreducible polynomials, first addressing the issue of addition, and only later dealing with multiplication.

10.2 The Rigorous Construction of Fields

If $P(x)$ is any polynomial over the field F, then we denote by $(P(x))$ the set of all the polynomials of $F[x]$ that are divisible by $P(x)$. The set $(P(x))$ is a normal subgroup of $(F[x], +)$, and hence it determines a quotient group $F[x]/(P(x))$. The cosets of this group all have the unwieldy form $M(x) + (P(x))$, and it is convenient to replace this clumsy expression by

$$[M(x)] = M(x) + (P(x))$$

It may be easily verified that in terms of this notation the addition of cosets assumes the form

$$[M(x)] + [N(x)] = [M(x) + N(x)] \tag{4}$$

The following proposition makes it easy to visualize the additive group $F[x]/(P(x))$:

■ **PROPOSITION 10.8.** *Let F be a field, and let $P(x)$ be a polynomial of degree d over F. Then*

$$F[x]/(P(x)) = \{[R(x)] \mid R(x) \in F[x] \text{ and either degree}(R(x)) < d \text{ or } R(x) = 0\}$$

Proof. By definition, every element of $F[x]/(P(x))$ has the form $[M(x)]$ for some polynomial $M(x)$ over F. Now $M(x)$ divided by $P(x)$ yields

$$M(x) = Q(x)P(x) + R(x)$$

with $R(x)$ being either the zero polynomial or else degree$(R(x)) < d$. Since

$$M(x) - R(x) = Q(x)P(x)$$

which is divisible by $P(x)$, it follows that $[M(x)] = [R(x)]$. In other words, every element of the quotient structure $F[x]/(P(x))$ can be represented in the form $[R(x)]$, where $R(x)$ is either the zero polynomial or else has degree $< d$.

Suppose that $R(x)$ and $R'(x)$ are two polynomials of degree $< d$ such that

$$[R(x)] = [R'(x)]$$

Then the difference $R(x) - R'(x)$ is a polynomial of degree $< d$ that is divisible by the polynomial $P(x)$ of degree d. Clearly $R(x) - R'(x)$ must be the zero polynomial; in other words,

$$R(x) = R'(x)$$

Hence we have shown that the nonzero elements of $F[x]/(P(x))$ are in a one-to-one correspondence with the polynomials of degree $<d$ over F. ∎

Accordingly

$$\mathbb{Z}_2[x]/(x^2 + x + 1) = \{[0], [1], [x], [1 + x]\}$$
$$\mathbb{Z}_2[x]/(x^2 + 1) = \{[0], [1], [x], [1 + x]\}$$
$$\mathbb{Z}_3[x]/(x^2 + 1) = \{[0], [1], [2], [x], [1+x], [2+x], [2x], [1+2x], [2+2x]\}$$
$$\mathbb{R}[x]/(x^2 + 1) = \{[a + bx] \mid a, b \in \mathbb{R}\}$$

Some puzzlement may be caused by the similarity between $\mathbb{Z}_2[x]/(x^2 + x + 1)$ and $\mathbb{Z}_2[x]/(x^2 + 1)$, since both have Table 10.3 as their addition tables. Indeed, as additive groups, these two structures are isomorphic. It is only when multiplication is also brought into play that the difference between them becomes evident. The additive group $F[x]/(P(x))$ is endowed with an operation of multiplication by the following definition:

$$(Q(x) + (P(x))) \cdot (R(x) + (P(x))) = Q(x)R(x) + (P(x))$$

or, in terms of the bracket notation for cosets,

$$[Q(x)] \cdot [R(x)] = [Q(x)R(x)] \quad (5)$$

The fact that this muliplication is unambiguous follows from an argument analogous to that used in the last paragraph of the previous section (Exercise 24). Thus, in $\mathbb{Z}_2[x]/(x^2 + x + 1)$,

$$[1 + x] \cdot [1 + x] = [(1 + x)^2] = [1 + 2x + x^2] = [x]$$

whereas, in $\mathbb{Z}_2[x]/(x^2 + 1)$,

$$[1 + x] \cdot [1 + x] = [(1 + x)^2] = [1 + 2x + x^2] = [0]$$

One more concept is necessary for the formulation of this section's main theorem. If F is a field and E is a subset of F that also constitutes a field with respect to the arithmetical operations it inherits from F, then E is a *subfield* of F and F is an *extension* of E. Thus the complex numbers are an extension of the reals which, in turn, are an extension of the rational numbers. Similarly every Galois field $GF(p, P(x))$ is an extension of $\mathbb{Z}_p$. We will use this terminology even when F only contains a subfield that is isomorphic to E rather than E itself. In general, isomorphic fields will be identified.

■ **THEOREM 10.9.** *If F is any field and $P(x)$ is any irreducible polynomial over F, then $F[x]/(P(x))$ is a field extension of F that contains a zero of $P(x)$.*

Proof. We first demonstrate that $F[x]/(P(x))$ is indeed a field. Since the addition and multiplication of the elements of $F[x]/(P(x))$ are given by (4) and (5), it follows that these operations satisfy all but the last of the requirements in Section 6.1 simply because the usual addition and multiplication of polynomials also satisfy those requirements. Thus the additive and multiplicative identities of $F[x]/(P(x))$ are

$$[0] \quad \text{and} \quad [1]$$

and, by way of example, commutativity holds for multiplication in $F[x]/(P(x))$ because

$$[Q(x)] \cdot [R(x)] = [Q(x)R(x)] = [R(x)Q(x)] = [R(x)] \cdot [Q(x)]$$

However, since the multiplicative inverse of a polynomial is not a polynomial, more work is required to demonstrate that the nonzero elements of $F[x]/(P(x))$ possess multiplicative inverses. The argument we give here is a slight modification of the proof of Lemma 7.1.

Let $Q(x)$ be any polynomial over F that is not in $(P(x))$; that is, $Q(x)$ is a polynomial that is not divisible by $P(x)$. Since $P(x)$ is irreducible, this is tantamount to saying that $P(x)$ and $Q(x)$ are relatively prime. Hence there exist $A(x), B(x) \in F[x]$ such that

$$A(x)Q(x) + B(x)P(x) = 1$$

Passing on to cosets, we conclude that

$$[A(x)] \cdot [Q(x)] = [A(x)Q(x)] = [1 - B(x)P(x)] = [1]$$

the justification for the last equality being that $1 - B(x)P(x)$ and 1 differ by an element of $(P(x))$, and hence they belong to the same coset of $(P(x))$. Since $[1]$ is the multiplicative identity of $F[x]/(P(x))$, it follows that $[A(x)]$ is the multiplicative inverse of $[Q(x)]$, and $F[x]/(P(x))$ is indeed a field.

It is clear from the definitions of addition and multiplication in $F[x]/(P(x))$ that the collection of cosets

$$F' = \{[r] \mid r \in F\}$$

constitutes a subfield of $F[x]/(P(x))$ that is isomorphic to F, the isomorphism $f: F \to F'$ being defined by

$$f(r) = [r]$$

Hence $F[x]/(P(x))$ is indeed an extension of F. Finally, we show that the coset $[x]$ is a zero of the polynomial $P(x)$ over the field $F[x]/(P(x))$. If

$$P(x) = a_0 x^n + a_1 x^{n-1} + \cdots + a_n$$

is a polynomial over F, then as a polynomial over $F[x]/(P(x))$, it should be written as

$$P(y) = [a_0] y^n + [a_1] y^{n-1} + \cdots + [a_n]$$

The reason the variable x of $P(x)$ needs to be replaced by y is that $[x]$ has become an element of the field $F[x]/(P(x))$. If the variable y of $P(y)$ is now replaced by this element $[x]$ of $F[x]/(P(x))$, then

$$\begin{aligned} P([x]) &= [a_0] \cdot [x]^n + [a_1] \cdot [x]^{n-1} + \cdots + [a_n] \\ &= [a_0 x^n] + [a_1 x^{n-1}] + \cdots + [a_n] \\ &= [a_0 x^n + a_1 x^{n-1} + \cdots + a_n] \\ &= [P(x)] = [0] \end{aligned}$$

In other words, the element $[x]$ of $F[x]/(P(x))$ is a zero of the polynomial $P(y)$, which of course is identical with the polynomial $P(x)$. ∎

Among other things, this theorem justifies Galois's assertion (Section 7.1) that every polynomial that is irreducible over $\mathbb{Z}_p$ has a zero which we called its Galois imaginary. Being mortals ourselves, we will not comment on why Galois could make such an assertion without falling into the pits where most unjustified assumptions lead.

Before this theorem is applied to the creation of some new fields, we will show how it can be used to give a rigorous construction of the complex numbers. The polynomial $x^2 + 1$ is irreducible over $\mathbb{R}$, and hence, by the above theorem, $\mathbb{R}[x]/(x^2 + 1)$ is a field that contains a subfield isomorphic to $\mathbb{R}$ and in which $[x]$ satisfies the equation

$$[x]^2 + [1] = [x^2 + 1] = [0]$$

Since the function $f(r) = [r]$ is an isomorphism of $\mathbb{R}$ onto a subfield of $\mathbb{R}[x]/(x^2 + 1)$, we may identify $[r]$ with r. In addition let us label $[x]$ by i, so the typical element of $\mathbb{R}[x]/(x^2 + 1)$ is

$$[a + bx] = [a] + [b] \cdot [x] = a + bi, \qquad a, b \in \mathbb{R}$$

where

$$i^2 = [x]^2 = -[1] = -1$$

It is now clear that $\mathbb{R}[x]/(x^2 + 1)$ is isomorphic to the complex number system. This might be the place to note that this application is in fact the origin of Theorem 10.9. In his paper of 1847 Cauchy used this very approach to justify the existence of complex numbers.

The Galois fields of Chapter 7 can also be regarded as a special case of this construction. If $P(x)$ is an irreducible polynomial of degree v over Z_p, then, according to Proposition 10.8, the elements of the field $\mathbb{Z}_p[x]/(P(x))$ all have the form

$$a_0 + a_1[x] + a_2[x]^2 + \cdots + a_{v-1}[x]^{v-1}$$

which, when $[x]$ is interpreted as the Galois imaginary i, is identical with expression (2) of Chapter 7. When this identification between the elements of $\mathbb{Z}_p[x]/(P(x))$ and $\text{GF}(p, P(x))$ is carried out, the arithmetical operations of the one are also identical with those of the other. In other words, $\mathbb{Z}_p[x]/(P(x))$ and $\text{GF}(p, P(x))$ are isomorphic fields. This observation is a special case of the following very general and very strong theorem whose proof falls outside the scope of this text. The *order* of a field is the number of elements it contains.

■ **THEOREM 10.10.** *Every finite field has order p^n for some prime p and some positive integer n. Given any such p and n, there is a field of order p^n, and any two fields of the same finite order are isomorphic.*

Consider next the field $F = \text{GF}(2, x^2 + x + 1)$, where α is the Galois imaginary associated with $x^2 + x + 1$:

$$F = \{0, 1, \alpha, 1 + \alpha\} \qquad \text{where } \alpha^2 = \alpha + 1$$

The quadratic polynomial $x^2 + x + \alpha$ has no zeros in F and is therefore irreducible over it. In accordance with Theorem 10.9,

$$F[x]/(x^2 + x + \alpha)$$

is a field. If we write λ for $[x]$ and replace $[r]$ by r for all $r \in F$, the elements of this field are

$[0] = 0$ $\quad\quad$ $[1] = 1$ $\quad\quad$ $[\alpha] = \alpha$
$[1+\alpha] = 1+\alpha$ $\quad\quad$ $[x] = \lambda$ $\quad\quad$ $[1+x] = 1+\lambda$
$[\alpha+x] = \alpha+\lambda$ $\quad\quad$ $[1+\alpha+x] = 1+\alpha+\lambda$ $\quad\quad$ $[\alpha x] = \alpha\lambda$
$[1+\alpha x] = 1+\alpha\lambda$ $\quad\quad$ $[\alpha+\alpha x] = \alpha+\alpha\lambda$ $\quad\quad$ $[1+\alpha+\alpha x] = 1+\alpha+\alpha\lambda$
$[(1+\alpha)x] = \lambda+\alpha\lambda$ $\quad\quad$ $[1+(1+\alpha)x] = 1+\lambda+\alpha\lambda$
$[\alpha+(1+\alpha)x] = \alpha+\lambda+\alpha\lambda$ $\quad\quad$ $[1+\alpha+(1+\alpha)x] = 1+\alpha+\lambda+\alpha\lambda$

The elements of this field are to be added and multiplied as indicated by the following examples:

$$(1 + \alpha\lambda) + (1 + \alpha + \lambda + \alpha\lambda) = 2 + \alpha + \lambda + 2\alpha\lambda = \alpha + \lambda$$

Bearing in mind that $\alpha^2 = \alpha + 1$ and $\lambda^2 = \lambda + \alpha$, we obtain

$$(1 + \alpha\lambda)(1 + \alpha + \lambda + \alpha\lambda) = 1 + \alpha + \lambda + \alpha\lambda + \alpha\lambda + \alpha^2\lambda + \alpha\lambda^2 + \alpha^2\lambda^2$$
$$= 1 + \alpha + \lambda + (1 + \alpha)\lambda$$
$$\quad + \alpha(\lambda + \alpha) + (1 + \alpha)(\lambda + \alpha)$$
$$= 1 + \alpha + \lambda + \lambda + \alpha\lambda + \alpha\lambda$$
$$\quad + \alpha^2 + \lambda + \alpha + \alpha\lambda + \alpha^2$$
$$= 1 + \lambda + \alpha\lambda$$

Since this field has the same number of elements as $GF(2, x^4 + x + 1)$, it follows from the unproved Theorem 10.10 that they should be isomorphic. This is borne out by the computation

$$\lambda^4 = (\lambda^2)^2 = (\lambda + \alpha)^2 = \lambda^2 + \alpha^2 = \lambda + \alpha + \alpha + 1 = \lambda + 1$$

which shows that λ is a zero of the polynomial $x^4 + x + 1$ over $\mathbb{Z}_2$. In other words, we can think of λ as the Galois imaginary of the field $GF(2, x^4 + x + 1)$.

Let us examine some extensions of the rationals $\mathbb{Q}$. Since there is no rational number whose square equals 2, it follows that the polynomial $x^2 - 2$ is irreducible over $\mathbb{Q}$. Consequently Theorem 10.9 yields

$$\mathbb{Q}[x]/(x^2 - 2)$$

as a new field. In accordance with Proposition 10.8, the elements of this field all have the form

$$[a] + [b][x]$$

10.2 The Rigorous Construction of Fields

where a and b are rational numbers and $[x]$ is a quantity such that

$$[x]^2 - [2] = [x^2 - 2] = [0]$$

Since $[x]$ behaves just like a square root of 2, we denote it by the formal symbol $\sqrt{2}$. This formalism notwithstanding, note that

$$(\sqrt{2})^2 = [x]^2 = [2] = 2$$

in this field. If we persist in identifying $[r]$ with r for each rational number r, then the elements of this new field have the form

$$a + b\sqrt{2}, \qquad a, b \in \mathbb{Q}$$

The addition and multiplication of the elements of this field are given by

$$(a_1 + b_1\sqrt{2}) + (a_2 + b_2\sqrt{2}) = (a_1 + a_2) + (b_1 + b_2)\sqrt{2}$$

and

$$(a_1 + b_1\sqrt{2})(a_2 + b_2\sqrt{2}) = (a_1 a_2 + 2b_1 b_2) + (a_1 b_2 + a_2 b_1)\sqrt{2}$$

Note that

$$(a + b\sqrt{2})^{-1} = \frac{a}{a^2 - 2b^2} + \frac{-b}{a^2 - 2b^2}\sqrt{2}$$

Let us denote this new field $\mathbb{Q}[x]/(x^2 - 2)$ by F_1. The field F_1 can itself serve as the ground field for the construction of another field. Consider, for example, the polynomial $x^2 - 3$. This polynomial is irreducible over F_1. To justify this claim, it suffices to show that the equation

$$(x + y\sqrt{2})^2 - 3 = 0$$

has no solution wherein both x and y are rational. However, this equation simplifies to

$$x^2 + 2y^2 + 2xy\sqrt{2} = 3$$

or

$$\sqrt{2} = \frac{3 - x^2 - 2y^2}{2xy}$$

which cannot have rational solutions, since $\sqrt{2}$ is known to be an irrational number.

The quadratic $x^2 - 3$ being irreducible over F_1, it yields yet a new field

$$F_1[x]/(x^2 - 3)$$

whose typical element, when $[x]$ is symbolized by $\sqrt{3}$ is

$$(a_1 + b_1\sqrt{2}) + (a_2 + b_2\sqrt{2})\sqrt{3}$$

If we abbreviate $\sqrt{2}\sqrt{3}$ to the symbol $\sqrt{6}$, then all the elements of $F_1[x]/(x^2 - 3)$ can be written in the form

$$a + b\sqrt{2} + c\sqrt{3} + d\sqrt{6}, \qquad a, b, c, d \in \mathbb{Q}$$

It is clear that Theorem 10.9 can be used to construct a myriad of new fields. The general theory of these fields and their classification falls outside the scope of this book.

We conclude this section with a warning about a possible source of confusion. While the powerful Theorem 10.9 resembles the Fundamental Theorem of Algebra, which asserts the existence of complex zeros to every complex polynomial, the two theorems are distinct. The zeros whose existence is guaranteed by Theorem 10.9 need not belong to the ground field, as is exemplified by the polynomial $x^2 + x + 1$ over $\mathbb{Z}_2$. The Fundamental Theorem of Algebra, above and beyond asserting the mere existence of zeros of complex polynomials also places them back in the ground field, which Theorem 10.9 does not do.

EXERCISES 10.2

For each pair F and $P(x)$ in Exercises 1–8, describe a subfield of the complex numbers that is isomorphic to $F[x]/(P(x))$.

1. $F = \mathbb{Q}$, $P(x) = x^2 - 5$
2. $F = \mathbb{Q}$, $P(x) = x^3 - 2$
3. $F = \mathbb{Q}$, $P(x) = x^2 + 1$
4. $F = \mathbb{Q}$, $P(x) = x^2 + 25$
5. $F = \mathbb{Q}$, $P(x) = x^2 + x + 1$
6. $F = \mathbb{Q}[x]/(x^2 - 2)$, $P(x) = x^2 - 5$
7. $F = \mathbb{Q}[x]/(x^2 - 2)$, $P(x) = x^2 + 1$
8. $F = \mathbb{R}$, $P(x) = x^2 + 5$

For each of the fields in Exercises 9–12, find a field that is described in Chapter 7 and is isomorphic to it.

9. $\mathbb{Z}_2[x]/(x^2 + x + 1)$
10. $\mathbb{Z}_2[x]/(x^3 + x^2 + 1)$
11. $\mathbb{Z}_2[x]/(x^4 + x^3 + x^2 + x + 1)$
12. $\mathbb{Z}_3[x]/(x^2 + x + 2)$

Explain why each of the fields in Exercises 13–16 is finite.

13. $\mathbb{Z}_2[x]/(x^2 + x + 1)$
14. $\mathbb{Z}_2[x]/(x^3 + x + 1)$
15. $\mathbb{Z}_2[x]/(x^4 + x^3 + x^2 + x + 1)$
16. $\mathbb{Z}_5[x]/(x^2 + 4x + 2)$

17. Prove that the set of real numbers

$$\{a + b\sqrt{2} + c\sqrt{3} \mid a, b, c \in \mathbb{Q}\}$$

does not constitute a subfield of the real numbers (with respect to the usual arithmetical operations).

18. Prove that the set of real numbers of the form

$$\{a + b\sqrt{7} + c\sqrt{11} \mid a, b, c \in \mathbb{Q}\}$$

does not constitute a subfield of the real numbers (with respect to the usual arithmetical operations).

19. What is the multiplicative inverse of the element $2 + 3[x]$ in the field $\mathbb{Q}[x]/(x^2 - 5)$?
20. What is the multiplicative inverse of the element $3 + 2[x]$ in the field $\mathbb{Q}[x]/(x^2 - 7)$?
21. What is the multiplicative inverse of the element $1 + [x]$ in the field $\mathbb{Q}[x]/(x^2 + 5)$?
22. Let r be any rational number such that $\sqrt{r}$ is not rational. Find a formula for the multiplicative inverse of $a + b[x]$ in $Q[x]/(x^2 - r)$.
23. Suppose that $P(x)$ is a polynomial of degree d over a field F. Prove that there exists an extention F' of F such that, counting multiplicities, $P(x)$ has d zeros in F'.
24. Prove that the multiplication of the elements of $F[x]/(P(x))$ defined in (5) is unambiguous.
25. Show that if $P(x)$ is a reducible element of $F[x]$, then the multiplication defined in (5) does not yield a field.

10.3 GALOIS GROUPS AND THE RESOLVABILITY OF EQUATIONS (an informal discussion)

We would like to conclude this chapter with a brief account of how the young Galois settled the question of the algebraic resolvability of equations. Because of the introductory nature of this text, such an account must of necessity be superficial, and a more complete exposition of the theory can be found in many graduate texts.

Briefly put, if $P(x)$ is an irreducible polynomial, with either real or complex coefficients, then Galois associated a certain group of permutations with the equation $P(x) = 0$ and then proved that an appropriate analysis of the group yields the answer as to whether or not this equation is algebraically resolvable. We will now discuss both the group and its analysis.

Let $P(x)$ be an irreducible polynomial with integer coefficients. The Galois group of the polynomial equation $P(x) = 0$ is a group of permutations of the roots of the equation (recall that the existence of these roots is guaranteed by the Fundamental Theorem of Algebra of Section 3.3) that enjoys two properties:

1. Every rational expression in the roots that is invariant under all the permutations in the group has a rational expression in the coefficients of the equation.
2. Conversely, every rational expression in the roots that is also a rational expression in the coefficients is necessarily invariant under all the permutations of the group.

Consider, for example, the cyclotomic equation $x^4 + x^3 + x^2 + x + 1 = 0$. Since $x^5 - 1 = (x - 1)(x^4 + x^3 + x^2 + x + 1)$, it follows that the roots of this equation are $\varepsilon, \varepsilon^2, \varepsilon^3, \varepsilon^4$, where ε is the first fifth root of unity. It is known that the Galois group of this equation is the permutation group of order 4 generated by the cycle $\sigma = (\varepsilon \; \varepsilon^2 \; \varepsilon^4 \; \varepsilon^3)$. We will now illustrate the meaning of properties 1 and 2 above by investigating this group's effect on some rational expressions in these roots. Suppose first that a, b, c, d are integers such that the expression

$$\phi = (\varepsilon)^a (\varepsilon^2)^b (\varepsilon^4)^c (\varepsilon^3)^d$$

is invariant under σ (and so it is necessarily also invariant under all the permutations in the Galois group, since in this case they are all powers of σ). This means that

$$(\varepsilon)^a (\varepsilon^2)^b (\varepsilon^4)^c (\varepsilon^3)^d = (\varepsilon^2)^a (\varepsilon^4)^b (\varepsilon^3)^c (\varepsilon)^d \tag{6}$$

and hence

$$a + 2b + 4c + 3d \equiv 2a + 4b + 3c + d \quad (\text{mod } 5) \tag{7}$$

Subtraction surprisingly yields

$$0 \equiv a + 2b + 4c + 3d \quad (\text{mod } 5) \tag{8}$$

From this it immediately follows that

$$\phi = \varepsilon^{a + 2b + 4c + 3d} = 1$$

Thus, as required by property 1, the invariance of ϕ under the permutations of the Galois group was sufficient to guarantee its rationality. Suppose now that we only know ϕ to be rational. Since ϕ is necessarily a fifth root of unity, it follows that $\phi = 1$, and so (8) holds. Equation (8) entails equation (7), and this one implies (6). In other words, from the mere assumption of the rationality of ϕ, it was possible to prove its invariance under σ and all the elements of the Galois group.

10.3 Galois Groups and the Resolvability of Equations

Galois himself gives two examples of these groups. If $x_1, x_2, \ldots, x_n$ denote the roots of the general equation

$$x^n + a_1 x^{n-1} + \cdots + a_{n-1} x + a_n = 0$$

then the Galois group of this equation consists of all the permutations of these roots. The Fundamental Theorem of Symmetric Polynomials, briefly mentioned in Section 6.4, asserts that every symmetric rational polynomial of the variables $x_1, x_2, \ldots, x_n$ can be expressed as a function of the elementary symmetric polynomials of these variables. By Theorem 6.12, when these variables denote the roots of the above equation, their elementary symmetric functions equal $(-1)^k a_k$, $k = 1, 2, \ldots, n$. Thus condition 1 is satisfied. That condition 2 is satisfied is harder to show, and we will not do so here.

The second example Galois gives is that of the cyclotomic equation $x^p - 1 = 0$, where p is a prime number. Since the polynomial $x^n - 1$ is never irreducible, it is necessary to divide out the factor $x - 1$, after which we get

$$x^{p-1} + x^{p-2} + \cdots + x + 1 = 0 \tag{9}$$

which can be proved to be irreducible (for prime p). The Galois group of this equation is the cyclic group $\langle \sigma \rangle$, where

$$\sigma = (\zeta \ \zeta^k \ \zeta^{k^2} \ \zeta^{k^3} \cdots \zeta^{k^{p-2}})$$

ζ being any primitive pth root of unity and k being any primitive root modulo p. In the special case of $p = 17$ where k was taken as 3, we get

$$\sigma = (\zeta \ \zeta^3 \ \zeta^9 \ \zeta^{10} \ \zeta^{13} \ \zeta^5 \ \zeta^{15} \ \zeta^{11} \ \zeta^{16} \ \zeta^{14} \ \zeta^8 \ \zeta^7 \ \zeta^4 \ \zeta^{12} \ \zeta^2 \ \zeta^6)$$

Since σ is in general a cyclic permutation of $p - 1$ elements, it follows that the Galois group of equation (9) is isomorphic to $(\mathbb{Z}_{p-1}, +)$.

The analysis of the Galois group that determines the algebraic resolvability of its originating equation has the form of a recursive procedure. If the group G has prime order p (so that it is isomorphic to $(\mathbb{Z}_p, +)$), then the equation is algebraically resolvable. Next, if G has composite order and it contains no proper normal subgroups, then the originating equation is *not* algebraically resolvable. If neither of these conditions hold, then G is a group of a composite order with a proper normal subgroup, say, H. Now apply the same analysis to both H and G/H. If at any time we encounter a group of composite order without a proper normal subgroup, then the originating equation is not algebraically resolvable. Since at each stage the orders of H and G/H are smaller than the order of G, this process is bound to terminate. If all the groups of composite order encountered in this procedure have proper normal subgroups, the originating equation is algebraically resolvable. Otherwise, it is not so resolvable.

As an example we consider the cyclotomic equation $x^{13} - 1 = 0$, whose irreducible part is

$$x^{12} + x^{11} + x^{10} + \cdots + x + 1 = 0 \tag{10}$$

and whose Galois group is isomorphic to $(\mathbb{Z}_{12}, +)$. The order of $(\mathbb{Z}_{12}, +)$ is composite, and $H = \{0, 6\}$ is a subgroup of it. Since $(\mathbb{Z}_{12}, +)$ is an abelian, group H is necessarily normal, and since $(\mathbb{Z}_{12}, +)$ is cyclic, so is $G_1 = (\mathbb{Z}_{12}, +)/H$. Now H has order 2 which is a prime, and so we are done with it. On the other hand, G_1 has the composite order 6, and so, by Theorem 9.14, G_1 is isomorphic to $(\mathbb{Z}_6, +)$. Consequently G_1 itself has a (necessarily normal) subgroup G_2 of order 3. Thus G_2 has order 3, and G_1/G_2 has order 2, both being prime numbers. It follows from Galois's theory that equation (10) is indeed algebraically resolvable, which resolution was first investigated by Gauss.

Let us examine the general quintic equation. As noted above, its Galois group is isomorphic to S_5 which has the composite order $5! = 120$. The symmetric group S_5 has the group A_5 of all the even permutations of $\{1, 2, 3, 4, 5\}$ as a subgroup, and it follows from Proposition 10.3 that A_5 is a normal subgroup of S_5 such that S_5/A_5 has order 2. However, while A_5 has a plenitude of proper subgroups, we will now show that none of these subgroups is normal. Consequently the general quintic equation is not resolvable by radicals, a fact that had of course already been proved by Abel.

■ **PROPOSITION 10.11.** *The group A_5 has no proper normal subgroups.*

Proof. Suppose that H is a normal subgroup of A_5 that contains some nonidentity element. We will show that H necessarily equals A_5 by demonstrating the following two statements:

1. H contains a 3-cycle.
2. H contains all the 3-cycles of A_5.

Since we already know that every even permutation is expressible as the composition of 3-cycles (Exercise 8.4.13), it will follow that $H = A_5$.

Proof of Part 1. Since all the elements of A_5 are even permutations, it follows that their disjoint cycle decompositions consist either of a single 5-cycle, a single 3-cycle, or a pair of transpositions such as (1 2)(3 4). If H contains, a 5-cycle, say, (1 2 3 4 5), then, because H is a normal subgroup of A_5, it must also contain the element

$$(1\ 2\ 3\ 4\ 5)^{-1}[(1\ 2\ 3)(1\ 2\ 3\ 4\ 5)(1\ 2\ 3)^{-1}]$$
$$= (5\ 4\ 3\ 2\ 1)(1\ 2\ 3)(1\ 2\ 3\ 4\ 5)(3\ 2\ 1) = (1\ 3\ 5)$$

If H contains a pair of transpositions, say, (1 2)(3 4), then, because H is a normal subgroup, it must also contain the element

$$(1\ 2)(3\ 4)\{[(2\ 5)(3\ 4)](1\ 2)(3\ 4)[(2\ 5)(3\ 4)]^{-1}\}$$
$$= (1\ 2)(3\ 4)(2\ 5)(3\ 4)(1\ 2)(3\ 4)(3\ 4)(2\ 5) = (1\ 5\ 2)$$

Thus, if H contains any nonidentity elements, then it necessarily contains some 3-cycle.

Proof of Part 2. By part 1 we know that H contains some 3-cycle, say, (1 2 3). Moreover, if $(a\ b\ c)$ is any 3-cycle of A_5, then by Exercise 8.2.23,

$$(1\ 2\ 3\ 4\ 5)(a\ b\ c)(1\ 2\ 3\ 4\ 5)^{-1} = (a+1\ b+1\ c+1) \quad \text{(addition mod 5)}$$

Hence, since $(1\ 2\ 3\ 4\ 5) \in A_5$ and H is normal in A_5, it follows that H must contain the 3-cycles (1 2 3), (2 3 4), (3 4 5), (4 5 1), (5 1 2) as well as their inverses. Since

$$[(1\ 2)(3\ 4)](1\ 2\ 3)[(1\ 2)(3\ 4)]^{-1} = (1\ 2)(3\ 4)(1\ 2\ 3)(3\ 4)(1\ 2) = (1\ 4\ 2)$$

it follows for similar reasons that H must also contain the 3-cycles (1 4 2), (2 5 3), (3 1 4), (4 2 5), (5 3 1), and their inverses. As these exhaust all the 20 3-cycles of A_5, the proof is complete. ∎

The above proposition generalizes to the statement that A_n is simple for each $n \geqslant 5$. This concludes our attempt at an elementary description of Galois theory. It remains only to say a few words about the subsequent evolution of group theory.

We saw that groups caught the attention of mathematicians because they provided the key to the question of which polynomials equations are algebraically resolvable. Thus the Galois group of an equation contains all the information that is required to decide on its algebraic resolvability. Mathematicians subsequently went on to try to classify all these new structures, and eventually a very clear-cut and apparently also very difficult question crystallized. *Is every finite group necessarily isomorphic to the Galois group of some equation with integer coefficients?* As of the writing of this text, the answer to this question is still unknown.

Groups that contain no proper normal subgroups were seen to play a key role in Galois theory and are called *simple* groups. The commutative simple groups are the cyclic groups of prime order. The group A_5, which is the subject of Proposition 10.11, is the smallest of the noncommutative simple groups. The appellation "simple" is not to be taken literally. These groups have in fact very complicated structures. The joint efforts of hundreds of group theorists resulted recently in the complete classification of the finite simple groups. This monumental work occupies about 14,000 pages of mathematical publications. The last finite simple group to be identified is affectionately known as *the monster*. Its order is

$$2^{46} \cdot 3^{20} \cdot 5^9 \cdot 7^6 \cdot 11^2 \cdot 13^2 \cdot 17 \cdot 19 \cdot 23 \cdot 29 \cdot 31 \cdot 41 \cdot 47 \cdot 59 \cdot 71 \approx 10^{54}$$

The monster happens to be the group of vertex symmetries of a solid that resides in a space of 196,883 dimensions. Strangely enough, the number 196,884 features in some applications of non-Euclidean geometry to number theory, but that's another story.

EXERCISES 10.3

1. Prove that no commutative group of composite (or infinite) order is simple.
2. Suppose that G is a subgroup of S_n. Prove that if G is simple and $o(G) > 2$, then $G \subset A_n$.
3. Prove that the group of vertex symmetries of the regular octahedron (Figure 9.6) is not simple.
4. Prove that the group of vertex symmetries of the cube (Figure 9.5) is not simple.
5. Suppose that G is a finite simple group and $1_G \neq a \in G$. Prove that every element of G can be expressed as a product of elements of the conjugacy class $C(a)$. (Compare Exercises 10.1.24–30.)
6. Prove that A_6 is a simple group.

CHAPTER SUMMARY

We have shown that new groups can be obtained from old ones by the quotient operation. This operation was applied to additive groups of polynomials to produce a host of new and old fields. Finally, the notion of quotient groups permitted us to formulate Galois's criterion for the resolvability of algebraic equations.

Chapter Review Exercises

Mark the following true or false.

1. If K is the Klein 4-group, then KK contains more elements than K.
2. K is a normal subgroup of D_4.
3. $\mathbb{Z}_{59}^*$ has a subgroup that is isomorphic to K.
4. The number of distinct elements of $\mathbb{Z}_5[x]/(x^2 + 2x + 1)$ is 25.
5. $(\mathbb{Z}_5, +)$ is a simple group.
6. The complex numbers constitute an extention of the rational numbers.

New Terms

Coset multiplication	221
Extension of a field	236
Galois group	243
Isomorphism of fields	234
Monster	247
Normal subgroup	224
Quotient of groups	227
Simple group	247
Subfield	236

Supplementary Exercises

1. Write a computer script that will decide whether a given subgroup of some group is normal. If the answer is yes, write out a multiplication table for the quotient group.
2. Prove that the alternating group A_n is simple for $n \geqslant 5$.

CHAPTER 11

Topics in Elementary Group Theory

This chapter displays some of the methods and results of elementary group theory. Specifically, we demonstrate how many more groups can be constructed and classify, up to isomorphism, all the groups of orders $2p$ and p^2 for p prime.

11.1 THE DIRECT PRODUCT OF GROUPS

In this section we describe one of many methods for combining groups to produce new groups. If G and H are two groups, then their *direct product*, denoted by $G \times H$, has as its elements the set of all the ordered pairs

$$(g, h) \quad \text{where } g \in G, h \in H$$

The binary operation of $G \times H$ is defined by

$$(g, h)(g', h') = (gg', hh')$$

The associativity of this operation follows directly from the associativity of the group operations of G and H. The identity element of $G \times H$ is $(1_G, 1_H)$, and the inverse of (g, h) is the pair (g^{-1}, h^{-1}). Thus $\mathbb{Z}_2 \times \mathbb{Z}_2$ consists of the four pairs $\{(0,0), (1,0), (0,1), (1,1)\}$ where

$$(1, 1)(1, 0) = (1 + 1, 1 + 0) = (0, 1)$$

and

$$(1, 1)(1, 1) = (1 + 1, 1 + 1) = (0, 0)$$

Similarly $\mathbb{Z}_2 \times \mathbb{Z}_3$ consists of the six pairs $(0,0)$, $(0,1)$, $(0,2)$, $(1,0)$, $(1,1)$, and $(1,2)$ where

$$(1,1)(0,2) = (1+0, 1+2) = (1,0)$$

and

$$(1,2)(0,2) = (1+0, 2+2) = (1,1)$$

It is clear that if G and H are finite groups then $o(G \times H) = o(G)o(H)$. It is also easy to see that the function $f((g,h)) = (h,g)$ defines an isomorphism of $G \times H$ and $H \times G$. It is equally clear that the subgroups

$$G' = \{(g, 1_H) \mid g \in G\} \quad \text{and} \quad H' = \{(1_G, h) \mid h \in H\}$$

of $G \times H$ are isomorphic to G and H, respectively. In particular, $o((g, 1_H)) = o(g)$ and $o((1_G, h)) = o(h)$. The next proposition tells us how to determine the order of any element of $G \times H$.

■ **PROPOSITION 11.1.** *If g and h are elements of finite orders in the groups G and H, respectively, then $o((g,h))$ is the least common multiple of $o(g)$ and $o(h)$.*

Proof. Let k be the least common multiple of $o(g)$ and $o(h)$, and let $d = o((g,h))$. Since

$$(g,h)^k = (g^k, h^k) = (1_G, 1_H)$$

it follows that d divides k. Conversely, since

$$(1_G, 1_H) = 1_{G \times H} = (g,h)^d = (g^d, h^d)$$

it follows that $1_G = g^d$ and $1_H = h^d$. Thus d is divisible by both $o(g)$ and $o(h)$, and so, by Exercise 4.2.31, d is divisible by k. Hence $d = k$. ■

It follows from this proposition that $\mathbb{Z}_2 \times \mathbb{Z}_2$ is a group of order 4 in which every nonidentity element has order 2, and hence this group is isomorphic to K. Similarly the element $(1,2)$ of $\mathbb{Z}_2 \times \mathbb{Z}_3$ has order $2 \cdot 3 = 6$, and hence this group is isomorphic to $(\mathbb{Z}_6, +)$. On the other hand, the group $\mathbb{Z}_2 \times \mathbb{Z}_4$ has order 8, is commutative, contains no elements of order 8, and contains an element of order 4, namely $(0,1)$. This is enough information to justify the assertion that $\mathbb{Z}_2 \times \mathbb{Z}_4$ is not isomorphic to any of the previously encountered groups of order 8, namely the Quaternion group, D_4, $(\mathbb{Z}_8, +)$, and $\mathbb{Z}_2[x, \leqslant 2]$.

The preceding paragraph makes it clear that it would be useful to have some criteria for recognizing when a group is isomorphic to the direct product of some two other groups. This is now provided.

■ **PROPOSITION 11.2.** *Suppose that the finite group P contains two normal subgroups G and H such that*

$$G \cap H = \{1_p\} \quad \text{and} \quad o(P) = o(G)o(H)$$

then $P \cong G \times H$.

Proof. We begin by proving that the elements of G commute with elements of H. Thus suppose that $g \in G$ and that $h \in H$. Since G and H are normal in P, it follows that

$$ghg^{-1}h^{-1} = (ghg^{-1})h^{-1} \in HH = H$$

and

$$ghg^{-1}h^{-1} = g(hg^{-1}h^{-1}) \in GG = G$$

Since $G \cap H = \{1_p\}$, it follows that $ghg^{-1}h^{-1} = 1_p$, and hence

$$gh = hg \quad \text{for all } g \in G \text{ and } h \in H$$

We are now ready to prove the required isomorphism. Let $f((g, h)) = gh$. This is clearly a function from $G \times H$ into P. If $f((g, h)) = f((g', h'))$, then

$$gh = g'h' \quad \text{or} \quad g'^{-1}g = h'h^{-1}$$

However, $g'^{-1}g \in G$ and $h'h^{-1} \in H$, and hence, since $G \cap H = \{1_p\}$, we have

$$g'^{-1}g = h'h^{-1} = 1_p \quad \text{or} \quad g = g' \quad \text{and} \quad h = h'$$

It follows that f maps distinct elements of $G \times H$ to distinct elements of P. Since $o(P) = o(G)o(H) = o(G \times H)$, it follows that f does indeed match all the elements of $G \times H$ with those of P. Finally, making use of the above proved commutativity, we note that

$$f((g, h)(g', h')) = f((gg', hh')) = gg'hh' = ghg'h' = f((g, h))f((g', h'))$$

so that f is indeed an isomorphism. ■

As a consequence of Proposition 11.2, we show that if p and q are any two distinct prime numbers, then $(\mathbb{Z}_{pq}, +) \cong (\mathbb{Z}_p, +) \times (\mathbb{Z}_q, +)$. Let $P = (\mathbb{Z}_{pq}, +)$, $G = \langle p \rangle \cong (\mathbb{Z}_q, +)$, and $H = \langle q \rangle \cong (\mathbb{Z}_p, +)$. Then $P, G,$ and H satisfy the hypotheses of Proposition 11.2, and hence the desired conclusion follows. The next corollary illustrates a somewhat more complicated application of

Proposition 11.2. First, however, we note that it is clear that for any three groups G, H, J, $(G \times H) \times J \cong G \times (H \times J)$, this isomorphism being established by the function

$$f(((g, h), k)) = (g, (h, k))$$

Consequently we can unambigously write $G \times H \times J$ for $(G \times H) \times J$ and $G \times (H \times J)$. In particular, if k is any positive integer, then we denote the direct product of k copies of G by G^k.

■ **COROLLARY 11.3.** *If P is a finite group in which every nonidentity element has order 2, then there exists a nonnegative integer k such that $P \cong (\mathbb{Z}_2, +)^k$.*

Proof. We proceed by induction on the order of P, the conclusion being trivially valid when $o(P) \leq 2$. We therefore assume that $o(P) = n$ and that the proposition holds for all groups of order less than n.

Let H be a proper subgroup of P that is contained in no other proper subgroup of P, and let a be any element of P that is not in H. Then, since P is a commutative group (Exercise 9.2.29), $H' = H \cup (aH)$ is also a subgroup of P. Since H' contains both H and a, it follows from the maximality of H that $H' = P$. Thus H has index 2 in P, and so

$$o(P) = 2o(H) = o(\langle a \rangle)o(H).$$

Now from Proposition 11.2 we have $P \cong \langle a \rangle \times H \cong (\mathbb{Z}_2, +) \times H$. Since every element of H also has order 2 and H has order less than n, it follows from the induction hypothesis that $H \cong (\mathbb{Z}_2, +)^k$ for some k, and hence $P \cong (\mathbb{Z}_2, +) \times (\mathbb{Z}_2, +)^k = (\mathbb{Z}_2, +)^{k+1}$. ■

EXERCISES 11.1

1. Let G and H be any two finite groups. Prove that the sets

$$G' = \{(g, 1_H) \mid g \in G\} \quad \text{and} \quad H' = \{(1_G, h) \mid h \in H\}$$

   are normal subgroups of $G \times H$. Prove that $(G \times H)/G' \cong H$ and $(G \times H)/H' \cong G$.

2. Let p be a prime, and let G be a finite commutative group in which every nonidentity element has order p. Prove that $G \cong (\mathbb{Z}_p, +)^k$ for some nonnegative integer k.

3. Let p be a prime. Prove that every commutative group of order p^2 is isomorphic to either $(\mathbb{Z}_{p^2}, +)$ or $(\mathbb{Z}_p, +)^2$.

4. Suppose that m and n are relatively prime positive integers. Prove that $(\mathbb{Z}_{mn}, +) \cong (\mathbb{Z}_m, +) \times (\mathbb{Z}_n, +)$.

5. Prove that $(\mathbb{Z}_{209}, +) \times (\mathbb{Z}_{221}, +) \cong (\mathbb{Z}_{143}, +) \times (\mathbb{Z}_{323}, +)$.

A group is said to be decomposable if it is isomorphic to the direct product of two nontrivial groups. Otherwise, it is indecomposable.

6. Prove that $(\mathbb{Z}_n, +)$ is indecomposable if and only if $n = p^r$ for some prime p.
7. Prove that D_5 is indecomposable.
8. Prove that S_3 is indecomposable.
9. Prove that S_4 is indecomposable.
10. Prove that S_5 is indecomposable.
11. Prove that the Quaternion group is indecomposable.
12. Prove that if $1 \leqslant k \leqslant n$, then S_n contains a subgroup $P \cong S_k \times S_{n-k}$.
13. Prove that if G and H are finite groups, then $G \times H$ is cyclic if and only if both G and H are cyclic and their orders are relatively prime.
14. Let p be any prime. Prove that for every positive integer n there are at least n nonisomorphic groups of order p^n.
15. Prove that every finite group G of order greater than 2 has an automorphism that is distinct from the identity function.

11.2 MORE CLASSIFICATIONS

We begin by reviewing some information that was relegated to the exercises of Chapters 9 and 10.

Two elements a and b of the group G are said to be *conjugate* if there exists an element x such that $xax^{-1} = b$. The set of all the elements of G that are conjugate to a is denoted by $C(a)$ and is called the *conjugacy class of a*. Note that if a is conjugate to b, then

$$a = x^{-1}xax^{-1}x = x^{-1}bx = x^{-1}b(x^{-1})^{-1}$$

so that b is also conjugate to a. Moreover, if b is also conjugate to $c \in G$, say, $b = ycy^{-1}$, then

$$c = y^{-1}by = y^{-1}(xax^{-1})y = (y^{-1}x)a(y^{-1}x)^{-1}$$

so that a is also conjugate to c. These observations have the following consequence:

■ **LEMMA 11.4.** *If a and b are elements of the group G, then $C(a)$ and $C(b)$ are either identical or disjoint.*

Proof. Suppose that $C(a)$ and $C(b)$ are not disjoint, so that $c \in C(a) \cap C(b)$ for some $c \in G$. It then follows from the above observations that a and b, both being conjugate to c, are also conjugate to each other. However, if $d \in C(a)$, then d, being conjugate to a, is also conjugate to b so that $C(a) \subset C(b)$. By symmetry, $C(b) \subset C(a)$, and hence $C(a) = C(b)$. ■

The *centralizer* Z_a of the element a of the group G consists of all the elements x in G such that $xa = ax$. (See Exercises 9.4.43–51.) Note that Z_a always contains 1_G, a, as well as every power of a. If a commutes with every element of G so that $Z_a = G$, then $xax^{-1} = a$ for all $x \in G$ so that $C(a) = \{a\}$. In the Quaternion group (Table 9.3) $Z_1 = Z_d = G$, and $Z_x = \{1, d, x, x^{-1}\}$ for all other $x \in G$. The following proposition shows that there is a very strong relationship between centralizers and conjugacy classes:

■ **PROPOSITION 11.5.** *Let G be a group and $a \in G$. Then Z_a is a subgroup of G and $|C(a)| = [G:Z_a]$.*

Proof. If $x, y \in Z_a$, then

$$xya = xay = axy \quad \text{and} \quad x^{-1}a = x^{-1}axxx^{-1} = x^{-1}xax^{-1} = ax^{-1}$$

so that $xy, x^{-1} \in Z_a$. Hence Z_a is a subgroup of G.

To prove the second part, note that each of the following statements is equivalent to the next:

x and y belong to the same coset of Z_a in G,
$x^{-1}y \in Z_a$,
$x^{-1}ya = ax^{-1}y$,
$yay^{-1} = xax^{-1}$.

In other words, the two elements x and y belong to the same coset of Z_a if and only if they conjugate a to the same element. Equivalently, the elements x and y belong to distinct cosets of Z_a in G if and only if they conjugate a to distinct elements of $C(a)$. Either way, the cosets of Z_a have been matched up in a one-to-one fashion with the elements of $C(a)$ so that $|C(a)| = [G:Z_a]$. ■

Note that in the quaternion group $xax^{-1} = a$ or e according as $x \in \{1, a, d, e\}$ or not. Thus $|C(a)| = 2 = [G:Z_a]$, since we saw above that $Z_a = \{1, a, d, e\}$.

The *center* $Z(G)$ of the group G consists of all the elements of G that commute with every element of G. In other words,

$$Z(G) = \bigcap_{a \in G} Z_a$$

It is clear that $Z(G) = G$ if and only if G is a commutative group. The following theorem provides a very useful tool in the search for the classification of groups:

■ **THEOREM 11.6.** *Let G be a finite group. Then there exist elements $a_1, a_2, \ldots, a_k \notin Z(G)$ such that*

$$o(G) = o(Z(G)) + \sum_{i=1}^{k} [G:Z_{a_i}] \qquad (1)$$

where, for each i, $[G:Z_{a_i}] > 1$ and $[G:Z_{a_i}]o(Z(G))$ is a proper divisor of $o(G)$.

Proof. As observed just prior to Proposition 11.5, every element of $Z(G)$ constitutes a conjugacy class by itself. Let $C(a_1), \ldots, C(a_k)$ be a list of the other distinct conjugacy classes of G. Since each element of G belongs in some conjugacy class, we have

$$o(G) = o(Z(G)) + \sum_{i=1}^{k} |C(a_i)|$$

Equation (1) now follows from Proposition 11.5.

If $[G:Z_{a_i}] = 1$, then $G = Z_{a_i}$, implying that a_i commutes with all the elements of G and contradicting the fact that $a_i \notin Z(G)$. Thus $[G:Z_{a_i}] > 1$ for $i = 1, 2, \ldots, k$. Finally, since $a_i \notin Z(G)$, it follows that $Z(G)$ is a *proper* subgroup of Z_{a_i}, and hence $[G:Z_{a_i}]o(Z(G))$ is a proper divisor of $[G:Z_{a_i}]o(Z_{a_i}) = o(G)$. ∎

If G is the quaternion group, $Z(G) = \{1, d\}$, and we can use $a_1 = a, a_2 = b$, and $a_3 = e$, since $C(a) = \{a, e\}$, $C(b) = \{b, f\}$, and $C(c) = \{c, g\}$.

Equation (1) is called the *class equation* of G. It has a surprising number of implications for finite groups. We demonstrate this first by classifying all the groups of order p^2 up to isomorphism. Note that this corollary is a very strong generalization of Proposition 9.17.

■ **COROLLARY 11.7.** *Let G be a group of order p^2, where p is a prime. Then G is isomorphic to either $(\mathbb{Z}_{p^2}, +)$ or to $(\mathbb{Z}_p, +)^2$.*

Proof. If G is cyclic, it is isomorphic to $(\mathbb{Z}_{p^2}, +)$. Hence we may assume that every nonidentity element of G has order p. We first show that in the class equation of G, $k = 0$ so that $Z(G) = G$, and hence G is in fact commutative. To see this, assume that $k > 0$, and let $a_1, \ldots, a_k$ be as in Theorem 11.6. Since $o(G) = p^2$, it follows that

$$[G:Z_{a_i}] = p \quad \text{for each } i = 1, 2, \ldots, k$$

Hence $o(Z(G))$ must also be divisible by p. This, however, contradicts the fact that $[G:Z_{a_i}]o(Z(G))$ is a proper divisor of $o(G) = p^2$. Thus G is commutative.

Now that G is known to be a commutative group in which each nonidentity element has order p, let $a, b \in G$ be such that $\langle a \rangle$ and $\langle b \rangle$ are distinct subgroups of G. Since they both have prime order p, it follows that $\langle a \rangle \cap \langle b \rangle = \{1_G\}$, and so, by Proposition 11.2,

$$G \cong \langle a \rangle \times \langle b \rangle \cong (\mathbb{Z}_p, +) \times (\mathbb{Z}_p, +) = (\mathbb{Z}_p, +)^2 \quad ∎$$

We are already familiar with two noncyclic groups of order p^2 for each prime p. These are the groups $\mathbb{Z}_p[x, \leqslant 1]$ and $(GF(p, P(x)), +)$, where $P(x)$ is any irreducible quadratic over $\mathbb{Z}_p$. Since all the nonzero elements of these groups have order p, it follows that they are isomorphic to $(\mathbb{Z}_p, +)^2$. The next theorem, above and beyond its utility in the classification of finite groups, also provides a partial converse to Lagrange's Theorem 9.8 about the orders of subgroups. The actual converse to Lagrange's Theorem is false (Exercises 9.4.28–38).

■ **THEOREM 11.8 (Cauchy).** *If the order of the finite group G is divisible by the prime p, then G contains an element of order p.*

Proof. We proceed by induction on $o(G)$, the theorem being trivial for groups of order 1. Let the prime p be fixed, let n be a positive integer divisible by p, and assume the theorem to be true for all groups of order less than n. Let G be a group of order n, and assume that G has no element of order p. By Proposition 9.7.3, we can assume that the order of every element of G is relatively prime to p.

Let $a_1, a_2, \ldots, a_k$ be as in Theorem 11.6. Since Z_{a_i} is a subgroup of G, it contains no element of order p. Since $Z_{a_i} \ne G$, it follows that $o(Z_{a_i}) < n$, so, by the induction hypothesis, p cannot be a divisor of $o(Z_{a_i})$. However, $o(G) = [G : Z_{a_i}] o(Z_{a_i})$, and hence p must be a divisor of $[G : Z_{a_i}]$ for each $i = 1, 2, \ldots, k$. It now follows from the class equation (1) that p divides $o(Z(G))$.

Let z be any nonidentity element of $Z(G)$ (z exists because $o(Z(G)) > 1$), and let $H = \langle z \rangle$. Since $Z(G)$ is commutative, it follows that H is a normal subgroup of $Z(G)$. Since p does not divide $o(H)$, it must divide $o(Z(G)/H) < n$. By the induction hypothesis, $Z(G)/H$ has some element, say, bH, $b \notin H$, of order p. This means that $(bH)^p = H$, and hence $b^p \in H$. However, since $o(b)$ and p are relatively prime, it follows that there exist integers A and B such that $Ao(b) + Bp = 1$, and so

$$b = b^1 = b^{Ao(b) + Bp} = (1_G)^A (b^p)^B \in H$$

which is a contradiction. Thus G must contain an element of order p. ■

We now have sufficient information to classify all the groups of order $2p$ where p is prime:

■ **COROLLARY 11.9.** *If $p > 2$ is a prime and G is a group of order $2p$, then G is isomorphic to either $(\mathbb{Z}_{2p}, +)$ or D_p.*

Proof. By Theorem 11.8, G has elements a and b of orders p and 2, respectively. Since $H = \langle a \rangle$ has half the elements of G, it follows from Proposition 10.3 that H is a normal subgroup of G. Consequently

$$bab^{-1} \in bHb^{-1} = H = \langle a \rangle$$

and hence $bab^{-1} = a^k$ for some $k \in \mathbb{Z}_p$. However,

$$ba^k b^{-1} = (bab^{-1})(bab^{-1}) \cdots (bab^{-1}) = a^k a^k \cdots a^k = a^{k^2}$$

and so

$$a = b^2 a b^{-2} = b(bab^{-1})b^{-1} = ba^k b^{-1} = a^{k^2}$$

It follows that $k^2 \equiv 1 \pmod{p}$ or $k \equiv \pm 1 \pmod{p}$.

If $k \equiv 1$, then $bab^{-1} = a$ or $ba = ab$. Since $o(a) = p$ and $o(b) = 2$, it follows from Proposition 9.7 that $o(ab) = 2p$ so that $G \cong (\mathbb{Z}_{2p}, +)$. Otherwise, $k \equiv -1 \pmod{p}$ so that $bab^{-1} = a^{-1}$, or $ba = a^{-1}b = a^{p-1}b$. In this case G is necessarily isomorphic to D_p. To see this, observe that the $2p$ elements of G can be listed as $\{a^i b^j \mid i = 0, 1, \ldots, p-1, j = 0, 1\}$. On the other hand, in D_p let $\alpha = (1\ 2 \cdots p)$, and let $\beta = (2\ p)(3\ p-1) \cdots (\frac{p-1}{2}\ \frac{p+1}{2})$. Then $\beta\alpha\beta^{-1} = \alpha^{-1}$, or $\beta\alpha = \alpha^{-1}\beta = \alpha^{p-1}\beta$. It is now easily verified that the function $f(a^i b^j) = \alpha^i \beta^j$ defines an isomorphism of G and D_p (Exercise 12). ∎

EXERCISES 11.2

1. Let G be a commutative group of order $p_1 p_2 \cdots p_r$, where $p_1, p_2, \ldots, p_r$ are distinct primes. Prove that $G \cong (\mathbb{Z}_{p_1}, +) \times (\mathbb{Z}_{p_2}, +) \times \cdots \times (\mathbb{Z}_{p_r}, +)$.
2. Prove that a finite group has exactly one conjugacy class if and only if it is trivial.
3. Prove that a finite group has exactly two conjugacy classes if and only if it has order 2.
4. Prove that a finite group has exactly three conjugacy classes if and only if it has order 3.
5. Let p and q be distinct primes. Prove that every noncommutative group of order pq has a trivial center.
6. Let k, n be arbitrary positive integers, and let p be a prime such that $(k, p) = 1$ and $n/2 < p \leqslant n$. Let f be a function of n variables that have k distinct variants. Prove that there is a p-cycle σ in S_n such that $\sigma f = f$.
7. Suppose that G is a commutative group of order $p_1^{r_1} p_2^{r_2} \cdots p_h^{r_h}$, where $p_1, p_2, \ldots, p_h$ are distinct primes. Prove that G has subgroups $G_1, G_2, \ldots, G_h$ of orders $p_1^{r_1}, p_2^{r_2}, \ldots, p_h^{r_h}$, respectively, such that $G \cong G_1 \times G_2 \times \cdots \times G_h$.
8. Classify the following groups according to isomorphism type:
   (a) $(\mathbb{Z}_6, +)$
   (b) $\langle (1\ 2\ 3\ 4\ 5\ 6) \rangle$
   (c) $\langle (1\ 2\ 3)(4\ 5) \rangle$
   (d) S_3
   (e) $\sqrt[6]{1}$
   (f) $S_{3,f}$, where $f = x_1 x_2 x_3$
   (g) $(\mathbb{Z}_9^*, \cdot)$
9. Classify the following groups according to isomorphism type:
   (a) $(\mathbb{Z}_8, +)$
   (b) D_4
   (c) Quaternion group
   (d) $\sqrt[8]{1}$
   (e) $\langle (1\ 2\ 3\ 4\ 5\ 6\ 7\ 8) \rangle$
   (f) $(GF(2, x^3 + x^2 + 1), +)$
   (g) $(GF^*(3, x^2 + x + 2), \cdot)$
   (h) $(\mathbb{Z}_2, +) \times (\mathbb{Z}_4, +)$
   (i) $(\mathbb{Z}_2, +)^3$

10. Classify the following groups according to isomorphism type:
    (a) $(\mathbb{Z}_9, +)$
    (b) $\sqrt[9]{1}$
    (c) $(GF(3, x^2 + x + 2), +)$
    (d) $\langle (1\ 2\ 3\ 4\ 5\ 6\ 7\ 8\ 9) \rangle$
    (e) $S_{9,f}$, where $f = x_1^2 x_2 + x_2^2 x_3 + \cdots + x_9^2 x_1$
    (f) $(\mathbb{Z}_3, +)^2$

11. Classify the following groups according to isomorphism type:
    (a) $(\mathbb{Z}_{10}, +)$
    (b) D_5
    (c) $\sqrt[10]{1}$
    (d) $S_{5,f}$, where $f = x_1 x_2 + x_2 x_3 + x_3 x_4 + x_4 x_5 + x_5 x_1$
    (e) $(\mathbb{Z}_{11}^*, \cdot)$
    (f) $\langle (1\ 2\ 3\ 4\ 5)(6\ 7) \rangle$

12. Prove that the function f defined in the proof of Corollary 11.9 is indeed an isomorphism.

13. Find a group of order 16 which is isomorphic to neither $(\mathbb{Z}_{16}, +)$ nor $(\mathbb{Z}_4, +)^2$.

CHAPTER SUMMARY

The direct product of groups was introduced as a means for constructing a host of new groups. Cauchy's partial converse to Lagrange's Theorem and the class equation proved to be useful tools in investigating the isomorphism types of groups. It was shown that for any prime number p there are only two nonisomorphic groups of order p^2 and only two nonisomorphic groups of order $2p$.

Chapter Review Exercises

1. The direct product of $(\mathbb{Z}_8, +)$ with S_4 has 32 elements.
2. $G \times \{1_G\} \cong G$.
3. If G is a group, then the conjugacy class $C(a)$ of any element a of G is a subgroup of G.
4. If G is a group and $a \in G$, then $\mathbb{Z}_a$ is a commutative group.
5. If G is a commutative group, then $\mathbb{Z}_a = G$ for all a in G.
6. Every group of order 49 is commutative.
7. If H is a normal subgroup of index 100 in G, then G/H contains an element of order 25.
8. If p is a prime number, then every group of order $2p$ is commutative.

New Terms

Center	255
Centralizer	254
Class equation	256
Conjugacy class	254
Direct product	250

Supplementary Exercises

1. Prove that every finite commutative group is the direct product of cyclic groups.
2. Characterize all the finite groups that have exactly n conjugacy classes for as many positive integers n as possible.

APPENDIX A

Excerpts from Al-Khwarizmi's *Solution of the Quadratic Equation**

THE BOOK OF ALGEBRA AND ALMUCABOLA

Containing Demonstrations of the Rules of the Equations of Algebra

... Furthermore I discovered that the numbers of restoration and opposition are composed of these three kinds: namely, roots, squares, and numbers.[2] However, number alone is connected neither with roots nor with squares by any ratio. Of these, then, the root is anything composed of units which can be multiplied by itself, or any number greater than unity multiplied by itself: or that which is found to be diminished below unity when multiplied by itself. The square is that which results from the multiplication of a root by itself.

Of these three forms, then, two may be equal to each other, as for example:

Squares equal to roots,
Squares equal to numbers, and
Roots equal to numbers.[3]

---

*Cardano, G., *Ars Magna or The Rules of Algebra* (T. Richard Witmer translator). Copyright © 1968 by The Massachusetts Institute of Technology. Reprinted by permission of MIT Press.

[2] The term "roots" (*radices*) stands for multiples of the unknown, our x; the term "squares" (*substantiae*) stands for multiples of our x^2; "numbers" (*numeri*) are constants.

[3] In our notation, $x^2 = ax$, $x^2 = b$, $x = c$.

261

Chapter I. Concerning Squares Equal to Roots[4]

The following is an example of squares equal to roots: a square is equal to 5 roots. The root of the square then is 5, and 25 forms its square which, of course, equals five of its roots.

Another example: the third part of a square equals four roots. Then the root of the square is 12 and 144 designates its square. And similarly, five squares equal 10 roots. Therefore one square equals two roots and the root of the square is 2. Four represents the square.

In the same manner, then, that which involves more than one square, or is less than one, is reduced to one square. Likewise you perform the same operation upon the roots which accompany the squares.

Chapter II. Concerning Squares Equal to Numbers

Squares equal to numbers are illustrated in the following manner: a square is equal to nine. Then nine measures the square of which three represents one root.

Whether there are many or few squares, they will have to be reduced in the same manner to the form of one square. That is to say, if there are two or three or four squares, or even more, the equation formed by them with their roots is to be reduced to the form of one square with its root. Further, if there be less than one square, that is, if a third or a fourth or a fifth part of a square or root is proposed, this is treated in the same manner.

For example, five squares equal 80. Therefore one square equals the fifth part of the number 80 which, of course, is 16. Or, to take another example, half of a square equals 18. This square therefore equals 36. In like manner all squares, however many, are reduced to one square, or what is less than one is reduced to one square. The same operation must be performed upon the numbers which accompany the squares.

Chapter III. Concerning Roots Equal to Numbers

The following is an example of roots equal to numbers: a root is equal to 3. Therefore nine is the square of this root.

Another example: four roots equal 20. Therefore one root of this square is 5. Still another example: half a root is equal to ten. The whole root therefore equals 20, of which, of course, 400 represents the square.

Therefore roots and squares and pure numbers are, as we have shown, distinguished from one another. Whence also from these three kinds which we have just explained, three distinct types of equations are formed involving three elements, as

> A square and roots equal to numbers,
> A square and numbers equal to roots, and
> Roots and numbers equal to a square.

[4] Latin: *de substantiis numeros coaequantibus*. The examples are $x^2 = 5x$, $\frac{1}{3}x^2 = 4x$, $5x^2 = 10x$.

Chapter IV. Concerning Squares and Roots Equal to Numbers

The following is an example of squares and roots equal to numbers: a square and 10 roots are equal to 39 units.[5] The question therefore in this type of equation is about as follows: what is the square which combined with ten of its roots will give a sum total of 39? The manner of solving this type of equation is to take one-half of the roots just mentioned. Now the roots in the problem before us are 10. Therefore take 5, which multiplied by itself gives 25, an amount which you add to 39, giving 64. Having taken then the square root of this which is 8, subtract from it the half of the roots, 5, leaving 3. The number three therefore represents one root of this square, which itself, of course, is 9. Nine therefore gives that square.

Similarly, however many squares are proposed all are to be reduced to one square. Similarly also you may reduce whatever numbers or roots accompany them in the same way in which you have reduced the squares.

The following is an example of this reduction: two squares and ten roots equal 48 units. The question therefore in this type of equation is something like this: what are the two squares which when combined are such that if ten roots of them are added, the sum total equals 48? First of all it is necessary that the two squares be reduced to one. But since one square is the half of two, it is at once evident that you should divide by two all the given terms in this problem. This gives a square and 5 roots equal to 24 units. The meaning of this is about as follows: what is the square which amounts to 24 when you add to it 5 of its roots? At the outset it is necessary, recalling the rule above given, that you take one-half of the roots. This gives two and one-half which multiplied by itself gives $6\frac{1}{4}$. Add this to 24, giving $30\frac{1}{4}$. Take then of this total the square root, which is, of course, $5\frac{1}{2}$. From this subtract half of the roots, $2\frac{1}{2}$, leaving 3, which expresses one root of the square, which itself is 9.

. . .

Chapter VI. Geometrical Demonstrations[6]

We have said enough, says Al-Khowarizmi, so far as numbers are concerned, about the six types of equations. Now, however, it is necessary that we should demonstrate geometrically the truth of the same problems which we have explained in numbers. Therefore our first proposition is this, that a square and 10 roots equal 39 units.

---

[5] This example, $x^2 + 10x = 39$, answer $x = 3$, "runs," as Karpinski notices in his introduction to this translation, "like a thread of gold through the algebras for several centuries, appearing in the algebras of Abu Kamil, Al-Karkhi and Omar al-Khayyami, and frequently in the works of Christian writers," and it still graces our present algebra texts. The solution of this type, $x^2 + ax = b$, is, as we can verify, based on the formula $x = \sqrt{(a/2)^2 + b} - a/2$.

[6] For these geometric demonstrations we must go back, as said, to Euclid's *Elements* (Book VI, Prop. 28, 29; see also Book II, Prop. 5, 6). See also on this subject the introduction to the *Principal works of Simon Stevin*, vol. IIB (Swets-Zeitlinger, Amsterdam, 1958), 464–467.

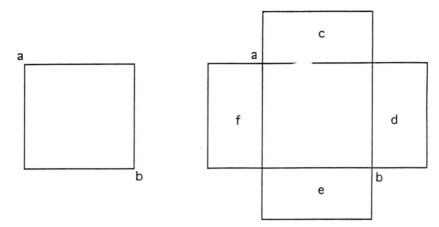

Figure A1

The proof is that we construct (Figure 1) a square of unknown sides, and let this square figure represent the square (second power of the unknown) which together with its root you wish to find. Let the square, then, be *ab*, of which any side represents one root. When we multiply any side of this by a number (or numbers) it is evident that that which results from the multiplication will be a number of roots equal to the root of the same number (of the square). Since then ten roots were proposed with the square, we take a fourth part of the number ten and apply to each side of the square an area of equidistant sides, of which the length should be the same as the length of the square first described and the breadth $2\frac{1}{2}$, which is a fourth part of 10. Therefore four areas of equidistant sides are applied to the first square, *ab*. Of each of these the length is the length of one root of the square *ab* and also the breadth of each is $2\frac{1}{2}$, as we have just said. These now are the areas *c, d, e, f*. Therefore it follows from what we have said that there will be four areas having sides of unequal length, which also are regarded as unknown. The size of the areas in each of the four corners, which is found by multiplying $2\frac{1}{2}$ by $2\frac{1}{2}$, completes that which is lacking in the larger or whole area. Whence it is that we complete the drawing of the larger area by the addition of the four products, each $2\frac{1}{2}$ by $2\frac{1}{2}$; the whole of this multiplication gives 25.

And now it is evident that the first square figure, which represents the square of the unknown [x^2], and the four surrounding areas [$10x$] make 39. When we add 25 to this, that is, the four smaller squares which indeed are placed at the four angles of the square *ab*, the drawing of the larger square, called *GH*, is completed (Figure 2). Whence also the sum total of this is 64, of which 8 is the root, and by this is designated one side of the completed figure. Therefore when we subtract from eight twice the fourth part of 10, which is placed at the extremities of the larger square *GH*, there will remain but 3. Five

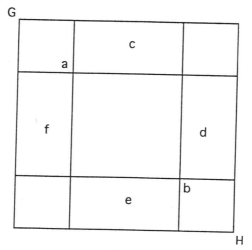

Figure A2

being subtracted from 8, 3 necessarily remains, which is equal to one side of the first square *ab*.

This three then expresses one root of the square figure, that is, one root of the proposed square of the unknown, and 9 the square itself. Hence we take half of ten and multiply this by itself. We then add the whole product of the multiplication to 39, that the drawing of the larger square *GH* may be completed; for the lack of the four corners rendered incomplete the drawing of the whole of this square. Now it is evident that the fourth part of any number multiplied by itself and then multiplied by four gives the same number as half of the number multiplied by itself. Therefore, if half of the root is multiplied by itself, the sum total of this multiplication will wipe out, equal, or cancel the multiplication of the fourth part by itself and then by four.

The remainder of the treatise deals with problems that can be reduced to one of the six types, for example, how to divide 10 into two parts in such a way that the sum of the products obtained by multiplying each part by itself is equal to 58: $x^2 + (10 - x)^2 = 58$, $x = 3$, $x = 7$. This is followed by a section on problems of inheritance.

APPENDIX B

Excerpts from Cardano's *Ars Magna**

CHAPTER XI. ON THE CUBE AND FIRST POWER EQUAL TO THE NUMBER

Scipio Ferro of Bologna well-nigh thirty years ago discovered this rule and handed it on to Antonio Maria Fior of Venice, whose contest with Niccolò Tartaglia of Brescia gave Niccolò occasion to discover it. He [Tartaglia] gave it to me in response to my entreaties, though withholding the demonstration. Armed with this assistance, I sought out its demonstration in [various] forms. This was very difficult. My version of it follows.

Demonstration

For example, let GH^3 plus six times its side GH equal 20, and let AE and CL be two cubes the difference between which is 20 and such that the product of AC, the side [of one], and CK, the side [of the other], is 2, namely one-third the coefficient of x. Marking off BC equal to CK, I say that, if this is done, the remaining line AB is equal to GH and is, therefore, the value of x, for GH has already been given as [equal to x].

In accordance with the first proposition of the sixth chapter of this book, I complete the bodies DA, DC, DE, and DF; and as DC represents BC^3, so DF represents AB^3, DA represents $3(BC \times AB^2)$ and DE represents $3(AB \times BC^2)$. Since, therefore, $AC \times CK$ equals 2, $AC \times 3CK$ will equal 6, the coefficient of x; therefore $AB \times 3(AC \times CK)$ makes $6x$ or $6AB$, wherefore three times the product of AB, BC, and AC is $6AB$. Now the difference between AC^3 and CK^3—manifesting itself as BC^3, which is equal to this by supposition—is 20, and from the first proposition of the sixth chapter is the sum of the bodies DA, DE, and DF. Therefore these three bodies equal 20.

*Cardano, G., *Ars Magna or The Rules of Algebra* (T. Richard Witmer, translator). Copyright © 1968 by The Massachusetts Institute of Technology, Reprinted by permission of MIT Press.

On the Cube and First Power Equal to the Number   267

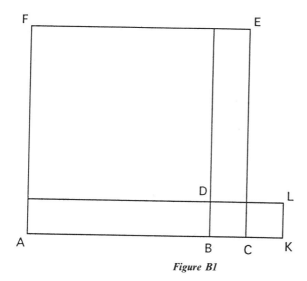

Figure B1

Now assume that BC is negative:

$$AB^3 = AC^3 + 3(AC \times CB^2) + (-BC^3) + 3(-BC \times AC^2),$$

by that demonstration. The difference between $3(BC \times AC^2)$ and $3(AC \times BC^2)$, however, is [three times] the product of AB, BC, and AC. Therefore, since this, as was demonstrated, is equal to $6AB$, add $6AB$ to the product of $3(AC \times BC^2)$, making $3(BC \times AC^2)$. But since BC is negative, it is now clear that $3(BC \times AC^2)$ is negative and the remainder which is equal to it is positive. Therefore,

$$3(CB \times AB^2) + 3(AC \times BC^2) + 6AB = 0.$$

It will be seen, therefore, that as much as is the difference between AC^3 and BC^3, so much is the sum of

$$AC^3 + 3(AC \times CB^2) + 3(-CB \times AC^2) + (-BC^3) + 6AB.$$

This, therefore, is 20 and, since the difference between AC^3 and BC^3 is 20, then, by the second proposition of the sixth chapter, assuming BC to be negative,

$$AB^3 = AC^3 + 3(AC \times BC^2) + (-BC^3) + 3(-BC \times AC^2).$$

Therefore, since we now agree that

$$AB^3 + 6AB = AC^3 + 3(AC \times BC^2) + 3(-BC \times AC^2) + (-BC^3) + 6AB,$$

which equals 20, as has been proved, they [i.e., $AB^3 + 6AB$] will equal 20. Since, therefore,

$$AB^3 + 6AB = 20,$$

and since

$$GH^3 + 6GH = 20,$$

it will be seen at once and from what is said in I, 35 and XI, 31 of the *Elements* that GH will equal AB. Therefore GH is the difference between AC and CB. AC and CB, or AC and CK, the coefficients, however, are lines containing a surface equal to one-third the coefficient of x and their cubes differ by the constant of the equation. Whence we have the rule:

Rule

Cube one-third the coefficient of x; add to it the square of one-half the constant of the equation; and take the square root of the whole. You will duplicate this, and to one of the two you add one-half the number you have already squared and from the other you subtract one-half the same. You will then have a *binomium* and its *apotome*. Then, subtracting the cube root of the *apotome* from the cube root of the *binomium*, the remainder [or] that which is left is the value of x.

For example,

$$x^3 + 6x = 20.$$

Cube 2, one-third of 6, making 8; square 10, one-half the constant; 100 results. Add 100 and 8, making 108, the square root of which is $\sqrt{108}$. This you will duplicate: to one add 10, one-half the constant, and from the other subtract the same. Thus you will obtain the *binomium* $\sqrt{108} + 10$ and its *apotome*

On the Cube and First Power Equal to the Number   269

$\sqrt{108} - 10$. Take the cube roots of these. Subtract [the cube root of the] *apotome* from that of the *binomium* and you will have the value of x:

$$\sqrt[3]{\sqrt{108} + 10} - \sqrt[3]{\sqrt{108} - 10}.$$

Again,

$$x^3 + 3x = 10.$$

Cube 1, one-third of 3, and 1 results; square 5, one-half of 10, and 25 results; add 25 and 1, making 26; add 5 to and subtract it from the square root of this. You will thus form the *binomium* $\sqrt{26} + 5$ and its *apotome* $\sqrt{26} - 5$; whence x equals $\sqrt[3]{\sqrt{26} + 5} - \sqrt[3]{\sqrt{26} - 5}$. Here you have proof:

	$\sqrt[3]{\sqrt{26} + 5}$	$-\sqrt[3]{\sqrt{26} - 5}$
The cubes of the parts: (As is evident, the sum of these is 10.)	$(\sqrt{26} + 5)$	$-(\sqrt{26} - 5)$
The squares of the parts:	$\sqrt[3]{51 + \sqrt{2600}}$	$\sqrt[3]{51 - \sqrt{2600}}$
Three times the squares of the parts:	$\sqrt[3]{1377 + \sqrt{1{,}895{,}400}}$	$\sqrt[3]{1377 - \sqrt{1{,}895{,}400}}$
The parts themselves:	$-\sqrt[3]{\sqrt{26} - 5}$	$+\sqrt[3]{\sqrt{26} + 5}$
The products of the parts and three times their squares:	$+\sqrt[3]{\sqrt{49{,}299{,}354} + 6885 - \sqrt{47{,}385{,}000} - 7020}$	$-\sqrt[3]{\sqrt{49{,}299{,}354} - 6885 - \sqrt{47{,}385{,}000} + 7020}$

Moreover, the cube roots contain four terms which can be reduced to two, for when 6885 is subtracted from 7020, the remainder is 135, and likewise when $\sqrt{47{,}385{,}000}$ is subtracted from $\sqrt{49{,}299{,}354}$ there is left $\sqrt{18{,}954}$. Therefore these products are

$$\sqrt[3]{\sqrt{18{,}954} - 135} - \sqrt[3]{\sqrt{18{,}954} + 135}.$$

The whole cube, then, from the demonstration in the third book is

$$10 + \sqrt[3]{\sqrt{18{,}954} - 135} - \sqrt[3]{\sqrt{18{,}954} + 135},$$

and three times the root, or $3x$, equals $\sqrt[3]{\sqrt{18{,}954} + 135} - \sqrt[3]{\sqrt{18{,}954} - 135}$. And, finally, having added all together, since the universal cube roots cancel each other, the whole becomes

$$x^3 + 3x = \text{exactly } 10.$$

A third example:

$$x^3 + 6x = 2.$$

Raise 2, one-third the coefficient of x, to the cube and 8 is the result; square 1, half of 2, making 1; add 8 to 1, and 9 is produced, the square root of which is 3. Now duplicate 3 and to one add 1, half the constant, thus making 4, and from the other subtract half the constant, thus making 2. Then subtract the cube root of the less from the cube root of the greater and you have $\sqrt[3]{4} - \sqrt[3]{2}$ as the value of x.

Remember what we said in the chapter in the third book on extracting cube roots whenever these universal cube roots are equivalent to a whole number or a fraction. Thus in the first example

$$\sqrt[3]{\sqrt{108} + 10} - \sqrt[3]{\sqrt{108} - 10}$$

is 2, as is indicated by the rule there given and as is perfectly clear if it is tried out.

APPENDIX C

A Demonstration of the Impossibility of the Algebraic Resolution of General Equations Whose Degree Exceeds Four[1]

by NIELS H. ABEL

As is well known, it is possible to resolve the general equations up to the fourth degree, but the equations of higher degree only in some special cases, and, if I am not mistaken, the question: "Is it possible to resolve in a general manner the equations whose degree exceeds four?" has yet to receive a satisfactory answer. It is the purpose of this memoir to respond to this question.

The algebraic resolution of an expression is an expression of its roots as algebraic functions of the coefficients. It is therefore necessary to first consider the general form of algebraic functions, and then to see whether it is possible to satisfy the given equation by replacing the unknown with an algebraic function.

. . .

---

[1] *Journal fur die reine und angewandte Mathematik* (Crelle), Vol 1, Berlin 1826.

III.

On the number of distinct values that a function of several variables can assume when these variables are interchanged amongst themselves.

Let v be a rational function of several independent variables $x_1, x_2, \ldots, x_n$. The number of different values to which this function is subjected upon interchanging the quantities on which it depends cannot exceed the product $1.2.3\ldots n$. Let μ be this product.

Now let

$$v \begin{pmatrix} \alpha & \beta & \gamma & \delta & \cdots \\ a & b & c & d & \cdots \end{pmatrix}$$

be the value that some function v assumes when x_a, x_b, x_c, x_d are substituted for $x_\alpha, x_\beta, x_\gamma, x_\delta$, etc. It is clear that when $A_1, A_2, \ldots, A_\mu$ denote the μ permutations that can be formed with the indices $1, 2, 3, \ldots, n$, the different values of v can be expressed as

$$v\begin{pmatrix} A_1 \\ A_1 \end{pmatrix}, v\begin{pmatrix} A_1 \\ A_2 \end{pmatrix}, v\begin{pmatrix} A_1 \\ A_3 \end{pmatrix}, \ldots, v\begin{pmatrix} A_1 \\ A_\mu \end{pmatrix}.$$

Suppose that the number of distinct values of v is less than μ, it is then necessary that some of the values of v be equal to each other, say,

$$v\begin{pmatrix} A_1 \\ A_1 \end{pmatrix} = v\begin{pmatrix} A_1 \\ A_2 \end{pmatrix} = v\begin{pmatrix} A_1 \\ A_3 \end{pmatrix} = \cdots = v\begin{pmatrix} A_1 \\ A_m \end{pmatrix}.$$

If the permutation denoted by $\begin{pmatrix} A_1 \\ A_{m+1} \end{pmatrix}$ is applied to these quantities we will have this new series of equal values

$$v\begin{pmatrix} A_1 \\ A_{m+1} \end{pmatrix} = v\begin{pmatrix} A_1 \\ A_{m+2} \end{pmatrix} = v\begin{pmatrix} A_1 \\ A_{m+3} \end{pmatrix} = \cdots = v\begin{pmatrix} A_1 \\ A_{2m} \end{pmatrix},$$

values which are different from the first ones, but the same in number. By changing these quantities by the substitution denoted by $\begin{pmatrix} A_1 \\ A_{2m+1} \end{pmatrix}$ we obtain a new system of equal quantities which are, however, different from the preceding ones. By continuing this process until all the permutations have been exhausted, the μ values of v will be partitioned into several systems, each of which will contain m equal values. It follows from this that if the number of distinct values of v is represented by ρ, a number that equals that of the systems, we have

$$\rho m = 1.2.3\ldots n,$$

that is:

The number of distinct values that a function of n quantities can assume under all the possible permutations of these quantities is necessarily a divisor of the product $1.2.3\ldots n$. This is known.

Now let $\begin{pmatrix} A_1 \\ A_m \end{pmatrix}$ be an arbitrary substitution. Suppose that in applying it several times successively to the function v we obtain the sequence of values

$$v, v_1, v_2, \ldots, v_{p-1}, v_p,$$

it is clear that v will be necessarily repeated several times. When v returns after p substitutions we say that $\begin{pmatrix} A_1 \\ A_m \end{pmatrix}$ is a *recurring substitution of order p*. We then have the following periodic series:

$$v, v_1, v_2, \ldots, v_{p-1}, v, v_1 v_2 \ldots$$

wherein, if $v\begin{pmatrix} A_1 \\ A_m \end{pmatrix}^r$ represents the value of v which is obtained after r repetitions of the substitution denoted by $\begin{pmatrix} A_1 \\ A_m \end{pmatrix}$, we obtain the series

$$v\begin{pmatrix} A_1 \\ A_m \end{pmatrix}^0, v\begin{pmatrix} A_1 \\ A_m \end{pmatrix}^1, v\begin{pmatrix} A_1 \\ A_m \end{pmatrix}^2, \ldots, v\begin{pmatrix} A_1 \\ A_m \end{pmatrix}^{p-1}, v\begin{pmatrix} A_1 \\ A_m \end{pmatrix}^0 \ldots$$

It follows that

$$v\begin{pmatrix} A_1 \\ A_m \end{pmatrix}^{\alpha p+r} = v\begin{pmatrix} A_1 \\ A_m \end{pmatrix}^r,$$

$$v\begin{pmatrix} A_1 \\ A_m \end{pmatrix}^{\alpha p} = v\begin{pmatrix} A_1 \\ A_m \end{pmatrix}^0 = v.$$

However, if p is the largest prime no greater than n, then if the number of distinct values of v is less than p, it must be the case that amongst p values some two must equal each other.

It therefore follows that of the p values

$$v\begin{pmatrix} A_1 \\ A_m \end{pmatrix}^0, v\begin{pmatrix} A_1 \\ A_m \end{pmatrix}^1, v\begin{pmatrix} A_1 \\ A_m \end{pmatrix}^2, \ldots, v\begin{pmatrix} A_1 \\ A_m \end{pmatrix}^{p-1},$$

some two must equal each other. Say

$$v\begin{pmatrix}A_1\\A_m\end{pmatrix}^r = v\begin{pmatrix}A_1\\A_m\end{pmatrix}^{r'},$$

then

$$v\begin{pmatrix}A_1\\A_m\end{pmatrix}^{r+p-r} = v\begin{pmatrix}A_1\\A_m\end{pmatrix}^{r'+p-r}.$$

Writing r for $r' + p - r$ and noting that $v\begin{pmatrix}A_1\\A_m\end{pmatrix}^p = v$, we conclude that

$$v = v\begin{pmatrix}A_1\\A_m\end{pmatrix}^r,$$

where r is clearly not a multiple of p. The value of v is therefore not changed by the substitution $\begin{pmatrix}A_1\\A_m\end{pmatrix}^r$, nor, consequently, by the repetition of this substitution. We therefore have

$$v = v\begin{pmatrix}A_1\\A_m\end{pmatrix}^{r\alpha},$$

α being an integer. If p is a prime number then it is clearly always possible to find two integers α and β such that

$$r\alpha = p\beta + 1,$$

hence,

$$v = v\begin{pmatrix}A_1\\A_m\end{pmatrix}^{p\beta+1},$$

and since

$$v = v\begin{pmatrix}A_1\\A_m\end{pmatrix}^{p\beta},$$

it follows that

$$v = v\begin{pmatrix}A_1\\A_m\end{pmatrix}.$$

The value of v will therefore not be changed by the recurrent substitution $\begin{pmatrix} A_1 \\ A_m \end{pmatrix}$ of order p.

However, it is clear that

$$\begin{pmatrix} \alpha\beta\gamma\delta\cdots\zeta\eta \\ \beta\gamma\delta\varepsilon\cdots\eta\alpha \end{pmatrix} \quad \text{and} \quad \begin{pmatrix} \beta\gamma\delta\varepsilon\cdots\eta\alpha \\ \gamma\alpha\beta\delta\cdots\zeta\eta \end{pmatrix}$$

are recurrent substitutions of order p when p is the number of indices $\alpha, \beta, \gamma, \ldots, \eta$. The value of v will therefore not be changed by the combination of these two. These substitutions are clearly equivalent to this single one

$$\begin{pmatrix} \alpha & \beta & \gamma \\ \gamma & \alpha & \beta \end{pmatrix},$$

and this one to the following two, applied successively,

$$\begin{pmatrix} \alpha & \beta \\ \beta & \alpha \end{pmatrix} \quad \text{and} \quad \begin{pmatrix} \beta & \gamma \\ \gamma & \beta \end{pmatrix}.$$

The value of v will therefore not be changed by the combination of these two substitutions. Hence

$$v = v \begin{pmatrix} \alpha & \beta \\ \beta & \alpha \end{pmatrix}\begin{pmatrix} \beta & \gamma \\ \gamma & \beta \end{pmatrix};$$

and similarly

$$v = v \begin{pmatrix} \beta & \gamma \\ \gamma & \beta \end{pmatrix}\begin{pmatrix} \gamma & \delta \\ \delta & \gamma \end{pmatrix},$$

from which it follows that

$$v = v \begin{pmatrix} \alpha & \beta \\ \beta & \alpha \end{pmatrix}\begin{pmatrix} \gamma & \delta \\ \delta & \gamma \end{pmatrix}.$$

We see from this that the function v is unchanged by two successive substitutions of the form $\begin{pmatrix} \alpha & \beta \\ \beta & \alpha \end{pmatrix}$, α and β being any two indices. If such a substitution is called a *transposition*, it may be concluded that any value of v will not be changed by an even number of transpositions, and that consequently all the values of v which result from an even number of substitutions are equal. Every exchange of the elements of a function can be effected by means of a certain number of transpositions; hence the function v can have only two values. The following theorem follows from this:

The number of different values that a function of n quantities can assume is not less than the largest prime not exceeding n, unless it is either 2 or 1.

It is therefore impossible to find a function of 5 quantities which has 3 or 4 different values.

The demonstration of this theorem is taken from a memoir of Mr. Cauchy which appears in the 17'th volume of the *Journal de l'école polytechnique*, p. 1.

APPENDIX D

On The Theory of Numbers[1]

by ÉVARISTE GALOIS

When it is agreed to consider as zero all the quantities which are the multiples of a given prime number p, and, subject to this convention, one looks for solutions to the polynomial equation $Fx = 0$, i.e., the equations that Mr. Gauss denotes by $Fx \equiv 0$, it is customary to consider only integer solutions to these sorts of questions. Having been led by some specific researches to consider their irrational solutions, I have arrived at some results that I believe to be new.

Let there be given such an equation or congruence, $Fx = 0$, and let p be the modulus. Suppose first that the congruence in question admits no rational factors, that is, there exist no three polynomials ϕx, ψx, χx such that

$$\phi x . \psi x = Fx + p . \chi x.$$

In that case the congruence has no integer roots, nor any irrational root of smaller degree. One should therefore regard the roots of this congruence as some kind of imaginary symbols (since they do not satisfy the same questions as integers), symbols whose employment, in calculations, will often prove as useful as that of the imaginary $\sqrt{-1}$ in ordinary analysis.

We are concerned here with the classification of these imaginaries and their reduction to the smallest possible number.

Let i denote one of the roots of the congruence $Fx = 0$, which can be supposed to have degree v.

---

[1] *Bulletin des Sciences mathematiques* de M. Ferussac, Vol. 13, June 1830; with the following note: "This memoire forms part of the research of Mr. Galois on the theory of permutations and algebraic equations."

Consider the general expression

$$a + a_1 i + a_2 i^2 + \cdots + a_{v-1} i^{v-1} \tag{A}$$

where $a, a_1, a_2, \ldots, a_{v-1}$ represent integers. When these numbers are assigned all their possible values, expression (A) runs through p^v values which possess, as I shall demonstrate, the same properties as the natural numbers in the *theory of residues of powers*.

Of the expressions (A) we shall only take the $p^v - 1$ values obtained when $a, a_1, a_2, \ldots, a_{v-1}$ are not all zero; let α be one of these expressions.

If α is successively raised to the second, third,... powers, a sequence of quantities all of which have the same form is obtained (since every function of i is reducible to the $(v-1)$th degree). Hence it must be that $\alpha^n = 1$ for some n; let n be the smallest number such that $\alpha^n = 1$. Then the set of numbers

$$1, \alpha, \alpha^2, \alpha^3, \ldots, \alpha^{n-1}$$

are all distinct. Multiply these n numbers by another expression of the same form. We then obtain another new group of quantities all different from the first group as well as from each other. If the quantities (A) have not been exhausted yet, the powers of α can be multiplied by a new expression γ, and so on. Consequently the number n necessarily divides the total number of quantities of type (A). Since this number is $p^v - 1$, we see that n divides $p^v - 1$. From this it also follows that

$$\alpha^{p^v - 1} = 1, \quad \text{or} \quad \alpha^{p^v} = \alpha.$$

Next it can be proven, just as is done in the theory of numbers, that there exist primitive roots α for which $p^v - 1 = n$, and which consequently reproduce, by their powers, the complete sequence of all the other roots.

And any one of these primitive roots depends only on a congruence of degree v, a congruence which must be irreducible, since otherwise the equation of i could not be irreducible either, as the roots of the congruence in i are all powers of the primitive root.

We note here the remarkable result that all the algebraic quantities that arise in this theory are roots of equations of the form

$$x^{p^v} = x.$$

This proposition is stated algebraically as follows: Given a function Fx and an integer p, one can write

$$fx \cdot Fx = x^{p^v} - x + p\phi x,$$

fx and ϕx being entire functions, whenever the congruence $Fx \equiv 0 \pmod{p}$ is irreducible.

If it is desired to express all the roots of such a congruence in terms of one, it suffices to note that in general

$$(Fx)^{p^n} = F(x^{p^n})$$

and that consequently, if one of the roots is x then the others are

$$x^p, x^{p^2}, \ldots, x^{p^{\nu-1}} \quad {}^2.$$

We now show that, conversely, the roots of the equation or of the congruence $x^{p^\nu} = x$ all depend on a single congruence of degree ν.

Let i be a root of an irreducible congruence and such that all the roots of the congruence $x^{p^\nu} = x$ are rational functions of i (as is the case for ordinary equations, it is clear that this property holds here as well)[3].

It is clear that the degree μ of the congruence in i cannot be less than ν, since otherwise the congruence

$$x^{p^{\nu-1}} - 1 = 0 \tag{ν}$$

would share all of its roots with the congruence

$$x^{p^{\mu-1}} - 1 = 0,$$

which is impossible, since the congruence (ν) has no repeated roots, as is seen

---

[2] It would be wrong to conclude from the fact that the roots of the irreducible congruence of degree ν

$$Fx = 0$$

are expressible as the sequence

$$x, x^p, x^{p^2}, \ldots, x^{p^{\nu-1}},$$

that these roots are always expressible by radicals. Here is an example to the contrary:
The irreducible congruence

$$x^2 + x + 1 = 0 \pmod{2}$$

yields

$$x = \frac{-1 + \sqrt{-3}}{2}$$

which reduces to

$$\frac{0}{0} \pmod{2},$$

from which formula we learn nothing.

[3] The general proposition in question here can be stated as follows: Given an algebraic equation, it is possible to find a rational function θ of all of its roots such that, reciprocally, each of the roots is rationally expressible in θ. This theorem was known to Abel as one can see in the first part of the memoire on eliptic functions which this celebrated geometer left.

by taking the derivative of the first part. I claim that neither can μ exceed ν. In fact, if that were the case, all the roots of the congruence

$$x^{p^\mu} = x$$

would depend rationally on those of the congruence

$$x^{p^\nu} = x.$$

But it is easily seen that if

$$i^{p^\nu} = i,$$

then every rational function $h = f(i)$ would yield

$$(fi)^{p^\nu} = f(i^{p^\nu}) = fi, \qquad \text{from which } h^{p^\nu} = h.$$

Hence all the roots of the congruence $x^{p^\mu} = x$ would also be roots of the equation $x^{p^\nu} = x$, which is impossible.

We therefore now know that all the roots of the equation $x^{p^\nu} = x$ necessarily depend on only *one irreducible* congruence of degree ν.

Now, the most general method for obtaining this irreducible congruence, on which the roots of the congruence $x^{p^\nu} = x$ depend, is to extract first out of this congruence all the factors that it shares with congruences of smaller degree and of the form

$$x^{p^\mu} = x.$$

One thus obtains a congruence which must factor into irreducible congruences of degree ν. And, since it is known how to express all the roots of each of these irreducible congruences in tems of a single one, it will be easy to obtain all of them by Mr. Gauss's method.

Most frequently, however, it will be easy to find by trial and error an irreducible congruence of a given degree ν, and the rest must be derived from it.

For example, let $p = 7$ and $\nu = 3$. Let us look for the roots of the congruence

$$x^{7^3} = x \pmod{7}. \tag{1}$$

I note first that since the congruence

$$i^3 = 2 \pmod{7} \tag{2}$$

being irreducible, and of degree 3, all the roots of congruence (1) depend rationally on those of congruence (2), so that all the roots of (1) have the form

$$a + a_1 i + a_2 i^2 \quad \text{or} \quad a + a_1 \sqrt[3]{2} + a_2 \sqrt[3]{4}. \tag{3}$$

It is now necessary to find a primitive root, that is, a form of expression (3) which, when raised to all possible powers, gives all the roots of the congruence

$$x^{7^3 - 1} = 1, \quad \text{or} \quad x^{2^1 \cdot 3^2 \cdot 19} = 1 \pmod{7},$$

and to accomplish this we only need a primitive root of each of the congruences

$$x^2 = 1, \quad x^{3^2} = 1, \quad x^{19} = 1.$$

The primitive root of the first is -1; those of $x^{3^2} - 1 = 0$ are given by the equations

$$x^3 = 2, \quad x^3 = 4,$$

so that i is a primitive root of $x^{3^2} = 1$.

It only remains to find a root of $x^{19} - 1 = 0$, or rather of

$$\frac{x^{19} - 1}{x - 1} = 0,$$

and we first try to see whether the requirements can be satisfied by taking $x = a + a_1 i$ rather than $a + a_1 i + a_2 i^2$; we must have

$$(a + a_1 i)^{19} = 1;$$

which, when developed by Newton's formula, after reducing the powers of $a, a_1,$ and i by applying the formulas

$$a^{m(p-1)} = 1, \quad a_1^{m(p-1)} = 1, \quad i^3 = 2,$$

reduces to

$$3[a - a^4 a_1^3 + (a^5 a_1^2 + a^2 a_1^5) i^2] = 1,$$

from which, by separation

$$3a - 3a^4 a_1^3 = 1, \quad a^5 a_1^2 + a^2 a_1^5 = 0.$$

These last two equations are satisfied by setting $a = -1$, $a_1 = 1$. Hence

$$-1 + i$$

is a primitive root of $x^{19} = 1$. We found above that the values -1 and i are primitive roots of $x^2 = 1$ and $x^{3^2} = 1$; it only remains to multiply the three quantities

$$-1, \quad i, \quad -1 + i,$$

and the product $i - i^2$ will be a primitive root of the congruence

$$x^{7^3 - 1} = 1.$$

Thus here the expression $i - i^2$ possesses the property that, in raising it to all powers, $7^3 - 1$ distinct expressions of the form

$$a + a_1 i + a_2 i^2$$

are obtained.

If we wish to find the lowest degree congruence on which our primitive root depends, it is necessary to eliminate i from the two equations

$$i^3 = 2, \quad \alpha = i - i^2.$$

One then obtains

$$\alpha^3 - \alpha + 2 = 0.$$

We agree to take imaginaries as a basis and to denote by i the root of this equation, so that

$$i^3 - i + 2 = 0, \tag{i}$$

and we will have all the imaginaries of the form

$$a + a_1 i + a_2 i^2$$

when i is raised to all of its powers and they are reduced by equation (i).

The main advantage of this new theory that is propounded here is that it restores to congruences the property (so useful for ordinary equations) that they possess as many roots as there are units in their degree.

The method of obtaining all of these roots is very simple. First, it is always possible to modify the given congruence $Fx = 0$ so that it does not have equal roots, or, in other words, so that it does not possess a common factor with

$F'x = 0$, and the means for doing so are evidently the same as those for ordinary equations.

Next, in order to obtain the integral solutions, it will suffice, as Mr. Libri seems to have been the first to remark, to look for the greatest common factor of $Fx = 0$ and $x^{p-1} = 1$.

To find the imaginary solutions of the second degree, it is necessary to look for the greatest common factor of $Fx = 0$ and $x^{p^2-1} = 1$, and, in general, the solutions of order v are given by the greatest common factor of $Fx = 0$ and $x^{p^v-1} = 1$.

It is above all in the theory of permutations, where it is often necessary to vary the form of the indices, that the consideration of imaginary roots of congruences seems indispensible. It provides a simple and easy method for recognizing in what case a primitive equation is solvable by radicals, as I will attempt to describe in a few words.

Let $fx = 0$ be an algebraic equation of degree p^v; suppose that the p^v roots are denoted by x_k, where the index k assumes the p^v values determined by the congruence $k^{p^v} = k \pmod{p}$.

Let V be an arbitrary rational function of the p^v roots x_k. Transform this function by replacing each index k by the index $(ak + b)^{p^r}$, a, b, r being arbitrary constants satisfying the requirements $a^{p^v-1} = 1$, $b^{p^v} = b \pmod{p}$ and r integral.

By assigning to the constants a, b, r all their admissible values, we obtain a total of $p^v(p^v - 1)v$ ways of permuting the roots amongst themselves by means of permutations of the form $[x_k, x_{(ak+b)}p^r]$, and the function V will in general assume $p^v(p^v - 1)v$ different forms as a result of these substitutions.

Assume now that the proposed equation $fx = 0$ is such that every function of its roots that is invariant under the action of the $p^v(p^v - 1)v$ permutations that were constructed above has a rational numerical value.

Under these circumstances the equation $fx = 0$ is solvable by radicals, and, to prove this result, it suffices to note that the value substituted for k, in each index, can be expressed in the three forms

$$(ak + b)^{p^r} = [a(k + b^1)]^{p^r} = a'k^{p^r} + b'' = a'(k + b')^{p^r}.$$

Those who are familiar with the theory of equations will see this easily.

This remark would have had little significance had I not succeeded in showing that, conversely, every primitive equation that is solvable by radicals must satisfy the conditions I have just stated. (The equations of the ninth and twenty fifth degrees are excepted from this rule.)

Thus, for each number of the form p^v it is possible to form a group of permutations such that every function of the roots that is invariant under the action of these permutations must admit a rational value when the equation of degree p^v is primitive and solvable by radicals.

Moreover, only equations of such degree p^v are simultaneously both primitive and solvable by radicals.

The general theorem I have just announced makes precise and develops the conditions that I specified in the *Bulletin* of the month of April. It indicates the means for forming a function of the roots whose value will be rational, whenever the primitive equation of degree p^v is solvable by radicals, and consequently it leads to a characterization of the solvability of these equations by means of calculations which, while perhaps not feasible in practice, are at least theoretically possible.

Note that in the case $v = 1$ the various values of k consist of the sequence of integers. There are then $p(p-1)$ substitutions of the form (x_k, x_{ak+b}).

The function which, in the case of equations that are solvable by radicals, has a rational value depends, in general, on an equation of degree $1.2.3\ldots(p-2)$, to which it is necessary, consequently, to apply the method of rational roots.

APPENDIX E

The Theory of Groups[1]

by ARTHUR CAYLEY

No. 1 — The Theory of Groups

SUBSTITUTIONS, and (in connexion therewith) groups, have been a good deal studied; but only a little has been done towards the solution of the general problem of groups. I give the theory so far as is necessary for the purpose of pointing out what appears to me to be wanting.

Let $\alpha, \beta, \ldots$ be functional symbols, each operating upon one and the same number of letters and producing as its result the same number of functions of these letters; for instance, $\alpha(x, y, z) = (X, Y, Z)$, where the capitals denote each of them a given function of (x, y, z).

Such symbols are susceptible of repetition and of combination; $\alpha^2(x, y, z) = \alpha(X, Y, Z)$, or $\beta\alpha(x, y, z) = \beta(X, Y, Z)$, = in each case three given functions of (x, y, z), and similarly $\alpha^3, \alpha^2\beta$, &c.

The symbols are not in general commutative, $\alpha\beta$ not $= \beta\alpha$; but they are associative, $\alpha\beta \cdot \gamma = \alpha \cdot \beta\gamma$, each $= \alpha\beta\gamma$, which has thus a determinate signification.

[The associativeness of such symbols arises from the circumstance that the definitions of $\alpha, \beta, \gamma, \ldots$ determine the meanings of $\alpha\beta, \alpha\gamma$, &c.: if $\alpha, \beta, \gamma \ldots$ were quasi-quantitative symbols such as the quaternion imaginaries i, j, k, then $\alpha\beta$ and $\beta\gamma$ might have by definition values δ and ε such that $\alpha\beta \cdot \gamma$ and $\alpha \cdot \beta\gamma$ ($= \delta\gamma$ and $\alpha\varepsilon$ respectively) have unequal values].

Unity as a functional symbol denotes that the letters are unaltered, $1(x, y, z) = (x, y, z)$; whence $1\alpha = \alpha 1 = \alpha$.

The functional symbols *may* be substitutions; $\alpha(x, y, z) = (y, z, x)$, the same letters in a different order: substitutions can be represented by the notation

---

[1] *The American Journal of Mathematics*, 1 (1878), 50–52.

$\alpha = \dfrac{yzx}{xyz}$, the substitution which changes xyz into yzx, or as products of cyclical substitutions, $\alpha = \dfrac{yzx\ wu}{xyz\ uw}$, $= (xyz)(uw)$, the product of the cyclical interchanges x into y, y into z, and z into x; and u into w, w into u.

A set of symbols $\alpha, \beta, \gamma \ldots$ such that the product $\alpha\beta$ of each two of them (in each order, $\alpha\beta$ or $\beta\alpha$,) is a symbol of the set, is a group. It is easily seen that 1 is a symbol of every group, and we may therefore give the definition in the form that a set of symbols, $1, \alpha, \beta, \gamma ..$ satisfying the foregoing condition is a group. When the number of the symbols (or terms) is $=n$, then the group is of the nth order; and each symbol α is such that $\alpha^n = 1$, so that a group of the order n is, in fact, a group of symbolical nth roots of unity.

A group is defined by means of the laws of combination of its symbols: for the statement of these we may either (by the introduction of powers and products) diminish as much as may be the number of independent functional symbols, or else, using distinct letters for the several terms of the group, employ a square diagram as presently mentioned.

Thus in the first mode, a group is $1, \beta, \beta^2, \alpha, \alpha\beta, \alpha\beta^2$ ($\alpha^2 = 1, \beta^3 = 1, \alpha\beta = \beta^2\alpha$); where observe that these conditions imply also $\alpha\beta^2 = \beta\alpha$:

Or in the second mode calling the same group $(1, \alpha, \beta, \gamma, \delta, \varepsilon)$, the laws of combination are given by the square diagram

	1	α	β	γ	δ	ε
1	1	α	β	γ	δ	ε
α	α	1	γ	β	ε	δ
β	β	ε	δ	α	1	γ
γ	γ	δ	ε	1	α	β
δ	δ	γ	1	ε	β	α
ε	ε	β	α	δ	γ	1

for the symbols $(1, \alpha, \beta, \gamma, \delta, \varepsilon)$ are in fact $= (1, \alpha, \beta, \alpha\beta, \beta^2, \alpha\beta^2)$.

The general problem is to find all the groups of a given order n; thus if $n = 2$, the only group is $1, \alpha$ ($\alpha^2 = 1$); $n = 3$, the only group is $1, \alpha, \alpha^2$ ($\alpha^3 = 1$); $n = 4$, the groups are $1, \alpha, \alpha^2, \alpha^3$ ($\alpha^4 = 1$), and $1, \alpha, \beta, \alpha\beta$ ($\alpha^2 = 1, \beta^2 = 1, \alpha\beta = \beta\alpha$);*
$n = 6$, there are three groups, a group $1, \alpha, \alpha^2, \alpha^3, \alpha^4, \alpha^5$ ($\alpha^6 = 1$); and two groups $1, \beta, \beta^2, \alpha, \alpha\beta, \alpha\beta^2$ ($\alpha^2 = 1, \beta^3 = 1$), viz: in the first of these $\alpha\beta = \beta\alpha$; while in the other of them (that mentioned above) we have $\alpha\beta = \beta^2\alpha, \alpha\beta^2 = \beta\alpha$.

*If $n = 5$, the only group is $1, \alpha, \alpha^2, \alpha^3, \alpha^4$ ($\alpha^5 = 1$). W. E. S.

But although the theory as above stated is a general one, including as a particular case the theory of substitutions, yet the general problem of finding all the groups of a given order n, is really identical with the apparently less general problem of finding all the groups of the same order n, which can be formed with the substitutions upon n letters; in fact, referring to the diagram, it appears that $1, \alpha, \beta, \gamma, \delta, \varepsilon$ may be regarded as substitutions performed upon the six letters $1, \alpha, \beta, \gamma, \delta, \varepsilon$, viz: 1 is the substitution unity which leaves the order unaltered, α the substitution which changes $1\alpha\beta\gamma\delta\varepsilon$ into $\alpha 1\gamma\beta\varepsilon\delta$, and so for $\beta, \gamma, \delta, \varepsilon$. This, however, does not in any wise show that the best or easiest mode of treating the general problem is thus to regard it as a problem of substitutions: and it seems clear that the better course is to consider the general problem in itself, and to deduce from it the theory of groups of substitutions.

CAMBRIDGE, *26th November, 1877.*

APPENDIX F

Mathematical Induction

The principle of mathematical induction is not a theorem. It is a powerful *method* for proving theorems. Other such methods are proof by contradiction, argument by symmetry, and the pigeonhole principle.

The following is probably the simplest form of this principle.

The Principle of Mathematical Induction — Version 1

Suppose that a set S of positive integers has the two properties:

1. $1 \in S$.
2. If $k \in S$, then also $k + 1 \in S$.

Then the set S consists of all the positive integers.

This is eminently reasonable. By the first property $1 \in S$. The second property therefore implies that

$$2 = 1 + 1 \in S$$

Now that $2 \in S$, it follows from the same second property that

$$3 = 2 + 1 \in S$$

and similarly $4 = 3 + 1$, $5 = 4 + 1, \ldots$ are all in S.

Let us see how this obvious principle can be used to prove a nonobvious fact.

■ **EXAMPLE F.1.** *If n is any positive integer, then*

$$1 + 2 + 3 + \cdots + n = \frac{n(n+1)}{2} \quad (1)$$

Proof. Let S be the set of all the positive integers n for which statement (1) is valid. The principle of mathematical induction will be used to demonstrate that S consists of all the positive integers, which is, of course, tantamount to proving the proposition.

It must first be shown that $1 \in S$. In other words, it must first be shown that when n is replaced by the integer 1 in equation (1), a true statement is obtained. This, however, is easily verified, since this replacement transforms (1) to the obviously true statement

$$1 = \frac{1(1+1)}{2}$$

To prove the validity of property 2 (in the definition of the induction principle), it must be shown that the assumption $k \in S$ leads to the conclusion $k + 1 \in S$. That $k \in S$ means that (1) is valid when n is replaced by k so that

$$1 + 2 + 3 + \cdots + k = \frac{k(k+1)}{2} \quad (2)$$

To conclude that $k + 1 \in S$, it must be shown that

$$1 + 2 + 3 + \cdots + k + (k+1) = \frac{(k+1)[(k+1)+1]}{2}$$

is also true. This is demonstrated, using equation (2), as follows:

$$1 + 2 + 3 + \cdots + k + (k+1) = \frac{k(k+1)}{2} + (k+1)$$

$$= \frac{k(k+1) + 2(k+1)}{2}$$

$$= \frac{(k+1)(k+2)}{2}$$

$$= \frac{(k+1)[(k+1)+1]}{2}$$

Thus the set S enjoys both of the properties required by the principle of mathematical induction, and so it consists of all the positive integers. In other words, statement (1) is valid for all the positive integers. ■

The same method is now used to prove a well-known formula that is often demonstrated by other means.

■ **EXAMPLE F.2.** *If r is any number distinct from 1 and n is any positive integer, then*

$$1 + r + r^2 + \cdots + r^n = \frac{1 - r^{n+1}}{1 - r} \tag{3}$$

Proof. Let S be the set of positive integers for which statement (3) is valid. As was the case above, our strategy will be to use the principle of mathematical induction to show that the set S consists of all the positive integers.

For $n = 1$, statement (3) reduces to

$$1 + r = \frac{1 - r^2}{1 - r}$$

which is true by virtue of the well known identity $1 - r^2 = (1 - r)(1 + r)$. Thus $1 \in S$.

Next, suppose that $k \in S$. In other words, suppose that

$$1 + r + r^2 + \cdots + r^k = \frac{1 - r^{k+1}}{1 - r} \tag{4}$$

Then, making use of (4),

$$\begin{aligned}
1 + r + r^2 + \cdots + r^k + r^{k+1} &= \frac{1 - r^{k+1}}{1 - r} + r^{k+1} \\
&= \frac{1 - r^{k+1} + (1 - r)r^{k+1}}{1 - r} \\
&= \frac{1 - r^{k+1} + r^{k+1} - r^{k+2}}{1 - r} \\
&= \frac{1 - r^{k+2}}{1 - r} = \frac{1 - r^{(k+1)+1}}{1 - r}
\end{aligned}$$

which means that statement (3) holds for $n = k + 1$ as well. In other words, $k + 1 \in S$.

Thus the set S enjoys both of the properties required by the principle of mathematical induction, and so it consists of all the positive integers. In other words, statement (3) is valid for all the positive integers. ■

Since the statement to be proved may involve several variables, it is advisable to begin a proof by mathematical induction by stating explicitly to which variable the process is applied. In any such proof, the verification of $1 \in S$ is referred to as *anchoring the induction*, the assumption that $k \in S$ is called the *induction hypothesis*, and the part of the proof that uses the induction hypothesis to prove that $k + 1 \in S$ is the *induction step*.

In all proofs by mathematical induction, the set S consists of the set of all the integers for which a certain statement is true, and it is customary to leave the set S implicit and to speak simply of the set of integers for which the statement holds. With this convention in mind, the principle of mathematical induction is restated in the following manner:

The Principle of Mathematical Induction — Version 2

Suppose that a statement about positive integers possesses the two properties:

1. The statement is true for 1.
2. If the statement holds for some integer $k \geq 1$, then it also holds for $k + 1$.

Then the statement in question holds for all the positive integers.

It is this new version that is employed in the next example.

■ **EXAMPLE F.3.** *If n is any positive integer, then*

$$1 + 3 + 5 + \cdots + (2n - 1) = n^2 \tag{5}$$

Proof. We proceed by induction on n. When n is replaced by 1, statement (5) becomes

$$1 = 1^2$$

which is clearly valid. Thus the induction process has been anchored.

Next suppose that (5) is valid for the integer k. In other words, the induction hypothesis is

$$1 + 3 + 5 + \cdots + (2k - 1) = k^2 \tag{6}$$

Then

$$1 + 3 + 5 + \cdots + (2k - 1) + [2(k + 1) - 1] = k^2 + [2(k + 1) - 1]$$
$$= k^2 + 2k + 2 - 1$$
$$= k^2 + 2k + 1$$
$$= (k + 1)^2$$

and so statement (5) also holds for $k + 1$. Thus statement (5) possesses property 2 above as well. It follows from version 2 of the principle of mathematical induction that statement (5) holds for all positive integers. ∎

The choice of 1 as the anchoring point for the principle of mathematical induction is merely a convention. Sometimes it is necessary to chose a different starting point, in which case this principle assumes a slightly different form.

The Principle of Mathematical Induction—Version 3

Suppose that a statement about integers possesses the two properties:

1. The statement is true for the integer a.
2. If the statement holds for some integer $k \geq a$, then it also holds for $k + 1$.

Then the statement in question holds for all integers that are greater than or equal to a.

■ **EXAMPLE F.4.** *Prove that if $n \geq 7$, then*

$$\left(\frac{3}{2}\right)^n > 2n + 1 \tag{7}$$

Proof. The statement of this proposition makes it clear that the induction should be anchored at $n = 7$. For this value of n, (7) reduces to

$$\left(\frac{3}{2}\right)^7 = 17.08\ldots > 2 \cdot 7 + 1$$

which is obviously true. Next assume the validity of (7) for some integer $k \geq 7$. In other words, assume that k is an integer such that $(3/2)^k > 2k + 1$ and $k \geq 7$. Then

$$\left(\frac{3}{2}\right)^{k+1} = \left(\frac{3}{2}\right)\left(\frac{3}{2}\right)^k > \frac{3}{2}(2k + 1) = 3k + \frac{3}{2}$$

$$= 2k + 3 + \left(k - \frac{3}{2}\right) > 2(k + 1) + 1$$

Notice that in verifying these steps, it was necessary to make use of both the induction hypothesis $(3/2)^k > 2k + 1$ and the assumption that $k \geq 7$. By version 3 of the principle of mathematical induction, inequality (7) is known to hold for all $n \geq 7$. ∎

The above examples are all algebraic in nature, but the following two examples come from calculus and geometry:

■ **EXAMPLE F.5.** *If $f(x)$ is any differentiable function of x, let $f'(x)$ denote its derivative with respect to x. Using the facts*

$$(x)' = 1 \qquad (8)$$

and

$$[u(x)v(x)]' = u'(x)v(x) + u(x)v'(x) \qquad \text{(the product rule)}$$

prove that

$$(x^n)' = nx^{n-1} \qquad \text{for every positive integer } n \qquad (9)$$

Proof. We proceed by induction on n. When $n = 1$, (9) reduces to (8), and so the induction is anchored at 1. Assume next that (9) holds for $n = k$. Then

$$(x^{k+1})' = [(x)(x^k)]' = (x)'(x^k) + (x)(x^k)' = 1 \cdot x^k + x \cdot kx^{k-1}$$
$$= x^k + kx^k = (k+1)x^k$$

Thus (9) holds for all positive integers n. ■

■ **EXAMPLE F.6.** *Prove that in a plane, any n straight lines, no two of which are parallel and no three of which are concurrent, divide the plane into $(n^2 + n + 2)/2$ regions.*

Proof. We proceed by induction on n. When $n = 1$,

$$\frac{n^2 + n + 2}{2} = \frac{(1^2 + 1 + 1)}{2} = 2$$

and it is clear that any 1 line divides the plane into 2 regions. Thus the induction is anchored at $n = 1$.

Assume that this proposition holds for $n = k$, and suppose that we are given $k + 1$ straight lines in a plane, no two of the lines being parallel and no three of them concurrent. Let one of these lines be labeled as q, and suppose that it is temporarily deleted. By the induction hypothesis, the remaining k lines divide the plane into $(k^2 + k + 2)/2$ regions. Restore the line q to its old position. Since q is not parallel to any of the other lines, and since it does not pass through any of their intersections, it follows that those lines cut q into $k + 1$ sections (two of which happen to be infinite). Each of these sections divides one of the old regions into two new regions. This means that the restoration of q raises the region count by

$$2(k+1) - (k+1) = k+1$$

Consequently the number of regions determined by the given $k+1$ lines is

$$\frac{k^2+k+2}{2} + k + 1 = \frac{k^2+k+2+2k+2}{2}$$

$$= \frac{(k^2+2k+1)+(k+1)+2}{2}$$

$$= \frac{(k+1)^2+(k+1)+2}{2}$$

Thus, by the principle of mathematical induction, we are done. ∎

Sometimes information about k does not transfer to information about $k+1$. For example, the prime factorization of k sheds no light whatsoever upon the prime factorization of $k+1$. For such cases we have yet another form of mathematical induction.

■ **The Principle of Mathematical Induction — Version 4**

Suppose that a statement about integers possesses the two properties:

1. The statement is true for the integer a.
2. For any integer $k \geq a$, if the statement is true for $a, a+1, \ldots, k-1$, then it is also true for k.

Then the statement in question holds for all integers that are greater than or equal to a.

■ **EXAMPLE F.7.** *Every positive integer $n \geq 2$ can be expressed as the product of a sequence of primes.*

Proof. We proceed by induction on n. Since 2 is a prime itself, the identity

$$2 = 2$$

anchors the induction at $n = 2$. Next let k be any positive integer ≥ 2 such that each of the integers $2, 3, \ldots, k-1$ can be factored into a product of primes. If k itself is a prime, then

$$k = k$$

is the required expression of k as the product of primes. If k is not a prime, then $k = ab$ for some integers $a, b \geq 2$. In that case we also have $a, b \leq k - 1$, so, by the above assumption, both a and b are expressible as products of primes. If

$$a = p_1 p_2 \cdots p_r \quad \text{and} \quad b = q_1 q_2 \cdots q_s$$

are such expressions, then

$$k = ab = p_1 p_2 \cdots p_r q_1 q_2 \cdots q_s$$

is the required expression of k as a product of primes. Hence, by version 4 of the principle of mathematical induction, each of the integers $n \geq 2$ is expressible as the products of primes. ∎

EXERCISES

Each of the examples in Exercises 1–17 is to be proved by mathematical induction on the integer n:

1. $1^2 + 2^2 + 3^2 + \cdots + n^2 = \dfrac{n(n + 1)(2n + 1)}{6}$

2. $1^3 + 2^3 + 3^3 + \cdots + n^3 = \dfrac{n^2(n + 1)^2}{4}$

3. $1 \cdot 2 + 2 \cdot 3 + 3 \cdot 4 + \cdots + n(n + 1) = \dfrac{n(n + 1)(n + 2)}{3}$

4. $\dfrac{1}{1 \cdot 2} + \dfrac{1}{2 \cdot 3} + \dfrac{1}{3 \cdot 4} + \cdots + \dfrac{1}{n(n + 1)} = \dfrac{n}{n + 1}$

5. $1 \cdot 2 \cdot 3 + 2 \cdot 3 \cdot 4 + 3 \cdot 4 \cdot 5 + \cdots + n(n + 1)(n + 2) = \dfrac{n(n + 1)(n + 2)(n + 3)}{4}$

6. $n^2 > 5n + 17$ for $n \geq 8$

7. $2^n > n^2$ for $n \geq 5$

8. $3^n > n^3$ for $n \geq 1$

9. $\left(\dfrac{4}{3}\right)^n > 1 + n$ for $n \geq 8$

10. $\lim\limits_{x \to \infty} x^n e^{-x} = 0$ for $n \geq 0$

11. $\int_0^\infty x^n e^{-x} \, dx = n!$ for $n \geq 0$

12. $n^3 + (n + 1)^3 + (n + 2)^3$ is divisible by 9 for all $n \geq 1$

13. $\cos \alpha \cos 2\alpha \cos 4\alpha \ldots \cos 2^n \alpha = \dfrac{\sin 2^{n+1} \alpha}{2^{n+1} \sin \alpha}$ for $n \geq 0$

14. The integer $11^{n+2} + 12^{2n+1}$ is divisible by 133 for each $n \geq 0$

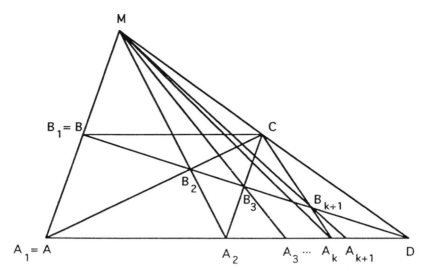

Figure F.1. A geometrical division method.

15. Given n planes, no two of which are parallel and no three of which contain the same straight line, they divide space into $(n^3 + 5n + 6)/6$ portions.

16. Every integer of the form $4n + 3$, $n \geq 0$, has a prime divisor of the same form.

17. Let $ABCD$ be a trapezoid in which the nonparallel sides AB and CD intersect in the point M (Figure F.1). Define $A_1 = A$, $B_1 = B$, and for each positive integer k, let B_{k+1} be the intersection of the lines CA_k and BD, and let A_{k+1} be the intersection of the lines MB_{k+1} and AD. Then

$$DA_n = \frac{DA}{n} \qquad \text{for } n \geq 1.$$

Biographies

CHAPTER 1

Archimedes (287–212 B.C.)

The Greek Archimedes was the greatest of the scientists, mathematicians, and engineers of antiquity. He lived in the city of Syracuse in Sicily. Among his accomplishments are the formulation of a precise theory of flotation, the computation of the volumes of many solids, including the sphere, and the construction of a variety of engines to defend his city against the besieging Romans.

Omar Khayyam (1048–1131)

Better known for his collection of poems *Rubaiyat*, the Persion Khayyam also wrote several mathematical, astronomical, and philosophical tracts. He tried to systematize the solution of the cubic equation and actually solved several special cases. He is also known for his work on Euclid's Parallel Postulate.

Muhammad ibn Musa al-Khwarezmi (ca. 780–850)

Al-Khwarezmi was on the faculty of the *House of Wisdom*, a scholarly institute in the city of Baghdad. The author of several mathematical and astronomical works, his name eventually gave rise to the term *algorithm*, and the first two syllables of his text *al-jabr wa'al-muqabala* mutated into the term *algebra*.

Rafael Bombelli (1526–1572)

An Italian engineer by profession, Bombelli also wrote a widely read teatise called *Algebra*. This book contains the first known attempt to systematize complex numbers.

Gerolamo Cardano (1507–1576)

A physician by profession, the Italian Cardano wrote the very influential text *Ars Magna* which included the solutions to the general cubic and quartic equations.

297

Leonhard Euler (1707–1783)

Together with the Frenchman Lagrange, the Swiss Euler completely dominated the mathematical developments of his time. He made fundamental contributions to all areas of mathematics, including some which did not begin to flourish until a century after his death.

Carl Friedrich Gauss (1777–1855)

The German Gauss is generally considered to be the greatest of the mathematicians to follow Newton. His profound work influenced the subsequent evolution of all the major areas of mathematics. In addition he also made important contributions to astronomy, physics, and geodesy.

Niels Henrik Abel (1802–1829)

Despite his early death from consumption, the Norwegian Abel exerted a major and lasting influence on the evolution of mathematics. He is best known for his work on the quintic equation and elliptic integrals.

Scipione del-Ferro (1465–1526)

The Italian del-Ferro was the first to succeed in solving cases of the cubic equation that had eluded both the Greek and the Islamic mathematicians.

Niccolo Tartaglia (1499–1557)

An Italian mathematician whose work on some cubic equations, together with that of his compatriots del-Ferro and Cardano, resulted in the complete solution of the general cubic equation.

Augustin-Louis Cauchy (1789–1857)

The Frenchman Cauchy was the most prolific of all the mathematicians of the nineteenth century. While he worked in many areas of mathematics, he is best remembered today as the founder of the theory of complex variables and also for his contributions to the rigorization of calculus.

CHAPTER 3

Lodovico Ferrari (1522–1565)

A student of Cardano's, Ferrari was the first to derive a formula for the solution of the general quartic equation.

Joseph Louis Lagrange (1736–1813)

Lagrange was born in Italy but spent most of his adult life in France, from where his ancestors had come. He made vast contributions to the fast evolving disciplines of calculus and differential equations. His work on algebraic equations, linear algebra, and number theory also proved very fecund in the long run.

A. T. Vandermonde (1735–1796)

A French mathematician whose pioneering work on the algebraic solution of equations was eclipsed by that of Lagrange.

Évariste Galois (1811–1832)

The Frenchman Galois possessed one of the most original mathematical minds of all times. Despite his untimely death at a duel, Galois's work on the solvability of equations eventually overshadowed that of his illustrious contemporaries Abel and Cauchy. He completely resolved the issue of solvability of equations and in the process created the modern mathematical discipline of group theory.

Isaac Newton (1642–1727)

An English scientist whose creativity influenced the evolution of both mathematics and physics more than that of any other individual. Among his many achievements are his theories of light and gravitation, and the development of calculus. His best-known book is the *Principia Mathematica*.

Joseph Raphson (1648–1715)

An Englishman whose tract *Analysis aequationum universalis* described Newton's numerical method for solving equations.

CHAPTER 4

Euclid (3rd century B.C.)

Euclid was a Greek who lived in Alexandria. He wrote several books of which the best known is *The Elements*. This textbook on geometry and number theory is arguably the most influential scientific tract of all times.

Adrien-Marie Legendre (1752–1833)

Legendre was a highly influential French mathematician. He made substantial contributions to geometry, analysis, and number theory, in each of which areas he also wrote a definitive text.

CHAPTER 5

Pierre de Fermat (1601–1655)

Known as the Prince of Mathematicians, the Frenchman Fermat was in fact a lawyer who regarded mathematics as a diversion. He did pioneering work in calculus and probability, and set number theory on a course it still follows.

Blaise Pascal (1623–1662)

Pascal was a French mathematician, philosopher and scientist. He invented the first adding machine and made major contributions toward the evolution of geometry and the theory of probability.

CHAPTER 8

Paolo Ruffini (1765–1822)

Ruffini was an Italian physician who published a proof of the unsolvability of the general quintic equation by radicals. While his proof was incomplete, it did contain some ideas that were eventually incorporated into Abel's proof of the same theorem.

CHAPTER 9

Felix Klein (1849–1925)

Klein was a highly influential German mathematician. His paper that came to be known as the *Erlanger Programm* focused the attention of mathematicians on the applications of group theory to geometry. He was also one of the pioneers of hyperbolic geometry and the calculus of complex variables.

Arthur Cayley (1821–1895)

Cayley was the most productive English mathematician to follow Newton. He made many contributions to geometry, algebra, and analysis but is best known for his work on invariant theory.

Camille Jordan (1838–1922)

A French mathematician who is best known for his pioneering work in group theory and linear algebra.

Bibliography

Abel, N. H., *Oeuvres complètes de Niels Henrik Abel*, Grondahl, Christiana, 1881.

Bell, E. T., *Men of Mathematics*, Simon and Schuster, New York, 1965.

Birkeland, B., Ludwig Sylow's Lectures on Algebraic Equations and Substitutions, Christiana (Oslo), 1862: An Introduction and a Summary, *Historia Mathematica*, 23 (1996), 182–199.

Birkhoff, G., and Mac Lane, S., *A Survey of Modern Algebra*, 4th ed., Macmillan, New York, 1977.

Bombelli, R., *Algebra*, Feltrinelli, Milan, 1966.

Borofsky, S., *Elementary Theory of Equations*, Macmillan, New York, 1959.

Cajori, F., *An Introduction to the Theory of Equations*, Macmillan, New York, 1969.

Cardano, G., *Ars Magna*, Dover, New York, 1993.

Cauchy, A. L., *Oeuvres complètes*, Gauthier-Villars, Paris, 1882.

Cauchy, A. L., Mémoire sur les premiers termes de la série des quantités qui sont propres a représenter le nombre des valeurs distinctes d'une fonction des n variables indépendantes, *Comptes Rendus Paris*, 21 (1845), 1093–1101.

Cauchy, A. L., Mémoire sur une nouvelle théorie des imaginaires, et sur les racines symboliques des équations et des équivalences. *Comptes Rendus Paris*, 24 (1847), 1120–1130.

Dickson, L. E., *New First Course in the Theory of Equations*, Wiley, New York, 1939.

Dickson, L. E., *Linear Groups*, Dover, New York, 1958.

Edwards, H. M., *Galois Theory*, Springer-Verlag, New York, 1984.

Euclid, *The Elements*, Dover, New York, 1956.

Fermat, P. de, *Oeuvres de Fermat*, ed. P. Tannery and C. Henry, Gauthier-Villars, Paris, 1922.

Gallian, J. A., *Contemporary Abstract Algebra*, 2nd ed., D. C. Heath, Lexington, MA, 1982.

Galois, E., *Écrits et mémoires mathematiques*, ed. R. Bourgne and J.-P. Azra, Gauthier-Villars, Paris, 1962.

Gillings, R. J., *Mathematics in the Time of the Pharaohs*, Dover, New York, 1972.

Hadlock, C. R., *Field Theory and Its Classical Problems*, The Mathematical Association of America, Washington, DC., 1978.

Hahn, L., *Complex Numbers and Geometry*, The Mathematical Association of America, Washington, DC., 1994.

Hall, H. S., and Knight, S. R., *Higher Algebra*, Macmillan, London, 1919.

Herstein, I. N., *Topics in Algebra*, 2nd ed., Wiley, New York, 1975.

Hungerford, T. W., *Algebra*, Springer-Verlag, New York, 1974.

Jordan, C. *Traité des substitutions et des equations algebriques*, Gauthier-Villars, Paris, 1870.

Katz, V., *A History of Mathematics: An Introduction*, HarperCollins, New York, 1993.

Kiernan, B. M., The development of Galois theory from Lagrange to Artin, *Arch. Hist. Exact Sci.*, 8 (1971), 40–154.

Kleiner, I., The evolution of group theory: A brief survey, *Math. Mag.*, 59 (1986), 195–215.

Kline, M., *Mathematical Thought from Ancient to Modern Times*, Oxford University Press, New York, 1990.

Knopp, K., *Problem Book in the Theory of Functions*, Dover, New York, 1948.

Lagrange, J. L., *Oeuvres de Lagrange*, Gauthier-Villars, Paris, 1867–1892.

MacWilliams, F. J. and Sloane, N. J. A., *The Theory of Error-Correcting Codes*, North-Holland, Amsterdam, 1977.

McCarthy, P. J., *Algebraic Extensions of Fields*, Dover, New York, 1991.

Needham, J., *Science and Civilization in China*, Cambridge University Press, Cambridge, 1959.

Serret, J. A., *Cours d'algebre superieure*, Gauthier-Villars, Paris, 1885.

Shanks, D., *Solved and Unsolved Problems in Number Theory*, Spartan Books, Washington, 1962.

Stauduhar, R. P., The Determination of Galois Groups, *Mathematics of Computation*, 27 (1973), 981–996.

Stewart, I., *Galois Theory*, Chapman and Hall, London, 1989.

Story, W. E., Note on the "15" Puzzle, *Amer. J. Math.*, 2 (1879), 399–404.

van der Waerden, B. L., *Modern Algebra*, F. Ungar, New York, 1953.

Wells, D., *The Penguin Dictionary of Curious and Interesting Numbers*, Penguin, Great Britain, 1986.

Wussing, H., *The Genesis of the Abstract Group Concept*, MIT Press, Cambridge, 1984.

Solutions to Selected Odd Exercises

EXERCISES 1.1

3. $\dfrac{1 \pm \sqrt{7}}{6}$ **9.** $\dfrac{3abc - b^3}{a^3}$ **11.** $\dfrac{b^2 - 4ac}{a^2}$

13. $-\dfrac{ab}{c^2}$ **15.** $x^2 + (p - q)x - pq = 0$ **17.** $\alpha \leqslant 0$

CHAPTER REVIEW EXERCISES

1. True   **3.** False

EXERCISES 2.1

1. Argument $= \arctan\left(\dfrac{3}{2}\right) \approx 56.3°$ and modulus $= \sqrt{2^2 + 3^2} = \sqrt{13} \approx 3.6$.

3. Argument $= 180° + \arctan\left(\dfrac{4}{3}\right) \approx 233.1°$ and modulus $= \sqrt{(-3)^2 + (-4)^2} = 5$.

5. $7 + 2i$ **7.** $-3 + 4i$ **9.** $13 + 13i$

11. $\dfrac{7}{26} + \dfrac{17}{26}i$ **13.** $-\dfrac{3\sqrt{3}}{7} + \dfrac{13}{7}i$ **15.** $-\dfrac{1}{2} - \dfrac{7}{2}i$

17. $-7 + 24i$ **19.** i **21.** $-i$

23. $z = -7 - 3i$ **25.** $w = z = i$

EXERCISES 2.2

1. $\left\{\dfrac{1}{\sqrt{2}} + \dfrac{1}{\sqrt{2}}i,\ i,\ -\dfrac{1}{\sqrt{2}} + \dfrac{1}{\sqrt{2}}i,\ -1,\ -\dfrac{1}{\sqrt{2}} - \dfrac{1}{\sqrt{2}}i,\ -i,\ \dfrac{1}{\sqrt{2}} - \dfrac{1}{\sqrt{2}}i,\ 1\right\}$

3. $\left\{\dfrac{\sqrt{3}}{2} + \dfrac{i}{2},\ i,\ -\dfrac{\sqrt{3}}{2} + \dfrac{i}{2},\ -\dfrac{\sqrt{3}}{2} - \dfrac{i}{2},\ -i,\ \dfrac{\sqrt{3}}{2} - \dfrac{i}{2}\right\}$

5. $\pm(2 - i)$

7. $\{1.08 + .29i,\ -.79 + .79i,\ -.29 - 1.08i\}$

9. $\{i, -\frac{\sqrt{3}}{2} - \frac{i}{2}, \frac{\sqrt{3}}{2} - \frac{i}{2}\}$ **11.** $\pm(c + i)$

15. $\{-i, \frac{-2+i}{5}\}$ **17.** $\{-i, -3 - 2i\}$

23. ± 1 **25.** ζ^{357}

EXERCISES 2.3

1. Rational **3.** Degree 2 algebraic
5. Rational **7.** Not algebraic
9. Degree 2 algebraic **11.** Not algebraic

EXERCISES 2.4

7. 102, 120, 128, 136, 160, 170, 192
13. Constructible **15.** Constructible
17. Constructible **19.** Not known to be constructible

EXERCISES 2.5

1. $\sqrt[4]{1} = \{i, -1, -i, 1\}$, with orders $= \{4, 2, 4, 1\}$
3. $\sqrt[6]{1} = \{-\omega^2, \omega, -1, \omega^2, -\omega, 1\}$ with orders $= \{6, 3, 2, 3, 6, 1\}$
5. $\sqrt[8]{1} = \{(1 + i)/\sqrt{2}, i, (-1 + i)/\sqrt{2}, -1, (-1 - i)/\sqrt{2}, -i, (1 - i)/\sqrt{2}, 1\}$ with orders $\{8, 4, 8, 2, 8, 4, 8, 1\}$.
7. Let $\zeta = \cos 2\pi/10 + i \sin 2\pi/10$. Then $\sqrt[10]{1} = \{\zeta, \zeta^2, \zeta^3, \zeta^4, \zeta^5, \zeta^6, \zeta^7, \zeta^8, \zeta^9, 1\}$ with orders $\{10, 5, 10, 5, 2, 5, 10, 5, 10, 1\}$.
9. $\{-\omega^2, -\omega\}$
11. $\{\zeta, \zeta^5, \zeta^7, \zeta^{11}\}$ where $\zeta = \cos 2\pi/12 + i \sin 2\pi/12$.

CHAPTER REVIEW EXERCISES

1. True **3.** False **5.** True
7. False **9.** True **11.** False
13. True

EXERCISES 3.1

1. $x_1 = 3^{2/3} - \dfrac{9}{3 \cdot 3^{2/3}} = 3^{2/3} - 3^{1/3}$

$x_2 = \omega 3^{2/3} - \omega^2 3^{1/3}$ and $x_3 = \omega^2 3^{2/3} - \omega 3^{1/3}$

3. $x_1 = 6^{2/3} - \dfrac{18}{3 \cdot 6^{2/3}} = 6^{2/3} - 6^{1/3}$

$x_2 = \omega 6^{2/3} - \omega^2 6^{1/3}$ and $x_3 = \omega^2 6^{2/3} - \omega 6^{1/3}$

5. $x_1 = y_1 - a/3 = 2^{2/3} - \dfrac{6}{3 \cdot 2^{2/3}} - 1 = 2^{2/3} - 2^{1/3} - 1$

$x_2 = \omega 2^{2/3} - \omega^2 2^{1/3} - 1$ and $x_3 = \omega^2 2^{2/3} - \omega 2^{1/3} - 1$

7. $x_1 = y_1 - a/3 = -2^{1/3}i\omega - \dfrac{-6}{3\cdot(-2^{1/3}i\omega)} - 1 = -2^{1/3}i\omega + 2^{2/3}i\omega^2 - 1$

$x_2 = -2^{1/3}i\omega^2 + 2^{2/3}i\omega - 1$ and $x_3 = -2^{1/3}i + 2^{1/3}i - 1$

9. $x_1 = y_1 - \dfrac{1}{2} = \dfrac{1}{2} - \dfrac{3/4}{3\cdot 1/2} - \dfrac{1}{2} = -\dfrac{1}{2}$

$x_2 = \omega\dfrac{1}{2} - \omega^2\dfrac{3/4}{3\cdot 1/2} - \dfrac{1}{2} = \dfrac{\omega - \omega^2 - 1}{2} = \omega$

$x_3 = \omega^2\dfrac{1}{2} - \omega\dfrac{3/4}{3\cdot 1/2} - \dfrac{1}{2} = \dfrac{\omega^2 - \omega - 1}{2} = \omega^2$

11. $-1 - i$

EXERCISES 3.3

1. $x_1 = 0$, $x_2 = -.6667$, $x_3 = -.5968$, $x_4 = x_5 = -.5958$
3. $x_1 = 0$, $x_2 = -.85$, $x_3 = x_4 = -.8449$
5. $x_1 = 0$, $x_2 = .5667$, $x_3 = x_4 = .5686$
7. $x_1 = -1$, $x_2 = -2.8305$, $x_3 = -2.0496$, $x_4 = -1.9387$, $x_5 = -1.9346 = x_6$
9. $x_1 = 2$, $x_2 = 2.1667$, $x_3 = 2.1545$, $x_3 = 2.1544 = x_4$
11. $x_1 = 3$, $x_2 = 4.5311$, $x_3 = 4.0488$, $x_4 = 3.7947$, $x_5 = 3.7311$, $x_6 = 3.7276 = x_7$

CHAPTER REVIEW EXERCISES

1. False   3. False   5. True

EXERCISES 4.1

1. $x \equiv 1$ 3. $x \equiv 0, 1$ 5. $x \equiv 0$
7. (1) $x \equiv 4$ (2) $x \equiv 2, 3$ (3) $x \equiv 0, 2, 3$
   (4) $x \equiv 4$ (5) $x \equiv 0$
9. No solutions   11. $y \equiv 3$, $x \equiv 6$ 13. $y \equiv 3$, $z \equiv 2$, $x \equiv 3$
15. $x \equiv 1$ 17. (11) $x \equiv 0$, $y \equiv 4$, (12) $y \equiv 3$, $x \equiv 11$.
19. (14) $x \equiv 7$, (15) No solutions in $\mathbb{Z}_{13}$.   21. 501
23. 4

EXERCISES 4.2

1. 365   3. 12   5. 1   7. 59
9. 18   11. 34   13. 72 15. 31
17. 65,521   19. $1 \equiv 1^{-1}$, $11 \equiv 11^{-1}$, $5 \equiv 5^{-1}$, $7 \equiv 7^{-1}$
21. $x = -42$, $y = 24$

EXERCISES 4.3

1. Squares $= \{0, 1, 2, 4\}$ 3. Squares $= \{0, 1, 3, 4, 9, 10, 12\}$
5. $\{0, 1, 6\}$ 7. $\{0, 1, 5, 8, 12\}$

EXERCISES 4.4

1. $2^6 \cdot 5^6$ **3.** $641 \cdot 6{,}700{,}417$
5. 2^{20} **15.** 325

CHAPTER REVIEW EXERCISES

1. True **3.** True
5. False **7.** True

EXERCISES 5.1

1. $128x^7 + 1344x^6y^2 + 6048x^5y^4 + 15{,}120x^4y^6 + 22{,}680x^3y^8 + 20{,}412x^2y^{10} + 10{,}206xy^{12} + 2187y^{14}$
3. $3x^{10} + 4y^5z^{15}$ **5.** z^{12}
7. $7{,}015{,}680a^{26}b^8c^{12}$ **9.** 489,888

EXERCISES 5.2

1. 0, 1, 2, 4, 4, 2, 1 **3.** $x = 1$
5. $x = 1$ **7.** $x = 7$
15. The primitive roots (mod 11) are the powers 2^n where n is relatively prime to 10; that is 2, 8, 7, 6.
17. $p = 2$, root = 1; $p = 3$, root = 2; $p = 5$, roots = 2, 4; $p = 7$, roots = 3, 5; $p = 11$, roots = 2, 6, 7, 8; $p = 13$, roots = 2, 6, 7, 11; $p = 17$, roots = 3, 5, 6, 7, 10, 11, 12, 14; $p = 19$, roots = 2, 3, 10, 13, 14, 15.
27. $a = 2, n = 4$

EXERCISES 5.3

1. $x^2 + y^2 + z^2 + 2xy + 2xz + 2yz$
3. $x^4 + y^4 + z^4 + 4x^3y + 4xy^3 + 4x^3z + 4xz^3 + 4y^3z + 4yz^3 + 6x^2y^2 + 6x^2z^2 + 6y^2z^2 + 12x^2yz + 12xy^2z + 12xyz^2$
5. $x^6 + y^9 + x^3y^3 + 3x^4y^3 + 3x^2y^6 + 3x^5y + 3x^4y^2 + 3xy^7 + 3x^2y^5 + 6x^3y^4$
7. 887,040
9. 2355

EXERCISES 5.4

1. 8, 48, 400

CHAPTER REVIEW EXERCISES

1. True **3.** False **5.** True
7. False **9.** True

EXERCISES 6.1

1. Quotient $= x^4 + x^3 + x^2 + x$, remainder $= x^2 + 1$

Solutions to Selected Odd Exercises   307

3. Quotient $= x^4 + 4x^3 + x^2 + 4x + 2$; remainder $= 2x^2 + 2x + 4$
11. $x^6 + 2x^3 + 1$ **13.** $x^6 + 2x^5 + 4x^4 + x^3 + 2x^2 + 4x + 1$

EXERCISES 6.2

1. 1
x^2
$x^2 + x = x(x + 1)$
x^3
$x^3 + x = x(x + 1)^2$
$x^3 + x^2 = x^2(x + 1)$
$x^3 + x^2 + x = x(x^2 + x + 1)$
x^4
$x^4 + x = x(x + 1)(x^2 + x + 1)$
$x^4 + x^2 = x^2(x + 1)^2$
$x^4 + x^2 + x = x(x^3 + x + 1)$
$x^4 + x^3 = x^3(x + 1)$
$x^4 + x^3 + x = x(x^3 + x^2 + 1)$
$x^4 + x^3 + x^2 = x^2(x^2 + x + 1)$
$x^4 + x^3 + x^2 + x = x(x + 1)^3$

x
$x + 1$
$x^2 + 1 = (x + 1)^2$
$x^2 + x + 1$
$x^3 + 1 = (x + 1)(x^2 + x + 1)$
$x^3 + x + 1$
$x^3 + x^2 + 1$
$x^3 + x^2 + x + 1 = (x + 1)^3$
$x^4 + 1 = (x + 1)^4$
$x^4 + x + 1$
$x^4 + x^2 + 1 = (x^2 + x + 1)^2$
$x^4 + x^2 + x + 1 = (x + 1)(x^3 + x^2 + 1)$
$x^4 + x^3 + 1$
$x^4 + x^3 + x + 1 = (x + 1)^2(x^2 + x + 1)$
$x^4 + x^3 + x^2 + 1 = (x + 1)(x^3 + x + 1)$
$x^4 + x^3 + x^2 + x + 1$

3. x^2
$x^2 + 2 = (x + 1)(x + 2)$
$x^2 + x + 1 = (x + 2)^2$
$x^2 + 2x = x(x + 2)$
$x^2 + 2x + 2$

$x^2 + 1$
$x^2 + x = x(x + 1)$
$x^2 + x + 2$
$x^2 + 2x + 1 = (x + 1)^2$

5. $x^3 + 1 = (x + 1)(x^2 + 4x + 1)$
$x^3 + 2x + 1$
$x^3 + 4x + 1 = (x + 2)(x^2 + 3x + 3)$

$x^3 + x + 1$
$x^3 + 3x + 1 = (x + 4)(x + 3)^2$

7. $\dfrac{p(p^2 - 1)}{3}$ **13.** $3x - 1$

15. $3x + 1$ **17.** $a \equiv 8, b \equiv 0 \pmod{11}$

EXERCISES 6.3

1. $x^2 + x + 1$ **3.** 1   **5.** 1
9. No   **11.** No
27. $(x^3 + 1) = x^2(x^7 + x^4 + x^3 + 1) + (x^3 + x + 1)(x^6 + x^4 + x + 1)$

EXERCISES 6.4

1. $a^2 - 2b$ **3.** $c - ab$ **5.** $-\dfrac{b}{c}$

7. $\dfrac{a^2 + b}{c - ab}$ **9.** $x^3 + 23x - 1$ **11.** $x^3 - 3x^2 + 26x - 23$

13. $x^3 + 23kx^2 + k^3$ **15.** 5   **17.** $\pm\sqrt{5}, \pm\sqrt{7}\,i, 4$

19. $\dfrac{5 \pm \sqrt{13}}{2}$ **27.** $-\dfrac{a_{n-1}}{a_n}$ **29.** $\dfrac{a_{n-2}}{a_n}$

EXERCISES 6.5

1. $\pm i, \pm 1$ 3. $\omega, \omega^2, 0, 1$

CHAPTER REVIEW EXERCISES

1. False   3. True   5. False
7. True   9. True

EXERCISES 7.1

1. $\tau, \tau^2, \tau^3 = \tau + 1, \tau^4 = \tau(\tau + 1) = \tau^2 + \tau,$
   $\tau^5 = \tau(\tau^2 + \tau) = \tau^3 + \tau^2 = \tau^2 + \tau + 1,$
   $\tau^6 = \tau(\tau^2 + \tau + 1) = \tau^3 + \tau^2 + \tau = \tau + 1 + \tau^2 + \tau = \tau^2 + 1,$
   $\tau^7 = \tau(\tau^2 + 1) = \tau^3 + \tau = \tau + 1 + \tau = 1.$

3. $\eta, \eta^2, \eta^3, \eta^4, \eta^5 = \eta^2 + 1, \eta^6 = \eta^3 + \eta,$
   $\eta^7 = \eta^4 + \eta^2, \eta^8 = \eta^5 + \eta^3 = \eta^3 + \eta^2 + 1,$
   $\eta^9 = \eta^4 + \eta^3 + \eta, \eta^{10} = \eta^5 + \eta^4 + \eta^2 = \eta^4 + 1,$
   $\eta^{11} = \eta^5 + \eta = \eta^2 + \eta + 1, \eta^{12} = \eta^3 + \eta^2 + \eta, \eta^{13} = \eta^4 + \eta^3 + \eta^2,$
   $\eta^{14} = \eta^5 + \eta^4 + \eta^3 = \eta^4 + \eta^3 + \eta^2 + 1,$
   $\eta^{15} = \eta^5 + \eta^4 + \eta^3 + \eta = \eta^4 + \eta^3 + \eta^2 + \eta + 1,$
   $\eta^{16} = \eta^5 + \eta^4 + \eta^3 + \eta^2 + \eta = \eta^4 + \eta^3 + \eta + 1,$
   $\eta^{17} = \eta^5 + \eta^4 + \eta^2 + \eta = \eta^4 + \eta + 1,$
   $\eta^{18} = \eta^5 + \eta^2 + \eta = \eta + 1, \eta^{19} = \eta^2 + \eta, \eta^{20} = \eta^3 + \eta^2,$
   $\eta^{21} = \eta^4 + \eta^3, \eta^{22} = \eta^5 + \eta^4 = \eta^4 + \eta^2 + 1,$
   $\eta^{23} = \eta^5 + \eta^3 + \eta = \eta^3 + \eta^2 + \eta + 1, \eta^{24} = \eta^4 + \eta^3 + \eta^2 + \eta,$
   $\eta^{25} = \eta^5 + \eta^4 + \eta^3 + \eta^2 = \eta^4 + \eta^3 + 1,$
   $\eta^{26} = \eta^5 + \eta^4 + \eta = \eta^4 + \eta^2 + \eta + 1,$
   $\eta^{27} = \eta^5 + \eta^3 + \eta^2 + \eta = \eta^3 + \eta + 1, \eta^{28} = \eta^4 + \eta^2 + \eta,$
   $\eta^{29} = \eta^5 + \eta^3 + \eta^2 = \eta^3 + 1, \eta^{30} = \eta^4 + \eta,$
   $\eta^{31} = \eta^5 + \eta = 1.$

5. $\xi, \xi^2 = \xi + 1, \xi^3 = \xi^2 + \xi = 2\xi + 1,$
   $\xi^4 = 2\xi^2 + \xi = 2\xi + 2 + \xi = 2, \xi^5 = 2\xi,$
   $\xi^6 = 2\xi^2 = 2\xi + 2, \xi^7 = 2\xi^2 + 2\xi = 2\xi + 2 + 2\xi = \xi + 2,$
   $\xi^8 = \xi^2 + 2\xi = \xi + 1 + 2\xi = 1.$

7. $\theta, \theta^2, \theta^3 = \theta^2 + 2, \theta^4 = \theta^3 + 2\theta = \theta^2 + 2\theta + 2,$
   $\theta^5 = \theta^3 + 2\theta^2 + 2\theta = \theta^2 + 2 + 2\theta^2 + 2\theta = 2\theta + 2,$
   $\theta^6 = 2\theta^2 + 2\theta, \theta^7 = 2\theta^3 + 2\theta^2 = 2\theta^2 + 4 + 2\theta^2 = \theta^2 + 1,$
   $\theta^8 = \theta^3 + \theta = \theta^2 + \theta + 2,$
   $\theta^9 = \theta^3 + \theta^2 + 2\theta = \theta^2 + 2 + \theta^2 + 2\theta = 2\theta^2 + 2\theta + 2,$
   $\theta^{10} = 2\theta^3 + 2\theta^2 + 2\theta = 2\theta^2 + 4 + 2\theta^2 + 2\theta = \theta^2 + 2\theta + 1,$
   $\theta^{11} = \theta^3 + 2\theta^2 + \theta = \theta^2 + 2 + 2\theta^2 + \theta = \theta + 2, \theta^{12} = \theta^2 + 2\theta,$
   $\theta^{13} = \theta^3 + 2\theta^2 = 2, \theta^{14} = 2\theta, \theta^{15} = 2\theta^2,$
   $\theta^{16} = 2\theta^3 = 2\theta^2 + 1, \theta^{17} = 2\theta^3 + \theta = 2\theta^2 + \theta + 1,$
   $\theta^{18} = 2\theta^3 + \theta^2 + \theta = 2\theta^2 + 4 + \theta^2 + \theta = \theta + 1, \theta^{19} = \theta^2 + \theta,$
   $\theta^{20} = \theta^3 + \theta^2 = 2\theta^2 + 2, \theta^{21} = 2\theta^3 + 2\theta = 2\theta^2 + 2\theta + 1,$
   $\theta^{21} = 2\theta^3 + 2\theta^2 + \theta = 2\theta^2 + 4 + 2\theta^2 + \theta = \theta^2 + \theta + 1,$
   $\theta^{22} = \theta^3 + \theta^2 + \theta = \theta^2 + 2 + \theta^2 + \theta = 2\theta^2 + \theta + 2,$
   $\theta^{23} = 2\theta^3 + \theta^2 + 2\theta = 2\theta^2 + 4 + \theta^2 + 2\theta = 2\theta + 1, \theta^{24} = 2\theta^2 + \theta,$
   $\theta^{25} = 2\theta^3 + \theta^2 = 2\theta^2 + 4 + \theta^2 = 1.$

9. $\alpha^2 = \alpha + 4$, $\alpha^3 = 5\alpha + 4$, $\alpha^4 = 2\alpha + 6$,
$\alpha^5 = \alpha + 1$, $\alpha^6 = 2\alpha + 4$, $\alpha^7 = 6\alpha + 1$,
$\alpha^8 = 3$, $\alpha^9 = 3\alpha$, $\alpha^{10} = 3\alpha + 5$,
$\alpha^{11} = \alpha + 5$, $\alpha^{12} = 6\alpha + 4$, $\alpha^{13} = 3\alpha + 3$,
$\alpha^{14} = 6\alpha + 5$, $\alpha^{15} = 4\alpha + 3$, $\alpha^{16} = 2$,
$\alpha^{17} = 2\alpha$, $\alpha^{18} = 2\alpha + 1$, $\alpha^{19} = 3\alpha + 1$,
$\alpha^{20} = 4\alpha + 5$, $\alpha^{21} = 2\alpha + 2$, $\alpha^{22} = 4\alpha + 1$,
$\alpha^{23} = 5\alpha + 2$, $\alpha^{24} = 6$, $\alpha^{25} = 6\alpha$,
$\alpha^{26} = 6\alpha + 3$, $\alpha^{27} = 2\alpha + 3$, $\alpha^{28} = 5\alpha + 1$,
$\alpha^{29} = 6\alpha + 6$, $\alpha^{30} = 5\alpha + 3$, $\alpha^{31} = \alpha + 6$,
$\alpha^{32} = 4$, $\alpha^{33} = 4\alpha$, $\alpha^{34} = 4\alpha + 2$,
$\alpha^{35} = 6\alpha + 2$, $\alpha^{36} = \alpha + 3$, $\alpha^{37} = 4\alpha + 4$,
$\alpha^{38} = \alpha + 2$, $\alpha^{39} = 3\alpha + 4$, $\alpha^{40} = 5$,
$\alpha^{41} = 5\alpha$, $\alpha^{42} = 5\alpha + 6$, $\alpha^{43} = 4\alpha + 6$,
$\alpha^{44} = 3\alpha + 2$, $\alpha^{45} = 5\alpha + 5$, $\alpha^{46} = 3\alpha + 6$,
$\alpha^{47} = 2\alpha + 5$, $\alpha^{48} = 1$.

11. $y = 1$, $x = 1 + \beta$ 13. $x = \beta^2$, $y = 1 + \beta + \beta^2$, $z = 1$

15. $y = 2$, $x = \beta - 2$ 17. $y = \beta + 2$, $z = \beta + 2$, $x = 0$

19. See Theorem 6.2.

EXERCISES 7.2

1. Each element except 0 and 1 has order 7.

3. Each element except 0 and 1 has order 31.

5. μ, μ^5, μ^7, μ^{11}, μ^{13}, μ^{17}, μ^{19}, μ^{23} have order 24;
μ^2, μ^{10}, μ^{14}, μ^{22} have order 12; μ^3, μ^9, μ^{15}, μ^{21} have order 8;
μ^4, μ^{20} have order 6; μ^6, μ^{18} have order 4; μ^8, μ^{16} have order 3;
μ^{12} has order 2. 1 has order 1.

7. $(x - \beta)(x - \beta^6)(x - \beta^2)(x - \beta^5)(x - \beta^3)(x - \beta^4)(x - 1)x =$
$(x + \beta)(x + \beta^6)(x + \beta^2)(x + \beta^5)(x + \beta^3)(x + \beta^4)(x + 1)x =$
$[x^2 + (\beta + \beta^6)x + 1][x^2 + (\beta^2 + \beta^5)x + 1][x^2 + (\beta^3 + \beta^4)x + 1](x^2 + x) =$
$(x^2 + \beta^2 x + 1)(x^2 + \beta^4 x + 1)(x^2 + \beta x + 1)(x^2 + x) =$
$[x^4 + (\beta^2 + \beta^4)x^3 + (1 + \beta^6 + 1)x^2 + (\beta^2 + \beta^4)x + 1][x^4 + (1 + \beta)x^3 + (1 + \beta)x^2 + x] =$
$(x^4 + \beta^5 x^3 + \beta^6 x^2 + \beta^5 x + 1)(x^4 + \beta^5 x^3 + \beta^5 x^2 + x) =$
$x^8 + (\beta^5 + \beta^5)x^7 + (\beta^6 + \beta^{10} + \beta^5)x^6 + (1 + \beta^{10} + \beta^{11} + \beta^5)x^5 +$
$(\beta^5 + \beta^{11} + \beta^{10} + 1)x^4 + (\beta^6 + \beta^{10} + \beta^5)x^3 + (\beta^5 + \beta^5)x^2 + x =$
$x^8 + x = x^8 - x$.

13. 0, except that when $p = 2$ and $v = 1$ the answer is 1.

EXERCISES 7.3

1. α and $\alpha^2 = 1 + \alpha$ 3. 2, 3

5. The primitive elements are σ, σ^2, $\sigma^4 = \sigma + 1$, $\sigma^7 = \sigma^3 + \sigma + 1$, $\sigma^8 = \sigma^2 + 1$,
$\sigma^{11} = \sigma^3 + \sigma^2 + \sigma$, $\sigma^{13} = \sigma^3 + \sigma^2 + 1$, $\sigma^{14} = \sigma^3 + 1$.

7. The primitive elements are σ, $\sigma^3 = 2\sigma + 2$, $\sigma^5 = 2\sigma$, $\sigma^7 = \sigma + 1$.

9. α, $\alpha^5 = \alpha + 1$, $\alpha^7 = 6\alpha + 1$, $\alpha^{11} = \alpha + 5$,
$\alpha^{13} = 3\alpha + 3$, $\alpha^{17} = 2\alpha$, $\alpha^{19} = 3\alpha + 1$,
$\alpha^{23} = 5\alpha + 2$, $\alpha^{25} = 6\alpha$, $\alpha^{29} = 6\alpha + 6$,

$\alpha^{31} = \alpha + 6, \alpha^{35} = 6\alpha + 2, \alpha^{37} = 4\alpha + 4,$
$\alpha^{41} = 5\alpha, \alpha^{43} = 4\alpha + 6, \alpha^{47} = 2\alpha + 5.$

EXERCISES 7.4

1. $x^2 + 4x + 2, x^2 + 3x + 3, x^2 + x + 2, x^2 + 2x + 3$
3. $x^3 + 2x + 1, x^3 + 2x^2 + x + 1, x^3 + x^2 + 2x + 1, x^3 + 2x^2 + 1$
11. $x^4 + x + 1$ is the minimal polynomial of $\sigma, \sigma^2, \sigma^4, \sigma^8$.
    $x^4 + x^3 + x^2 + x + 1$ is the minimal polynomial of $\sigma^3, \sigma^6, \sigma^9, \sigma^{12}$.
    $x^2 + x + 1$ is the minimal polynomial of σ^5, σ^{10}.
    $x^4 + x^3 + 1$ is the minimal polynomial $\sigma^7, \sigma^{14}, \sigma^{13}, \sigma^{11}$.
    x and $x + 1$ are the minimal polynomials of 0 and 1.
13. $GF(5, x^2 + 4x + 2)$. Employing the solution to Exercise 7.1.8,
    $x^2 + 4x + 2$ is the minimal polynomial for μ and μ^5.
    $(x - \mu^2)(x - \mu^{10}) = x^2 + 3x + 4, (x - \mu^3)(x - \mu^{15}) = x^2 + 3,$
    $(x - \mu^4)(x - \mu^{20}) = x^2 + 4x + 1, (x - \mu^7)(x - \mu^{11}) = x^2 + 3x + 3,$
    $(x - \mu^8)(x - \mu^{16}) = x^2 + x + 1, (x - \mu^9)(x - \mu^{21}) = x^2 + 2,$
    $(x - \mu^{13})(x - \mu^{17}) = x^2 + x + 2, (x - \mu^{14})(x - \mu^{22}) = x^2 + 2x + 4,$
    $(x - \mu^{19})(x - \mu^{23}) = x^2 + 2x + 3,$
    $x - a$ is the minimal polynomial for all $a \in \mathbb{Z}_5$.

CHAPTER REVIEW EXERCISES

1. True   3. False
5. False   7. True

EXERCISES 8.1

1. 1   3. 1   5. 1
7. 1   9. 2   11. 3
13. 1   15. 3   17. 6
19. 4   21. 6   23. 6
25. 24   27. 1   29. 1
31. 20   33. 15   35. 120
37. $x_1 x_2 \cdots x_n$ 39. $x_1^2 x_2^2 x_3 \cdots x_n$
41. $(x_1 x_2 + x_3 x_4) x_5 x_6 \cdots x_n$
43. $x_1 x_2 \cdots x_{n-k} x_n^2$

EXERCISES 8.2

1. (1 9 8 6 5)(2 3 4 7)
3. (1 9)(2 8)(3 7)(4 6)(5)
5. (1)(2), (1 2)
7. (1), (1 2), (1 3), (1 4), (2 3), (2 4), (3 4),
   (1 2 3), (1 3 2), (1 2 4), (1 4 2), (1 3 4), (1 4 3), (2 3 4), (2 4 3),
   (1 2)(3 4), (1 3)(2 4), (1 4)(2 3),
   (1 2 3 4), (1 2 4 3), (1 3 2 4), (1 3 4 2), (1 4 2 3), (1 4 3 2)

9. (1 3 7)(2 4 9 6)(5)(8), order = 12
11. (1 7)(8 3)(2 5 4)(6)(9), order = 6
13. (1 5 2)(3 4 7 9)(6 8), order = 12
15. (1 2)(2 3)(4 5)(5 6)(6 7)(8 9)
17. (1 9)(9 8)(8 6)(6 5)(2 3)(3 4)(4 7)

EXERCISES 8.3

1. Yes, x_1 3. Yes, $x_1 - x_2$
5. Yes, $\Delta_3 = (x_1 - x_2)(x_1 - x_3)(x_2 - x_3)$
7. Yes, $x_1 x_2 x_3 x_4$
9. Yes, $x_1 x_2 + x_3 x_4$ 11. Yes, $x_1 x_2 x_3 x_4 x_5$
13. No, by Corollary 8.9   15. Yes, x_3
17. Yes, $x_1 x_2 x_3 x_4 x_5 x_6$ 19. No, by Theorem 8.8, with $p = 5$
21. No, by Theorem 8.8, with $p = 7$ 23. Yes, $x_1 x_2$
25. Yes, $(x_1 x_2 + x_3 x_4) x_5 x_6 \cdots x_n$

EXERCISES 8.4

1. Odd   3. Even   5. (1)(2)
7. (1), (1 2 3), (1 3 2), (1 2 4), (1 4 2), (1 3 4), (1 4 3), (2 3 4), (2 4 3), (1 2)(3 4), (1 3)(2 4), (1 4)(2 3)
9. (1 2), (1 3), (2 3)   23. Yes, $x_n \Delta_{n-1}$
25. Yes, $x_{n-2} x_{n-1} x_n \Delta_{n-3}$ if $n \geqslant 5$; $x_4 \Delta_3$ if $n = 4$, Δ_3 if $n = 3$.

CHAPTER REVIEW EXERCISES

1. False   3. True   5. True
7. False   9. False

EXERCISES 9.1

1. Yes   3. No   5. No
7. Id, (1 2)   9. All S_3 11. Id, (2 3)
13. Id, (1 2)(3 4), (1 3)(2 4), (1 4)(2 3)
15. Id, (1 2), (3 4), (1 2)(3 4), (1 3)(2 4), (1 4)(2 3), (1 3 2 4), (1 4 2 3)
19. After Id, the rotations are grouped by the nature of their axes.
    Axis joining vertices: (5 4 6 2), (5 6)(2 4), (5 2 6 4), (1 6 3 5), (1 3)(5 6), (1 5 3 6), (1 2 3 4), (1 3)(2 4), (1 4 3 2).
    Axis joining midpoints of edges: (1 2)(3 4)(5 6), (1 4)(2 3)(5 6), (1 6)(3 5)(2 4), (1 5)(3 6)(2 4), (1 3)(2 6)(4 5), (1 3)(2 5)(4 6).
    Axis joining centers of opposite faces: (1 2 6)(5 3 4), (1 6 2)(5 4 3), (1 6 4)(2 3 5), (1 4 6)(2 5 3), (1 4 5)(2 6 3), (1 5 4)(2 3 6), (1 2 5)(4 6 3), (1 5 2)(4 3 6).
21. The icosahedron has 10 pairs of opposite faces. The line joining the center of such a pair of faces acts as the axis of 2 nontrivial rotations of angles $\pm 120°$. This

312 Solutions to Selected Odd Exercises

accounts for $10 \cdot 2 = 20$ symmetries. The icosahedron has 15 pairs of opposite edges. The line joining the midpoints of such a edge acts as the axis of a $180°$ rotation. This accounts for 15 symmetries. The icosahedron has 6 pairs of opposite vertices. The line joining each such pair of vertices acts as the axis of 4 nontrivial rotations of angles $72°$, $144°$, $216°$, and $288°$ respectively. This accounts for $6 \cdot 4 = 24$ symmetries. Including the identity, we thus have $24 + 15 + 20 + 1 = 60$ symmetries.

23. (1 3 4), $120°$ clockwise rotation about altitude from 2.
25. (1 2 3), $120°$ clockwise rotation about altitude from 4.
27. (1 4)(2 3), $180°$ rotation about line joining centers of edges 14 and 23.
29. (1 3 6)(4 7 5). A $120°$ rotation about the diagonal joining 2 and 8.
31. (1 7)(2 6)(3 5)(4 8). This is a $180°$ rotation about the axis joining the midpoints of the edges 26 and 48.
33. (1 4)(2 8)(3 5)(6 7). This is a $180°$ rotation about the axis joining the midpoints of the edges 14 and 67.
35. $2n$.

EXERCISES 9.2

1. Yes **3.** No **5.** No
7. No **9.** No **11.** Yes
13. Yes **15.** $\{0,\}, \{1,3\}, \{2\}$ **17.** $\{1\}, \{i, -i\}, \{-1\}$
19. Each element is its own inverse.
21. $\{0\}, \{1,4\}, \{2,3\}$ **23.** $\{1\}, \{5\}$ **25.** $\{0\}, \{1,5\}, \{2,4\}, \{3\}$
27. $\{1\}, \{a, e\}, \{b, f\}, \{c, g\}, \{d\}$

EXERCISES 9.3

3. No, K and $(\mathbb{Z}_4, +)$ are not isomorphic to each other.
11. No **21.** Define $f(z) = z^2$.

EXERCISES 9.4

1. $m = 1: \{0\}; m = 2: \{0\}, \mathbb{Z}_2; m = 3: \{0\}, \mathbb{Z}_3;$
  $m = 4: \{0\}, \{0, 2\}, \mathbb{Z}_4; m = 5: \{0\}, \mathbb{Z}_5;$
  $m = 6: \{0\}, \{0, 3\}, \{0, 2, 4\}, \mathbb{Z}_6; m = 7: \{0\}, \mathbb{Z}_7;$
  $m = 8: \{0\}, \{0, 4\}, \{0, 2, 4, 6\}, \mathbb{Z}_8; m = 9: \{0\}, \{0, 3, 6\}, \mathbb{Z}_9;$
  $m = 10: \{0\}, \{0, 5\}, \{0, 2, 4, 6, 8\}, \mathbb{Z}_{10}.$
3. $\{\text{Id}\}, \{\text{Id}, (1\ 2)\}, \{\text{Id}, (1\ 2\ 3), (1\ 3\ 2)\},$
  $\{\text{Id}, (1\ 2\ 3\ 4), (1\ 3)(2\ 4), (1\ 4\ 3\ 2)\}, S_3, D_4, A_4, S_4$, respectively.
5. $\{\text{Id}, (1\ 2)(3\ 4\ 5), (1\ 2)(3\ 5\ 4), (1\ 2), (3\ 4\ 5), (3\ 5\ 4)\}, D_4, D_5, A_4$, respectively.
7. $\{0\}, \{0, \beta^k\}, k = 0, 1, \ldots, 6; \{0, \beta^i, \beta^j, \beta^i + \beta^j\}$, where i and j are distinct elements of $\{0, 1, \ldots, 6\}$; $(\text{GF}(2, x^3 + x^2 + 1), +)$.
9. $\{1\}, \{1, d\}, \{1, a, d, e\}, \{1, b, d, f\}, \{1, c, d, g\}$, the whole group.
11. Also $\{1, 4, 7, 10, 13\}$ and $\{2, 5, 8, 11, 14\}$.
13. Also $\{1, 10\}, \{2, 11\}, \{3, 12\}, \{4, 13\}, \{5, 14\}, \{6, 15\}, \{7, 16\}, \{8, 17\}$.
15. Also $\{(1\ 3), (1\ 2\ 3)\}, \{(2\ 3), (1\ 3\ 2)\}$.

Solutions to Selected Odd Exercises 313

17. $\{\text{Id}, (1\ 2)(3\ 4), (1\ 3)(2\ 4), (1\ 4)(2\ 3)\}, \{(1\ 2\ 3), (1\ 3\ 4), (2\ 4\ 3), (1\ 4\ 2)\}$,
    $\{(1\ 3\ 2), (2\ 3\ 4), (1\ 2\ 4), (1\ 4\ 3)\}$.
19. Also $\{a, e\}, \{b, f\}, \{c, g\}$.
21. $H = \{1, \rho, \rho^2, \ldots, \rho^{n-1}\}$, where ρ is the counterclockwise rotation by the angle $2\pi/n$. There is only one other coset, namely $G - H$.
23. $k = 1$: $x_1 x_2 x_3$; $k = 2$: $(x_1 - x_2)(x_1 - x_3)(x_2 - x_3)$;
    $k = 3$: x_1; $k = 4, 5$: not possible; $k = 6$: $x_1 x_2^2 x_3^3$.
    No such functions exist for $k > 6 = 3!$.
25. (a) $k = 1$: $x_1 x_2 x_3 x_4 x_5$; $k = 2$: Δ_5, $k = 5$: x_1; $k = 3, 4$: no such function can exist.
    (b) $k = 10$: $x_1 x_2$; $k = 12$: let f be the function such that $D_5 = S_{5,f}$; $k = 11, 13, 14$ no such functions exist.
29. No   31. S_4 33. A_4 35. D_4
37. $\{\text{Id}, (1\ 2\ 3\ 4\ 5), (1\ 3\ 5\ 2\ 4), (1\ 4\ 2\ 5\ 3), (1\ 5\ 4\ 3\ 2)\}$
47. $\{1, d\}$ 49. A_4
51. S_n: if $n = 1$ or 2, then the center equals S_n. For $n > 2$, the center of S_n is $\{\text{Id}\}$.

EXERCISES 9.5

3. $(\mathbb{Z}_4, +)$ 5. $(\mathbb{Z}_4, +)$ 7. K 9. $(\mathbb{Z}_4, +)$
11. K 15. n 17. 10
19. 5   21. 21   23. 4

EXERCISES 9.6

1. $P_0 = \text{Id}, P_1 = (0\ 1)$
3. $P_0 = \text{Id}, P_1 = (0\ 1\ 2\ 3), P_2 = (0\ 2)(1\ 3), P_3 = (0\ 3\ 2\ 1)$.
5. $P_0 = \text{Id}, P_1 = (0\ 1\ 2\ 3\ 4), P_2 = (0\ 2\ 4\ 1\ 3), P_3 = (0\ 3\ 1\ 4\ 2)$,
   $P_4 = (0\ 4\ 3\ 2\ 1)$.
7. $P_0 = \text{Id}, P_1 = (0\ 1\ 2\ 3\ 4\ 5), P_2 = (0\ 2\ 4)(1\ 3\ 5)$,
   $P_3 = (0\ 3)(1\ 4)(2\ 5), P_4 = (0\ 4\ 2)(1\ 5\ 3), P_5 = (0\ 5\ 4\ 3\ 2\ 1)$.

CHAPTER REVIEW EXERCISES

1. True   3. False   5. True   7. True
9. True   11. False   13. True
15. True   17. True   19. True

EXERCISES 10.1

1. $(\mathbb{Z}_4, +)$ 3. $(\mathbb{Z}_5, +)$ 5. H is not normal in G.
7. $(\mathbb{Z}_3, +)$ 9. H is not normal in G.   11. $(\mathbb{Z}_4, +)$
13. $G/\langle(1\ 2\ 3\ 4)\rangle \cong (\mathbb{Z}_2, +)$, $G/\{\text{Id}, (1\ 3)(2\ 4)\} \cong K$.
15. $G/\langle x \rangle \cong K$ if $x \neq 0$.
    $G/\{0, x, y, x + y\} \cong (\mathbb{Z}_2, +)$ if x and y are distinct and nonzero.
17. The nontrivial subgroups of G are $\{1, \sigma^2, \sigma^4, \sigma^6\}$ and $\{1, \sigma^4\}$. The quotients are isomorphic to $(\mathbb{Z}_2, +)$ and $(\mathbb{Z}_4, +)$, respectively.
19. $G/K \cong (\mathbb{Z}_3, +)$.

29. It follows from Exercise 8.2.23 that each conjugacy class consists of a maximal set of permutations whose disjoint cycle decompositions all have the same number of k-cycles for each positive integer k.

EXERCISES 10.2

1. $\{a + b\sqrt{5} \mid a, b \in \mathbb{Q}\}$
3. $\{a + bi \mid a, b \in \mathbb{Q}\}$
5. $\{a + b\omega \mid a, b \in \mathbb{Q}\} = \{a + b\sqrt{-3} \mid a, b \in \mathbb{Q}\}$
7. $\{a + b\sqrt{2} + ci + di\sqrt{2} \mid a, b, c, d \in \mathbb{Q}\}$
9. $\mathrm{GF}(2, x^2 + x + 1)$
11. $\mathrm{GF}(2, x^4 + x^3 + x^2 + x + 1)$
13. This is $\mathrm{GF}(2, x^2 + x + 1)$ which has order 4.
15. This is $\mathrm{GF}(2, x^4 + x^3 + x^2 + x + 1)$ which has order 16.
19. $-\dfrac{2}{41} + \dfrac{3}{41}[x]$ **21.** $\dfrac{1}{6} - \dfrac{1}{6}[x]$

CHAPTER REVIEW EXERCISES

1. False   **3.** False   **5.** True

EXERCISES 11.2

9. (a), (d), (e), (g) are all cyclic of order 8. (f) has every element but the identity of order 2, so it is isomorphic to (i). (h) is commutative with an element of order 4 but none of order 8. (b) and (c) are noncommutative and also not isomorphic to each other.

11. (a), (c), (d), (e) are all cyclic and so isomorphic to each other. (i) and (d) are isomorphic to each other.

CHAPTER REVIEW EXERCISES

1. False   **3.** False
5. True   **7.** True

Symbols

$(\mathbb{C}, +)$	193	$\phi(n)$	93
$(F^*, \cdot)$	193	F^*	193
$(G, \cdot)$	191	$F[x, \leq n]$	193
(m, n)	62	$F[x]$	101
$\binom{n}{k}$	75	G/H	227
$(\mathbb{Q}, +)$	193	$\mathrm{GF}(p, P(x))$	133
$(\mathbb{R}, +)$	193	G^k	253
$(\mathbb{Z}_n^*, \cdot)$	193	i	9
$(\mathbb{Z}, +)$	193	i_c	43
$(\mathbb{Z}_n, +)$	193	ma	100
$(\sqrt[n]{1}, \cdot)$	193	$o(a)$	39, 88, 135, 201
0_c	42		
1_c	43	$\mathbb{Q}$	99
$\langle a \rangle$	212	$\mathbb{R}$	99
$[G:H]$	208	$\sqrt[n]{z}$	18
$a^{\#}$	191	S_n	182
$\frac{a}{b}$	102	$S_{n,f}$	182
a^m	100	ω	19
A_n	184	$\mathbb{Z}$	58
$\mathbb{C}$	99	$\mathbb{Z}_n^*$	193
$C(a)$	230	$\mathbb{Z}_n$	58
$c_{n,k}$	75	$\circ$	156
Δ_n	166	$|z|$	11
D_n	187	$\|\sigma\|$	162
		$\equiv$	57

Pronounciation Guide

a	ask, cat
ä	father, balm
ẹ	her, system
e	set, mess
ē	me, beet
g	get, rag
i	is, fit
KH	as in Scottish loch or in German ich
N	as in French bon or un
o	ox, not
ō	note, no
ô	thought, saw
oi	boil, boy
ōō	too, moon
R	rolled r as in French rouge or in German rot
u	but, rum
z	zebra, maze
zh	pleasure, age

Index

Abel, Niels Henrick (ä' bel), 7, 53, 154, 165, 181, 246, 271, 298
Abelian group, 194
Abstract group, 191
Addition:
 Cartesian numbers, 42
 complex numbers, 10
 in fields, 98
 modular, 58
 polynomials, 101
 table, 58
Additive:
 identity, 99
 inverse, 43, 60, 99
al-Khayyami, Umar (äl KHä yä mē), 1, 297
al-Khwarizmi, Muhammad (äl KHwär iz' mē), 3, 261, 297
Algebra, etymology, 3
Algebraic expression, 24
Algebraic resolvability, 25, 52, 243–246
Algebraic solution, 24
Algorithm, etymology, 3
Alternating group, 184
Archimedes (är ke mē' dēz), 1, 297
Argument, 12
Argument principle, 15
Ars Magna, 5, 266–270
Associativity:
 abstract group, 191
 Cartesian numbers, 42
 cosets, 222
 field, 98
 modular arithmetic, 58
 permutations, 156
 polynomials, 101
Automorphism, 202

Binet's formula, 91
Binomial coefficient, 75

Binomial Theorem, 76, 100
Bombelli, Rafael, 5, 297

Cardano, Gerolamo (kär dä' nō), 1, 5, 266, 297
Carmichael number, 90
Cartesian numbers, 41
 addition, 42
 associativity, 42
 unit, 43
 zero, 43
Cartesian plane, 10
Cartesian representation, 10
Cauchy, Augustin-Louis (kō shē'), 7, 154, 165, 239, 257, 298
Cauchy's Theorem, 164
Cayley, Arthur, 191, 200, 215, 217, 285, 300
Cayley's Theorem, 217
Center, 211, 255
Centralizer, 211, 254
Chinese Remainder Theorem, 68
Circle, 27, 37
Class equation, 256
Closure, 98
Common divisor, 62
Commutative group, 194
Commutativity:
 Cartesian numbers, 42, 44
 field, 98
 modular arithmetic, 58
 polynomials, 101
Complex number, 7, 10
 addition, 10
 conjugate, 16
 division, 10
 multiplication, 10, 14
 subtraction, 10
Composite number, 39
Composition of permutations, 156

319

Configuration:
  constructible, 28
  15-puzzle, 168
Congruence, 57
Conjugacy class, 230, 254
Conjugate:
  complex number, 16
  group element, 230, 254
Constant polynomial, 102
Constructible:
  configuration, 28
  number, 28
Continuous function, 52
Coset, 205
Coset multiplication, 221
Cube, 27, 37, 189
Cubic equation, 4, 47, 121
Cycle, 157
Cyclic:
  group, 213
  permutation, 157
  table, 134
Cyclotomic equation, 51, 181, 245
Cylindrical (k, n)-puzzle, 176

Decomposable group, 253
Degree 2 algebraic expression, 25
Degree of polynomial, 102
del Ferro, Scipione, 1, 298
Dihedral group, 187
Direct product, 250
Discriminant:
  of cubic equation, 51
  polynomial, 166
Disjoint cycle decomposition, 159
Disquisitiones Arithmeticae, 28, 52
Distinct variant, 153
Distributivity:
  Cartesian numbers, 42, 44
  field, 98
  modular arithmetic, 58
  polynomials, 101
Division:
  complex numbers, 10
  polynomials, 102
Divisor:
  integer, 39
  polynomial, 107
Dodecahedron, 190
Doubling the cube, 27, 37

Elementary symmetric polynomial, 120
The Elements, 62, 66

Equation:
  cubic, 4, 47, 121
  cyclotomic, 51, 181, 245
  fifth degree, 51, 53, 154, 246
  first degree, 2
  fourth degree, 51, 124, 152
  quadratic, 3
  quartic, 51, 124, 152
  quintic, 51, 53, 153, 246
Equilateral triangle, 19, 27
Erlanger Programm, 185
Euclid (yōō' klid), 62, 66, 299
Euclidean algorithm:
  integers, 63
  polynomials, 111
Euclidean geometry, 27
Euler, Leonhard (oi' lėr), 7, 69, 93, 144, 298
Even permutation, 167
Exponent, 203
Extension, 236

Factorization:
  integers, 70
  polynomials, 102, 106, 116
Fermat, Pierre de (feR mä'), 86, 214, 299
Fermat's Theorem, 86, 214
Ferrari, Lodovico, 51, 298
Ferro-Tartaglia-Cardano formula, 1, 5, 53
Fibonacci numbers, 84, 91, 97
Field:
  addition, 98
  associativity, 98
  closure, 98
  commutativity, 98
  distributivity, 98
  isomorphism, 234
  unit, 99
  zero, 99
15-puzzle, 168
Fifth degree equation, 51, 53, 154, 246
First degree equation, 2
First nth root of unity, 20
Fourth degree equation, 51, 124, 152
Fundamental Theorem of Algebra, 53, 110, 242
Fundamental Theorem of Arithmetic, 70
Fundamental Theorem of Symmetric Polynomials, 120, 245

Galois, Évariste (gal wä'), 52, 53, 129, 144, 181, 214, 238, 243, 277, 299
  field, 133
  group, 243

imaginary, 130, 238
   polynomial, 140
   Theorem, 139, 214
Gauss, Carl Friedrich (gous), 7, 28, 35, 52, 57, 69, 142, 144, 181, 298
Generating function, 80
Generator, 213
Geometric progression, 23
Greatest common divisor:
   integers, 62
   polynomials, 111
Ground field, 101
Group:
   Abelian, 194
   abstract, 191
   alternating, 184
   commutative, 194
   identity, 191
   inverse, 191
   multiplication table, 191
   of permutations, 182

i, 9, 101, 102, 130
Identity:
   additive, 99
   multiplicative, 99
   permutation, 155
Imaginary numbers, 6, 9, 130
Indecomposable group, 253
Index of subgroup, 208
Induction, 288–296
Invariant, 153
Inverse:
   additive, 43, 60, 99
   group element, 191
   multiplicative, 43, 60, 64, 99, 133
Irrational numbers, 9
Irreducible polynomial, 108
Isomorphism:
   field, 234
   group, 198

Jordan, Camille (zhôR däN), 204, 300

k-cycle, 158
(k, n)-puzzle, 176
   cylindrical, 176
   Klein bottle, 178
   Moebius, 177
   toroidal, 176
Khayyam, Omar. See al-Khayyami, Umar
Klein, Felix, 185, 300
Klein bottle, 178
Klein 4-group, 185, 223

Lagrange, Joseph Louis (la gRäNzh'), 51, 124, 152, 204, 209, 298
Latin square, 194
Law of quadratic reciprocity, 69
Least common multiple, 68
Legendre, Adrien-Marie (lę zhäN' drę), 69, 299
Legitimate move, 168
Lindemann, Charles, 37
Loop, 218

Method of false position, 2
Minimal polynomial, 150
Modular:
   addition, 58
   additive inverse, 60
   arithmetic, 57
   multiplication, 58
   multiplicative inverse, 60, 64, 65
   powers, 85
   radicals, 68
   roots of unity, 87
   subtraction, 60
Modulus, 11
Moebius (k, n)-puzzle, 177
Monic polynomial, 102
Monster, 247
Multinomial Theorem, 91
Multiple, 39
Multiplication:
   Cartesian numbers, 42
   complex numbers, 10
   cosets, 221
   field, 98
   modular arithmetic, 58
   polynomials, 101
   table, 191
Multiplicative:
   identity 43, 99
   inverse, 43, 60, 64, 99, 133
Multiplicatively perfect number, 71
Multiplicity of zero, 110

nth root of unit, 20, 39–41
Negative numbers, 9
Newton, Isaac, 80, 299
Newton–Raphson method, 53
Non-Euclidean geometry, 248
Normal subgroup, 224, 245, 252
Number of permutations, 156

Octahedron, 189
Odd permutation, 167

Order:
   coset, 228
   element, 39, 88, 135, 201
   field, 239
   group, 200

Parity of permutation, 168
Pascal, Blaise, 300
Pascal's identity, 76
Pascal's triangle, 77, 78
Perfect number, 71, 86
Permutation, 155
   even, 167
   odd, 167
Permutation group, 182
Polar form, 14
Polynomial, 100
   addition, 101
   degree, 102
   division, 102
   factorization, 102, 106, 116
   irreducible, 108
   multiplication, 101
Prime factorization, 70
Prime number, 39
Primitive:
   element, 134, 143, 213
   nth root of unity, 40, 57
   polynomial, 149
   root (mod p), 88
Primitive Element Theorem, 143, 213
Proper subgroup, 204

Quadratic equation, 3
Quartic equation, 51, 124, 152
Quaternion group, 194, 222
Quintic equation, 51, 53, 154, 246
Quotient of groups, 227

Raphson, Joseph, 53, 299
Rational expression, 25
Rational zero, 68
Real number, 10
Regular:
   hexagon, 27
   octagon, 27
   pentagon, 27
   polygon, 20, 27
   17-sided polygon, 28, 33, 57, 181
Relative prime:
   integers, 63
   polynomials, 115

Rhind mathematical papyrus, 2
Root, 17
Roots of unity:
   complex, 20
   Galois field, 138
   modular, 87
Rotation, 185
Ruffini, Paolo, 154, 300
Ruler and compass constructibility, 27

Simple group, 247
Solvability by radicals, 25, 51, 181, 243
Square, 27
Squaring the circle, 37
Standard configuration, 171
Subfield, 236
Subgroup, 203
Subtraction:
   complex numbers, 10
   modular, 60
Symmetric group, 182
Symmetry, 185

Tartaglia, Niccolo (täR ta' glē e), 1, 298
Tetrahedron, 187
Toroidal (k, n)-puzzle, 176
Transposition, 160
Triangular prism, 188
Trisecting the angle, 37
Trivial:
   group, 204
   symmetry, 186

Unchanged, 164
Uniqueness of factorization:
   integers, 70
   polynomials, 116
Unit segment, 28

Vandermonde, A.T., 51, 299
Variant, 153
Vector, 11
Vertex symmetry, 186

Wantzel, Pierre (vän' tsel), 38

Zero of polynomial, 106
Zero polynomial, 102